METALLIZED PLASTICS 1

Fundamental and Applied Aspects

METALLIZED PLASTICS 1

Fundamental and Applied Aspects

Edited by

K. L. Mittal
IBM US Technical Education
Thornwood, New York

and

J. R. Susko
IBM Corporation
Endicott, New York

PLENUM PRESS • NEW YORK AND LONDON

Library of Congress Cataloging in Publication Data

Electrochemical Society Symposium on Metallized Plastics (1988: Chicago, Ill.)
Metallized plastics 1: fundamental and applied aspects / edited by K. L. Mittal and J. R. Susko.
p. cm.
"Proceedings of the Electrochemical Society Symposium on Metallized Plastics, held October 9–12, 1988, in Chicago, Illinois – T.p. verso.
Includes bibliographical references.
ISBN 0-306-43389-3
1. Plastic coatings – Congresses. I. Mittal, K. L., 1945– . II. Susko, J. R. III. Title.
TP1175.M4E35 1988 89-28395
668.4 – dc20 CIP

Proceedings of the Electrochemical Society Symposium on
Metallized Plastics, held October 9–12, 1988, in Chicago, Illinois

A Division of Plenum Publishing Corporation
233 Spring Street, New York, N.Y. 10013

Printed in the United States of America

PREFACE

This volume chronicles the proceedings of the Symposium on Metallized Plastics: Fundamental and Applied Aspects held under the auspices of the Dielectrics and Insulation Division of the Electrochemical Society in Chicago, October 10-12, 1988. This was the premier symposium on this topic and if the comments from the attendees are any barometer of the success of a symposium then it was a grand success. Concomitantly, it has been decided to hold it on a regular basis (at intervals of 18 months) and the second event in this series is planned as a part of the Electrochemical Society meeting in Montreal, Canada, May 6-10, 1990.

Metallized plastics find a legion of applications ranging from mundane to very sophisticated. A complete catalog of the various technological applications of metallized plastics will be prohibitively long, so here some eclectic examples should suffice to show why there is such high tempo of R&D activity in the arena of metallized plastics, and all signals indicate that this high tempo will continue unabated. For example, polymeric films are metallized for packaging (food and other products) purposes, and the applications of metallized plastics in the automotive industry are quite obvious. In the field of microelectronics and computer technology, insulators are metallized for interconnection and other functional purposes. Also plastics are metallized to provide electromagnetic shielding.

This symposium was organized with the following objectives in mind: (1) to bring together researchers and technologists involved in the many facets of metallized plastics; (2) to provide a forum for discussion of latest research findings and applied aspects; (3) to provide an opportunity for cross-pollination of ideas; and (4) to highlight areas, within the broad purview of metallized plastics, which needed intensified R&D efforts. The scope of the symposium was quite comprehensive, and papers dealing with topics ranging from ways to metallize plastics to latest understanding in unravelling the metal/plastic interface using sophisticated spectroscopic techniques were solicited. Also there was a deliberate attempt to have a blend of invited overviews and original research contributions.

The final program contained a total of 39 papers covering a variety of topics ranging from very fundamental to applied. It should be recorded that there were brisk, lively and illuminating (not exothermic) discussions throughout the duration of the symposium, both formally and informally. The general consensus from the audience seemed to be that this symposium covered many important ramifications of metallized plastics, and our putative objectives were amply fulfilled.

The present volume contains a total of 18 papers divided into three parts as follows: Part I. Metallization Techniques and Properties of

Metal Deposits; Part II. Spectroscopic Investigation of Interfacial Interactions; and Part III. Surface Modification and Adhesion Aspects. The topics covered include: various ways to metallize plastics including laser stimulated chemical vapor deposition of metals, and UV laser-induced metallization; performance of silvered-polymer mirrors; stresses in metal coatings; metal interlayers in polymer films; graded metallization of polymers; application of various spectroscopic techniques (e.g., ESCA, SIMS, ISS, HREELS) in understanding interfacial interactions in metallized polymer systems; surface modification of polymers by various techniques before metallization; and adhesion aspects of thin metal films. It must be recorded here that all manuscripts were rigorously peer reviewed and were suitably revised before inclusion in this volume.

We certainly hope this volume will serve as a repository of latest information and thinking regarding metallized plastics and will provide a ready reference to veterans in this field and a guide to those who wish to make their maiden voyage into the wonderful world of metallized plastics.

Acknowledgements: Now we have the pleasant task of acknowledging those who contributed in many and varied ways. Thanks are due to the appropriate officials of the Dielectrics and Insulation Division of the Electrochemical Society for sponsoring this event. We are thankful to the appropriate management of IBM Corporation for permitting us to organize this symposium and to edit this volume. KLM would like to extend his appreciation to his wife, Usha, for her help in more ways than one. The time and efforts of the unsung heroes (reviewers) must be acknowledged as the comments from the peers are a desideratum to maintain the high standard of publications. Last, but most important, the enthusiasm, co-operation and contribution of authors is gratefully acknowledged without which this book would not have seen the light of day.

K.L. Mittal
IBM U.S. Technical Education
500 Columbus Ave.
Thornwood, NY 10594

J.R. Susko
IBM Corporation
Endicott, NY 13760

CONTENTS

PART I. METALLIZATION TECHNIQUES AND PROPERTIES OF METAL DEPOSITS

Laser Stimulated Chemical Vapor Deposition of Metals on Polyimides
S.J. Bezuk and C. Kryzak.. 1

Graded Metallization of Nonconducting Substrates by Glow Discharge Plasma Polymerization
B.K. Sun, D.L. Cho, T.J. O'Keefe and H. Yasuda.................. 9

Metallization of Plastics with Resistance Heated Sources
K.M. Anetsberger.. 29

High Rate Reactive Sputtering onto Flexible Polymer Sheet
R.P. Howson, A.G. Spencer, K. Oka and R.W. Lewin................ 45

UV Laser-Induced Metallization of Polyimide from Electroplating Solution
J. Zahavi, M. Rotel, D. Katz and S. Levine...................... 55

Simple Metal-on-Plastics Patterning by Pulsed Electron Beam in Soft Vacuum
J. Krishnaswamy, M. Eyolfson, L. Li, G.J. Collins, H. Hiraoka and M.A. Caolo.. 69

Optical Performance and Durability of Silvered Polymer Mirrors
G. Jorgensen and P. Schissel.................................... 79

Deposition of Adherent, Thick Copper Coatings on Glass
W.C. Cowden, T.G. Beat, T.A. Wash and J.W. Dini................. 93

Dependence of Stress on Deposition Conditions for Sputtered Copper Films onto Flexible Polyimide Substrates
A. Entenberg, V. Lindberg, L. Fendrock, S.K. Hong, T.S. Chen and R.S. Horwath.................................... 103

Metal Interlayers in Polymer Films: A Survey of Deposition Processes, Morphology, Patterning Methods and Physical Properties
S. Mazur, L.E. Manring, M. Levy, G.T. Dee, S. Reich and C.E. Jackson.. 115

PART II. SPECTROSCOPIC INVESTIGATION OF INTERFACIAL INTERACTIONS

Surface Spectroscopic Techniques Applied to Metallized Plastics
N.J. DiNardo 137

A. Static SIMS Study of Interfaces between Evaporated Metal Films and Polyimides
W.J. van Ooij, R.H.G. Brinkhuis and J.M. Park 171

XPS and HREELS Study of the Aluminum/Poly(ethylene terephthalate) Interface and the Influence of the Polymer Surface Pretreatment on Adhesion
Y. Novis, M. Chtaib, J. Vohs, J.J. Pireaux, R. Caudano, P. Lutgen and G. Feyder 193

Interfacial Chemistry of Metal Films on Polymers: Diffusion, Oxidation, Trace Components Studied by XPS, ISS, and SIMS
G.R. Sparrow and L. Homstad 205

PART III. SURFACE MODIFICATION AND ADHESION ASPECTS

Characterization of Surface Modifications during Metallization of Polyetherimide
M.C. Burrell, B.R. Karas, D.F. Foust, W.V. Dumas, E.J. Lamby, W.T. Grubb and J.J. Chera 223

Effect of Plasma Treatment and Ion-Plating on Adhesion of Metallized Polyimide and Poly(ethylene terephthalate)
K. Nakamae, S. Tanigawa and T. Matsumoto 235

Adhesion of Chromium, Nickel and Copper to Polymers Containing Basic Functional Groups
S.R. Cain, L.J. Matienzo and F. Emmi 247

Adhesion and Deformation Behavior of Thin Metal Films on Polyimide
Y.H. Jeng, F. Faupel, S.T. Chen and P.S. Ho 265

About the Contributors 271

Index 281

PART I. METALLIZATION TECHNIQUES AND PROPERTIES OF METAL DEPOSITS

LASER STIMULATED CHEMICAL VAPOR DEPOSITION OF METALS ON POLYIMIDES

S.J. Bezuk and C. Kryzak

UNISYS Corp.
10850 Via Frontera
San Diego, CA 92127

Polyimides are of great interest as insulators for microelectronic fabrication due to their low dielectric constant and planarizing properties. LCVD has been used to deposit material on a wide variety of substrates and for many microelectronic applications. We focus on the laser CVD of nickel on polyimide and the effects of this deposition on the electrical and mechanical surface properties of polyimides. We report on the effects of writing speed, laser power and reactant pressure on the deposition characteristics as well as the effects of the deposition on the underlying polyimide. A review of the microelectronic applications of metal depositions on polyimides will also be presented.

INTRODUCTION

The use of laser direct writing to deposit or etch metal interconnect introduces many possibilities for improving integrated circuit and circuit package manufacturing. These laser techniques exploit the use of focused laser beams for their highly localized processing capability. In addition, laser direct writing techniques do not require masks for the fabrication of patterns. Thus, laser direct writing can be used to repair defective circuits and photomasks, to manufacture small numbers of circuits economically, and to develop novel, yield improving IC packaging techniques[1-5].

The current trend in microelectronics toward increasing device performance and packaging density has raised strong interest in the study of polyimide materials[6]. The advantages of polyimides as interconnect dielectrics include their superior planarization properties and low dielectric constants. Therefore, the lower temperature and local metal deposition characteristics of laser direct writing in conjunction with the superior electrical and mechanical properties of polyimides offer strong incentives for investigating the compatibility of the two technologies.

APPLICATIONS

Applications using LCVD to deposit metal on polyimide are emerging within the areas of IC circuit and packaging manufacturing and repair.

The earliest deposition of a metal on polyimide was the deposition of cadmium from dimethyl cadmium by a photolytic decomposition[7]. The deposition was accomplished using a frequency doubled argon ion laser operating at 257 nm. The rates for photolytic laser depositions are less than for pyrolytic laser depositions (0.1-100 μm/sec) by a factor of 10^3-10^4. Therefore, photolytic depositions have limited usefulness. LCVD technology was extended to the deposition of nickel[8] which is discussed in greater detail below.

Tungsten LCVD deposition from WF_6 on polyimide has been demonstrated by several groups[9-11]. It has been used for high accuracy tuning of GaAs planar millimeterwave circuits[9]. Tungsten was added to an attenuator circuit to tune the attenuation frequency of the circuit after it was fabricated. Tuned GaAs microwave circuits are very difficult and expensive to fabricate. Tuning these circuits after fabrication can dramatically increase yield and lower costs.

Figure 1 shows modifications made to interdigitated capacitors for tuning GaAs microwave circuits. Nickel was deposited on the gold interconnect and across the polyimide dielectric to extend the interdigitated pattern and change the capacitance thereby allowing the tuned operational frequency of an amplifier to be changed.

LCVD tungsten has been used for supplemental interconnects for GaAs digital circuits with polyimide interlevel dielectrics[10]. Vias were excimer laser etched through the polyimide dielectric to expose first level metallization. Tungsten was then deposited across the polyimide dielectric to change the connections of the first level gold metallization to restructure the circuitry after manufacture. The ability to personalize or repair these circuits after manufacture is extremely beneficial.

The deposition of gold on polyimide has been accomplished from the pyrolytic decomposition of dimethyl gold acetylacetonate[12]. This has potential application for personalization of GaAs circuits, IC packaging, and X-ray photomasks.

We report on the laser CVD of nickel on polyimide and the effects of this decomposition on the electrical and mechanical surface properties of polyimides. Thermal and chemically induced surface degradation was electrically probed by measuring surface leakage current at various substrate temperatures after chemically removing the laser deposited

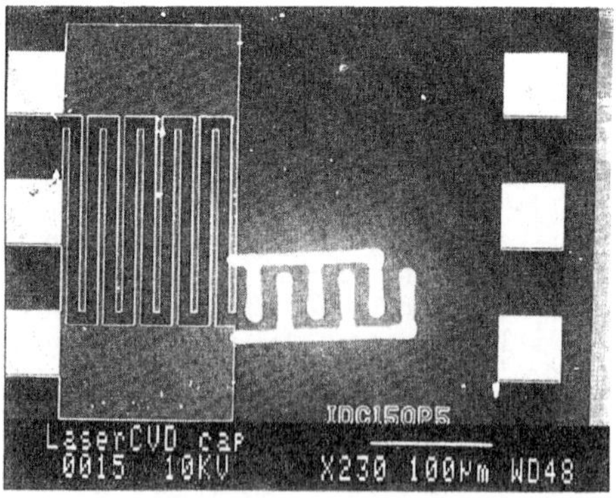

Figure 1. Direct write LCVD nickel modification of interdigitated capacitor for tuning GaAs Microwave circuits.

nickel lines. Emphasis was placed on the importance of tailoring thelaser processing in order to optimize the electrical performance of the desired metal interconnect. The effects of laser power, gas pressure, and scan rate on the aspect ratio of the deposited metal lines are also discussed and compared with similar depositions on silicon dioxide.

EXPERIMENTAL

The polyimides used in this experiment were Toray SP-710, Hitachi PIQ-13, and Ciba-Geigy XU284. The polyimides were used as received. Films of each sample were spun onto 4 inch silicon substrates. All substrates were coated with an adhesion promoter, gamma-aminopropyltriethoxysilane, prior to polyimide application. The polyimide films were cured by baking in a furnace at 420°C for 1 hour. Plasma deposited SiO_2 films were compared to polyimide films. The final thickness of all films tested was 1.2 μm. The experimental apparatus used for these studies is similar to that previously described[1,2]. An argon ion laser beam tuned to 514.5 nm was folded into the optical path of a microscope. The microscope's objective lens focused the beam to a spot size of approximately 12 μm as measured by a scanned pinhole-photodiode arrangement. A TV camera connected to a VCR was used to observe and record the processing in real time. Metal patterns were drawn by translating the substrate under the laser beam on computer controlled stepper motor driven stages. The beam was modulated with a galvanometer scanner with a rise time less than 0.5msec. The nickel was deposited by pyrolytic decomposition of neat $Ni(CO)_4$ (Strem Chemical, Newburyport, MA) at reduced pressure in a static chamber at room temperature. The nickel deposits easily nucleated on the surface of the polyimides used. Depositions were readily accomplished on polyimide films thicker than 50 μm. We believe that nucleation can occurs by the direct heating of the polyimide film by the absorption of laser light. Measurement of the optical absorption properties of the polyimide films showed a small (~1%) but measureable absorbance at 514 nm. This result has since been substantiated[12]. Resistivity measurements were made using four point probe techniques. The physical dimensions of the deposits were determined by stylus profilometry or by SEM.

In order to electrically monitor polyimide surface degradation, nickel lines were deposited at 1 Torr $Ni(CO)_4$ partial pressure at laser powers of 0.5, 0.75, and 1 watt. The Ni metallization was etched to completion in a solution of aqua regia and the samples were coated with 0.5 nm of resistively evaporated aluminum. A 400 μm long, 126 finger interdigitated capacitor pattern with 9 μm spaces was aligned over the laser deposited areas. The Al was wet etched to define the fingers, and the interdigitated patterns were electrically probed with a shielded HP 4061 test system. Variations in polyimide sheet resistivity were determined, for a given laser power, as a function of substrate temperature (25-120°C) and applied bias (0-100 volts).

RESULTS AND DISCUSSION

Figures 2 and 3 show SEM photos of a nickel line deposited on polyimide. The deposited line was characterized by a smooth surface without periodic structure and without physical damage to the underlying polyimide. Auger and electron microprobe spectroscopic measurements detected less than 5% carbon and oxygen in the nickel deposits, near the detection limits of the techniques.

Measurements of the deposit resistivity were approximately 2 times the resistivity of bulk nickel (20-30 $\mu\Omega\cdot$cm) without post deposition processing. Figure 3 shows the columnar grain structure in the center of the deposits. The size of the grains is a direct indication of the

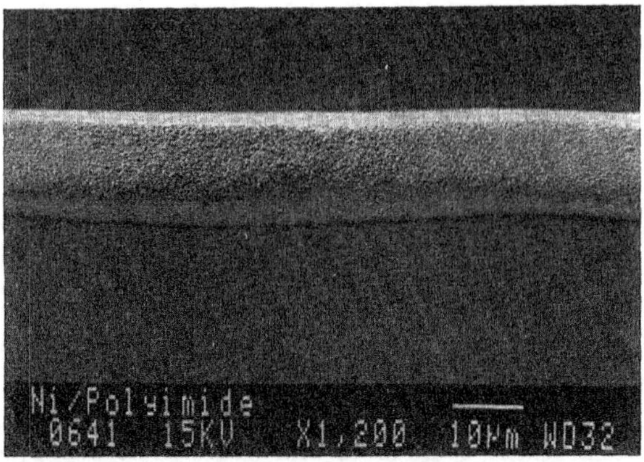

Figure 2. SEM micrograph of the top view of a laser deposited nickel line over 1.2 μm polyimide film.

Figure 3. SEM of the cross section of nickel line over 1.2 μm polyimide film.

temperature attained during deposition in that area of the deposit. The higher temperature in the center of the deposit leads to columnar deposits with 1lm cross sections and the lower temperature at the edges of the deposit leads to powdery deposits. These results are similar to those previously reported for depositions of nickel on inorganic materials[13-17]. The vertical growth rate for nickel under the studied conditions was found to be 10-100 μm/sec. Figure 4 shows the calculated growth rate of nickel as a function of temperature[18]. This shows that the temperature at the surface of the deposition should be 275-400°C. This surface temperature is just under the 450°C decomposition temperature of the polyimides used. To achieve a similar rate for the deposition of tungsten by the hydrogen reduction of tungsten hexafluoride[19] a temperature of between 800 and 1100°C would need to be attained on the surface.

Figure 5 shows a typical comparison of the deposition of nickel as a function of the scan rate for polyimide and silicon dioxide. Rate-limiting induced by diffusion-limited material transport was not found for the conditions investigated[20]. However, an enhancement of the deposition rate by a factor of 3 to 10 was found on polyimide compared to

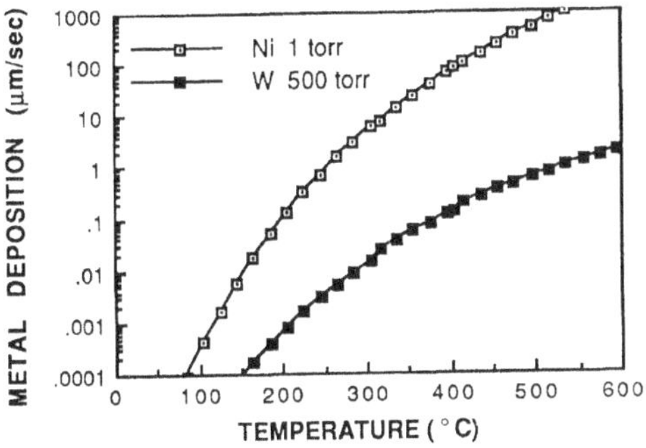

Figure 4. Calculated rate of nickel and tungsten deposition as a function of temperature.

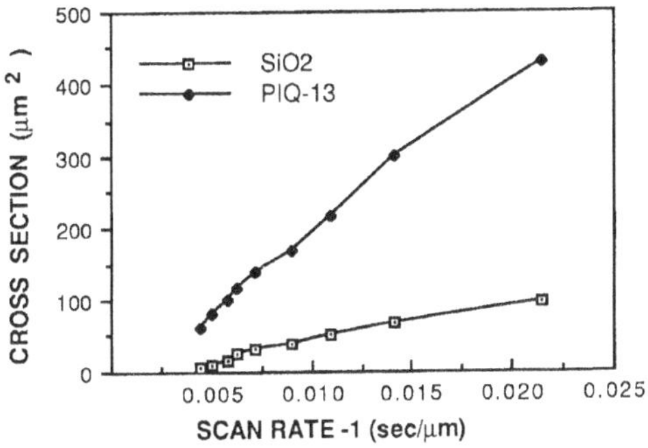

Figure 5. Cross section of nickel lines as a function of 1/scan rate, for a laser power of 0.94 W and $Ni(CO)_4$ pressure of 1 torr.

SiO_2. The temperature rise at the surface at the laser spot and the distribution of that heat are determined in this case by the thermal properties of silicon dioxide and polyimide[13]. The thermal conductivity of silicon dioxide (3.9E-3 cal/cm-sec°K) is a factor of 8.5 greater than polyimide (3.9E-4 cal/cm-sec°K). This translates to a higher peak temperature and more localized temperature distribution for polyimide and therefore higher deposition rate for the same incident power. Deposits with aspect ratios of greater than 0.5 were obtained.

Figure 6 shows the power dependence of the pyrolytic deposition of nickel on polyimide and silicon dioxide substrates. The depositions were not diffusion-limited in this power regime. The deposition rates for the polyimides were higher than silicon dioxide but did not increase as quickly with increasing power. At the upper range of powers used, the polyimide began to discolor and decompose. This indicates that a portion of the input power was being used to decompose the polymer.

The shape of deposited metal lines on inorganic materials and their dependence on power has been documented [13,14,16]. As power is increased, the transition from rounded peaks to flat-topped mesa-like structures and then to double-humped structures as power is increased further has been attributed to reactant gas depletion or absorption-desorption effects (i.e., nonlinear kinetics)[21].

Figure 7 shows results for line shapes obtained for nickel deposits on PIQ-13 as a function of scan speed. At low scan speeds, line shapes tended toward rounded peaks, flattening at intermediate speeds and at the

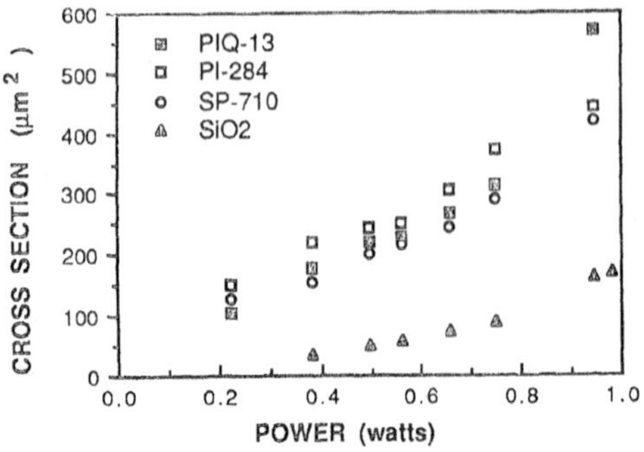

Figure 6. Cross section of nickel lines as a function of laser power, for a scan speed of 80lm/sec and $Ni(CO)_4$ pressure of 1 torr.

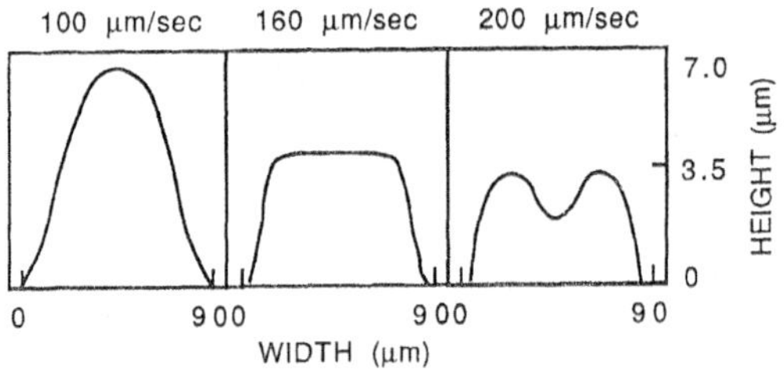

Figure 7. Cross section of nickel lines as a function of scan speed for a laser power of 0.65 W and $Ni(CO)_4$ pressure of 1 torr.

highest speeds studied became double-humped. The results can be explained by an increase in temperature at the depositing edge of the line[13-17].

Figure 8 illustrates this point. At low scan speed, the observed laser beam is on the top of the nickel deposit which allows incident power to be dissipated by the nickel conductor resulting in a low temperature at the deposits leading edge. At higher scan speeds the laser reflection is observed to move towards the leading edge of the deposit and more power is dissipated at the thinner leading edge of the deposit and substrate resulting in a higher temperature at the deposition point. This is confirmed by the observation that for a given input power the decomposition of polyimide films is a direct function of the scan speed. As the scan speed is increased, the amount of decomposition in the polyimide also increases.

I-V measurements were made on an interdigitated capacitor which was deposited over a region of the polyimide exposed to laser-induced nickel deposition with laser powers of 0.5 to 1.0 watts and a $Ni(CO)_4$ pressure of 1.4 torr. The capacitor was biased from 0 to 100 volts and the temperature was varied from 25 C to 125°C. The current measured was less than 0.1 picoamps under all conditions. With the described interdigitated structure and film thickness of 1.2 μm, a film resistivity of approximately 5.0E15 $\mu\Omega\cdot cm$ (i.e., 7.4E19 Ω/square) was found. Similar resistivity values were found at laser powers of 0.75 and 1 watt. As a control, identical interdigitated structures as well as dot capacitor structures were placed on sections of polyimide film which were not

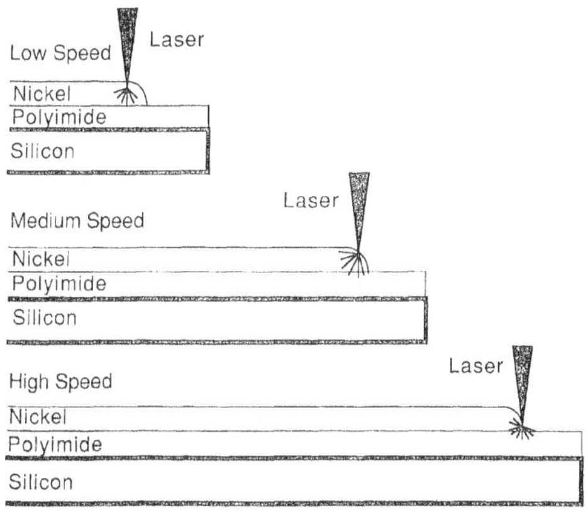

Figure 8. Position of laser beam focus as a function of scan speed.

exposed to any laser processing. The measured film resistivity averaged 1.0E15 $\mu\Omega\cdot$cm. These values are identical within the error limits of the measurement equipment and typical of the bulk resistivities of polyimide. They indicate the surface resistivity of the polyimide films were not measurably altered by the laser deposition reaction.

SUMMARY

The deposition of metals by LCVD is an expanding field. Important applications have been demonstrated for the technology in the area of microelectronic circuit manufacturing. We have shown the pyrolytic deposition of nickel over polyimide. The surface electrical properties of the polyimide are unaffected by the deposition at low power levels and flat-topped microelectronically ideal nickel lines can be deposited under a variety of conditions.

REFERENCES

1. R. M. Osgood, Jr., Ann. Rev. Phys. Chem. 34, 77 (1983).
2. D.J. Ehrlich and J.Y. Tsao, "VLSI Electronics: Microstructure Science", Vol. 7,p. 129 Academic Press, New York, 1983.
3. D.J. Ehrlich and J.Y. Tsao, J. Vac. Sci. Technol. B1(4), 969 (1983).
4. Irving P. Herman, in "Laser Processing and Diagnostics", p. 396.,Springer-Verlag, New York, 1984.
5. D. Bauerle, "Chemical Processing With Lasers", D. Bauerle, editor, (Springer-Verlag, New York, 1986).
6. G. Samuelson and S. Lytle, in "Polyimides: Synthesis, Characterization and Applications", K.L.Mittal, editor, Vol. 2, pp. 751-766, Plenum Press, New York, 1984.
7. H.H. Gilgen, C.J.Chen, R. Krchnavek, and R.M. Osgood, in "Laser Processing and Diagnostics", D. Bauerle, editor, p. 225 (Springer-Verlag, New York, 1984).
8. S.J. Bezuk, R.J. Baseman, C. Kryzak, K. Warner, and G. Thomes, in "Materials Research Society Symposia Proceedings", V.M. Donnelly, Irving P. Herman, and M. Hirose, editors, Vol. 75, , p. 75 (Elsevier Science, Inc., 1987).
9. D.J. Ehrlich, D.F. Williams, J.H. Sedlacek, M. Rothchild, and Steven E. Schwarz, IEEE Electron Device Letters 8, 110 (1987).
10. J.G. Black, S.P. Doran, M. Rothschild, and D.J. Ehrlich, Appl.Phys. Lett. 50, 1016 (1987)

11. W.M. Grossman and M. Karnezos, J. Vac. Sci. Technol. B5, 843 (1987).
12. T.H. Baum, J. Electrochem. Soc., 134, 2616 (1987).
13. Dieter Bauerle, in "Laser Processing and Diagnostics", D. Bauerle, editor, p. 166, (Springer-Verlag, New York, 1984).
14. W. Krauter, D. Bauerle, and F. Fimberger, Appl. Phys. A31, 13 (1983).
15. Klaus Piglmayer, Josef Doppelbauer and Dieter Bauerle, in "Materials Research Society Symposia Proceedings", A.W. Johnson, D.J. Ehrlich, and H.R. Schlossberg, editors, Vol. 29, p. 47, (Elsevier, 1984).
16. Dieter Bauerle, in "Materials Research Society Symposia Proceedings", R.M. Osgood, S.R.J. Bruek, and H.R. Schlossberg, editors, Vol. 17, p. 19, (Elsevier Science, Inc., 1983).
17. Irving P. Herman, Roderick A. Hyde, Bruce M. McWilliams, Andrew H. Weisberg and Lowell L. Wood, in "Materials Research Society Symposia Proceedings", R.M. Osgood, S.R.J. Bruek, and H.R. Schlossberg, editors, Vol. 17, p. 9, (Elsevier, 1983).
18. H.E. Carlton and J.H. Oxley, AIChE J.13,86 (1967).
19. E.K. Broadbent and C.L. Ramiller, J. Electrochem. Soc., 131, 1427 (1984).
20. D.J. Ehrlich and J.Y. Tsao, in "Materials Research Society Symposia Proceedings", R.M. Osgood, S.R.J. Bruek, and H.R. Schlossberg, editors, Vol. 17, p. 3, (Elsevier, 1983).
21. D.C. Skouby and K.J. Jensen, J. Appl. Phys. 63, 198 (1988).

GRADED METALLIZATION OF NONCONDUCTING SUBSTRATES BY GLOW DISCHARGE PLASMA POLYMERIZATION

Byung K. Sun, Dong L. Cho, Thomas J. O'Keefe, and H. Yasuda

Graduate Center for Materials Research
University of Missouri-Rolla
Rolla, MO 65401

The research described concerns the application of a composition graded film for use as a preplate for metallizing nonconducting materials. The composition of the graded film, which changes from 100% organic to nearly pure metallic over the thickness of the film, eliminates sudden changes in material structure and chemical composition between two dissimilar materials, and assists in establishing a stronger, more durable interface. The graded film was layered onto various polymeric substrates by glow discharge plasma polymerization, and the final metallizing material was plated onto the graded film electrolytically.

INTRODUCTION

Metallized nonconductors such as polymeric and ceramic materials are widely used in appliance, automobile, and electronics industries. But, metallization of nonconducting substrate is technically difficult because of the structural incompatibility between the substrate and metallizing material, in terms of both chemical bonding and properties. The abrupt mismatch at the interface between them has been blamed for the major portion of failures of metallized parts under operating conditions.

Lamination of a metal foil with an adhesive has been widely used, because of its simplicity to obtain a conductive layer on the surface of nonconducting material. One of the disadvantages of this process is that it has limits on the shape of a substrate and film thickness that can be used. Such limits were excluded when autocatalytic chemical metallization was developed. However, this process has certain limitations as well, even though it is being widely used in the field. Several pre-steps, each requiring precise handling and conditioning of the materials involved, are necessary and effluent treatment should be considered.

The continually growing demand for improved metallized nonconducting materials has created interest in developing new efficient and dependable processes. This paper presents a metallization technique using glow discharge plasma polymerization, which has been developed in an attempt to generate a film capable of minimizing some of the difficulties cited above. Usually, a lower degree of adherence between the metal and the insulator is obtained if little or no primary chemical bonding is present. Except for some special cases, the substrate-metal adhesion is believed to be obtained by physical adsorption bonds, although activated chemisorption, diffusion, chemical reaction, electrostatic attraction, mechanical interlocking, or some combination of these processes may be occurring, depending on the process condition.[1,2] Therefore, substrate surface cleaning, etching or appropriate surface treatment to achieve favorable surface roughness for mechanical interlocking, and surface modification by oxygen plasma for forming chemical bonds such as organic-oxygen-metal bonds, etc. to improve adhesion between metal and nonconducting substrates, have been tried in the vacuum metallizing field, and great improvements in adhesion of metal to insulators have been reported.[3-5]

The establishment of chemical bonds between a nonconducting substrate and metallizing material, based on research carried out in the field to date, seems to be one of the most crucial factors to improve the performances of metallized insulators under operating conditions. To accomplish this, the present research was directed to applying a composition graded film for use as a preplate for metallizing nonconducting materials. The composition of the graded film, which changes from 100% organic to nearly pure metallic over the thickness of the film, eliminates sudden changes in physical and chemical structures between two dissimilar materials and assists in establishing a stronger, more durable interface. This increases the chances of obtaining improved chemical bonding between the polymer substrate surface and the plasma polymer layer at the bottom of the composition graded film as well as the metallic bonding between the nearly pure metallic layer at the top of the film and subsequent metallizing material. In this case, the process scheme consists of 1) layering a composition graded film by sputter deposition of a suitable electrode material with simultaneous plasma polymerization of methane, followed by 2) deposition of the desired metallizing material by electroless or electroplating. The details and performance of the process will be given in the following sections.

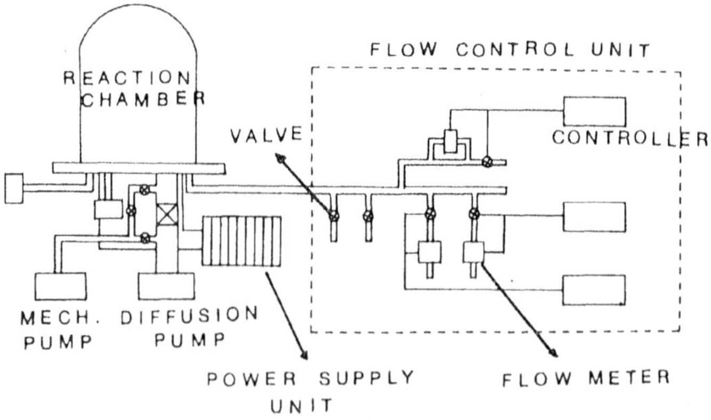

Figure 1. Plasma reactor system.

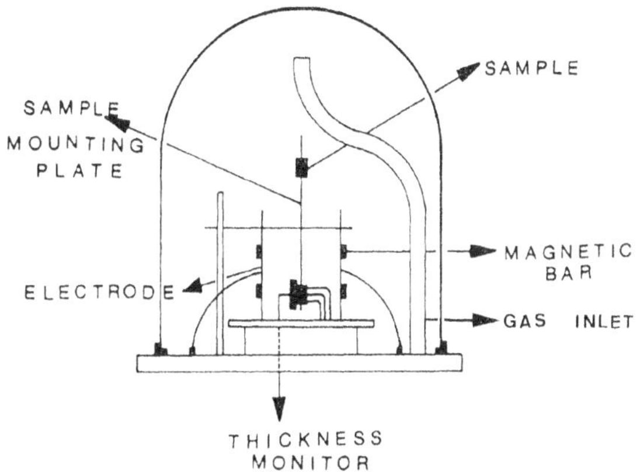

Figure 2. Reaction chamber.

EXPERIMENTAL

A capacitively coupled bell-jar type reactor as shown in Fig. 1 and Fig. 2, operating at a frequency of 10 kHz, was used for plasma polymerization. Either a single or double (Fig. 3) diode magnetron electrode system was used to create a plasma. Two copper plates (18 x 18 cm) were used as the electrodes for a single electrode system, and two titanium plates were added as the second set of electrodes for a double electrode system. A sample mounting disk was located in the middle of diode and rotated at 50 rpm for uniform coating of a plasma polymer film. Methane was used as a monomer for plasma polymerization.

Substrates used included fiber reinforced plastic [FRP, epoxy base], polytetrafluoroethylene [Teflon], poly(ethylene terephthalate) [PET], nylon 66, phenolic resin, and thermoplastic polyimide [ULTEM, GE].

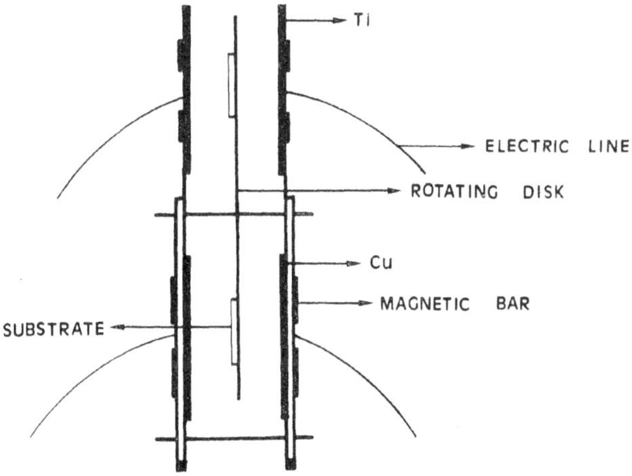

Figure 3. Electrode system for a double-graded process.

FRPs were the primary substrates used. Initially, they were cleaned with detergent in an ultrasonic bath followed by rinsing with deionized water and alcohol. For further cleaning, they were treated with oxygen plasma (1.33 sccm, 60 W, 5 min) followed by hydrogen plasma treatment (3 sccm, 60 W, 5 min).

Copper was selected as the metallizing material because of its wide industrial application. Copper was then plated onto the preplate layer either by electroless- or electroplating. A commercial solution was used for the electroless copper plating to obtain a 1.0 μm thick copper deposit. A 20 μm thick layer of copper was deposited electrolytically under the following conditions.

	$CuSO_4 \cdot 5H_2O$	199 g/l
	H_2SO_4	30 g/l
	With or Without Additives	
Current Density:	20 mA/cm^2	
Bath Temperature:	25°C	
Magnetic Stirring		

For semi-quantitative adhesion data, a tape test was carried out with a tape of peel strength 10 N/cm according to ASTM D3359-78. Tensile lap shear strength tests[6,7] were performed with an Instron floor model testing machine to obtain quantitative adhesion data. Circular copper films (1 cm^2 area) were plated onto plasma-processed 1.5 x 9.0 cm coupons. This film was glued onto another freshly cleaned unplated coupon surface of the same dimension with epoxy glue, completely overlap bonded, and cured for 48 hrs at room temperature. The bonds were sheared at a cross head rate of 0.025 cm/min. The structure of lap-shear test and dimension of the sample are depicted in Fig. 4.

The thicknesses of plasma films were estimated using a film thickness monitor (Kronos, Model QM301) located in the plasma zone. However, the thickness measured with a thickness monitor is not the thickness of the plasma film deposited on a substrate but that of the plasma film deposited on a quartz crystal of the thickness monitor. Therefore, the

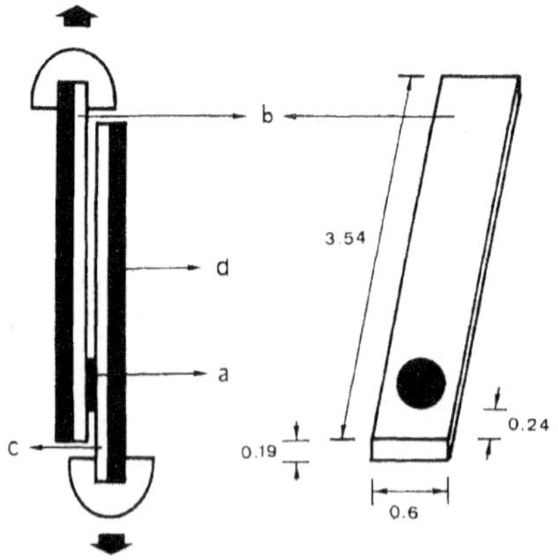

Figure 4. Schema of specimen for the measurement of adhesion: a, lap shear bond composed of metallizing coating and epoxy glue; b, metallized substrate; c, unplated fresh substrate; d, mechanical support (unit, inch).

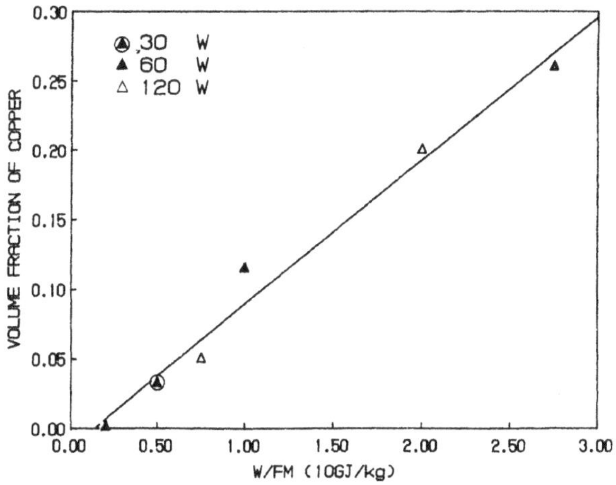

Figure 5. Dependence of the sputter deposited copper content in the plasma polymerized methane film on W/FM value of methane plasma.

film thickness estimated by the thickness monitor reading was calibrated by the thickness value of the film measured with an interferometer (Nomarski, Reichert).

RESULTS AND DISCUSSION

Deposition of Metal-Containing Polymer Film

Metal-containing polymer films were prepared by simultaneous sputter deposition of electrode material with plasma polymerization of methane. Since it was found in the earlier work[8] that sputtering of the electrode surface in a polymer-forming plasma was more efficient in the presence of a magnetic field, a magnetron glow discharge was employed in this work. The maximum flux density, perpendicular component to the electric field, on the electrode surface was 208 Gauss. Methane was chosen as a monomer because its low molecular weight allows a higher energy input, which is favored for sputtering of the electrode surface. In terms of energy per unit mass of a monomer, this is related to the expression, W/FM, where W is the discharge power, F the monomer flow rate, and M the molecular weight of the monomer. In addition, a methane plasma polymer was reported to provide an unique water insensitive adhesion and excellent lap-shear strength.[6,9]

In a magnetron discharge of methane with copper electrodes, copper particles were incorporated into a polymer film beyond the W/FM value of approximately 2 x 10^9 J/kg, as shown in Fig. 5. The volumetric fraction of copper in the polymer film is almost linearly dependent on W/FM. The copper content was estimated by XPS analysis assuming that the copper particles were uniformly distributed throughout the polymer film.

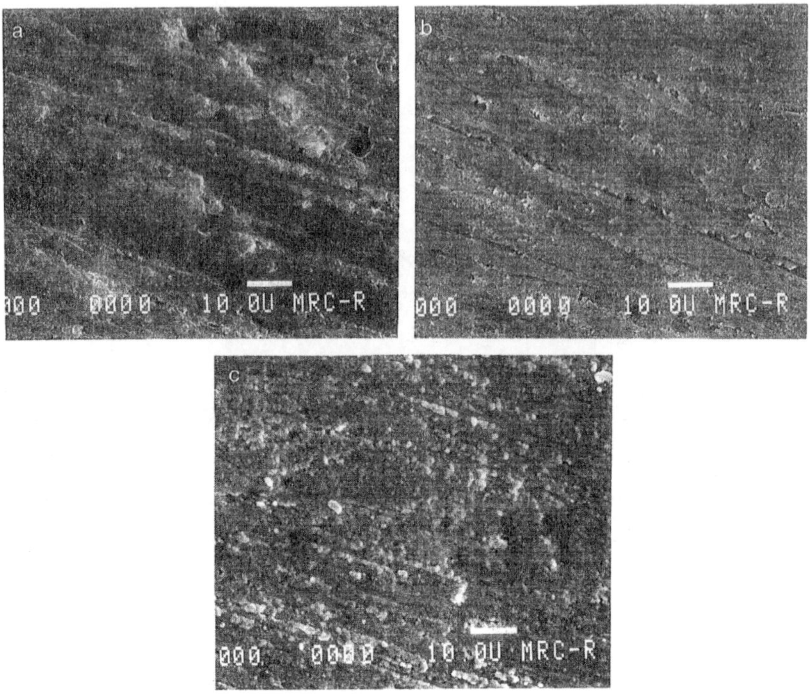

Figure 6. SEM micrographs of as-received FRP substrate (a) and plasma-activated (0.15 volume fraction copper) (b), and electroless copper plated (c) surfaces (X1000).

Activation of Nonconducting Surface for Electroless Plating

The activating effect of copper-containing polymer films for electroless plating of metal was investigated by varying the copper content in a polymer film. Electroless plating of copper was attempted after depositing a 200 Å thick copper-containing methane plasma polymer film on FRP (CH_4 0.5 sccm, 80 W). Visual plating of copper was observed indicating the activation capability of sputtered copper particles in a polymer film. For coatings of 0.15 volume fraction Cu, almost spontaneous electroless-plating with uniform distribution of copper particles throughout the film was observed. As the copper content decreased, the uniformity and coverage of electroless copper also decreased.

The surface morphologies of fresh substrate, plasma-activated and electroless copper plated surfaces are shown in Fig. 6. The figure shows uniform coverage of surface with electroless copper.

Adhesion of electroless-plated copper films was tape tested with an adhesive tape of peel strength of 10 N/cm. The samples produced without plasma preconditioning, such as oxygen or hydrogen treatment, passed the test only about 70% of the time. Surface contamination was suspected, and further cleaning with oxygen plasma treatment (1.3 sccm, 60 W, 5 min) was done before plasma polymerization. Even poorer results were obtained. All the samples failed the test. One interesting phenomenon in this case was that copper particles seemed to partially dissolve into the plating solution before the surface is covered entirely by a plated copper film. This fact became more evident when electroless plating was carried out onto sputter coated copper films in argon or helium plasma either with or without oxygen plasma treatment. Partially uncovered areas were observed and even the sputter coated copper disappeared in those areas. Since copper electrodes can be oxidized in air or by an oxygen plasma treatment, the particles sputtered from the electrode surface can consist of copper oxides instead of copper during initial plasma deposition. This could account for the dissolution of copper particles during electroless plating in highly basic solution (pH, 12-13).

To eliminate the possible incorporation of copper oxides into a polymer film, a hydrogen plasma treatment was carried out after the oxygen plasma treatment and before plasma polymerization. It was found in earlier work[10] that a hydrogen plasma can remove the oxide layer on the electrode surface. Adhesion of electroless-plated copper films improved considerably when the combined sequence (oxygen treatment/hydrogen treatment/plasma polymerization/electroless plating) was used. No failures were observed during the tape test when this treatment was used. The same process was applied to Nylon 66, phenolic resin, and ULTEM. In all cases, good adhesion was obtained.

The activation of nonconducting materials by the deposition of metal-containing polymer films could improve the autocatalytic metallization process by eliminating the aqueous etching and sensitizing steps. In addition, the substrates hard to etch and activate by the aqueous process could be plated by this technique.

Direct Electroplating onto Composition Graded Composite Films

If the film, layered by the plasma polymerization technique, is sufficiently conductive and adheres to the substrate, direct electroplating can be performed. This simplifies the conventional process even further by removing the necessity of several pre-steps such as etching, neutralization, catalyzing, acceleration, and electroless plating. Five different process schemes were tried to further investigate the effectiveness of composition graded films on the metallization of nonconducting substrates in terms of adhesion and serviceability.

Composition graded films were layered by simultaneous sputter deposition of electrode material with plasma polymerization using a methane (CH_4) monomer and argon carrier. The composition gradient through the film was accomplished by controlling discharge power and the gas flow rates of methane and argon.

Substrate/Evaporation Deposited (EVD) Copper/Electrocopper (EVD Process). EVD-copper was deposited at a pressure of 2.1×10^{-3} Pa as a preplate for electroplating. A commercial unit utilizing a tungsten basket to contain and heat the metal was used. A 1500 Å thick copper layer was obtained, and then followed by electrodeposition of copper.

Sputtered Copper/Electrocopper (Sputtering Process). The following AC plasma sputtering conditions were used for depositing sputtered copper layer as a preplate of thickness of 1500 Å.

Plasma carrier gas:	argon
Gas flow rate:	0.5 sccm
Power input:	140 W (419 V/336 mA)
Reactor pressure:	1.4 - 1.9 Pa

The copper was then electroplated onto the sputtered copper layer, as described previously.

Composition Ungraded Film/Electrocopper (Ungraded Process). For these runs, flow rates of 1.5 sccm and 1.0 sccm, methane and argon respectively, for the composition-ungraded step and 0.5 sccm as the flow rate of argon for the final sputtering step were used. The treatment scheme employed was as follows, giving a 1500 Å thick layer. The experimental reactor pressure ranges are given in brackets.

100 W (10 min) → 140 W (18 min) → 140 W (up to 1500 Å)
$F(CH_4)$ = 1.5 sccm ↑ (stop methane flow) ↑ (evacuation) F(Ar) = 0.5 sccm
F(Ar) = 1.0 sccm
[6.7 - 5.5 Pa] [5.6 - 2.7 Pa] [1.9 - 1.4 Pa]
stop methane flow evacuation

Before the methane flow was stopped, there was simultaneous sputter deposition of electrode material and plasma polymerization, depositing a constant composition (ungraded) composite layer of copper and methane. But, the methane to argon ratio decreased in the plasma reactor when the methane flow was stopped. The methane content in the reactor was nearly negligible after 4 minutes after stopping the flow according to pressure change in the reactor. Therefore, the deposition process at 140 W, after 4 minutes, can be considered to be just sputter deposition of copper.

By the process described above, a plasma film could be obtained which had high enough electrical conductivity to allow direct electrodeposition of copper. The bulk resistivity of film measured by a four point probe was 2.6×10^{-4} ohm.cm when deposition was stopped after 18 minutes at 140 W. This value is critical if a uniform electrolytic deposit is to be obtained. For safety, deposition was carried out until a total film thickness of 1500 Å was obtained, giving a nearly pure metallic layer on the top thick enough to allow subsequent electroplating. The final bulk resistivity was 7.8×10^{-5} ohm.cm, taking the total thickness into account for calculation.

The copper was then electroplated onto the composition ungraded plasma film.

Composition Graded Film/Electrocopper (Graded Process).

Single Electrode Process (Single-Graded Process). For these layers, the following parameters were used: 1.5 sccm or 1.0 sccm and 1.0 sccm as flow rates of methane and argon, respectively, for the composition-graded step, and 0.5 sccm as the flow rate of argon for the final sputtering step.

60 W (8 min) → 80 W (2 min) → 80 W (2 min) →
$F(CH_4)$ = 1.0 sccm or 1.5 sccm ↑
F(Ar) = 1.0 sccm
[5.3 - 4.5 Pa or 6.6 - 5.6 Pa]
stop methane flow

100 W (2 min) → 120 W (2 min) → 140 W (6 min) →
[4.5 - 2.8 Pa or 5.6 - 2.8 Pa] ↑
evacuation

140 W (up to 1500 Å)
F(Ar) = 0.5 sccm
[1.9 - 1.4 Pa]

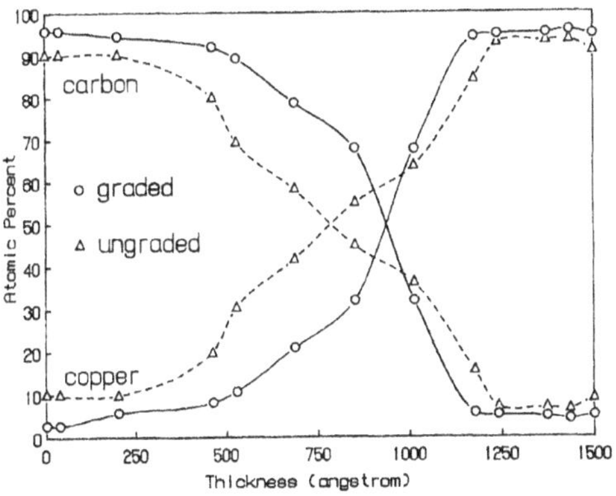

Figure 7. The composition depth profiles of the composition ungraded and graded (layered by single electrode process) films.

The combined gradual change in the methane to argon ratio and increase in discharge power resulted in the formation of a composition graded film. During this cycle, the deposition rate of the methane plasma polymer decreased while that of copper increased.

The composition depth profiles obtained from XPS data of the composition ungraded and graded interfaces are given in Fig. 7.

<u>Double Electrode Process (Double-Graded Process)</u>. Titanium plates, which have lower sputtering tendency, were used for the other set of electrodes during these experiments and were arranged as shown in Fig. 3. A discharge power of 80 W was supplied to the titanium plates to try to obtain higher degree of chemical bonding between the plasma polymer film and polymer substrate. This reduced the incorporation of metal in the very initial film, and kept the lower electrode set as clean as possible by avoiding deposition of the strongly adhering methane plasma polymer layer. Then, the supply of the discharge power was switched to the copper electrode plates from the titanium electrode plates to incorporate the copper into the film by sputter deposition.

The process is as follows: 1.5 sccm and 1.0 sccm as flow rates of methane and argon, respectively, for the composition-graded step, and 0.5 sccm as flow rate of argon for the final sputtering step were used.

80 W (5 min) ⟶ 60 W (3 min) ⟶ 80 W (2 min) ⟶
$F(CH_4)$ = 1.5 sccm ↑
F(Ar) = 1.0 sccm
[6.4 - 5.3 Pa] [5.3 - 5.1 Pa]
switch electrode from Ti to Cu ↑ stop methane flow

80 W (2 min) ⟶ 80 W (2 min) ⟶ 100 W (2 min) ⟶
120 W (2 min) ⟶ 140 W (6 min) ⟶ 140 W (up to 1500 Å)
[5.2 - 3.1 Pa] ↑ F(Ar)= 0.5 sccm
[1.9 - 1.4 Pa]
evacuation

Adhesion Strength

The Effects of Oxygen-Hydrogen Plasma Pretreatment. The results of tensile lap shear tests of coatings produced on FRP substrates for different process schemes with and without plasma pretreatment are shown in Table I. Examination of failed surfaces showed that separation occurred at the interface between the substrate and the bottom of the plasma film. No failure occurred between the plasma film and electrocopper layer.

The indication was that substrate condition is a crucial factor for attaining good film adhesion. Although detergent or solvent cleaning may provide a microscopically clean surface by removing certain contaminants such as scale, dust, or heavy grease, it is not satisfactory if a truly clean surface is needed. Also, the surface can be recontaminated in air during handling prior to the film deposition. These problems can be minimized or eliminated by conditioning the substrate with an oxygen plasma. Residual solvents or other contaminants from the air, which are normally hydrocarbons, can be easily removed by an oxygen plasma through ashing process. Recontamination after the oxygen plasma treatment can be eliminated by in situ deposition of a plasma polymer film. The oxygen plasma treatment may give undesirable results. Weak zones on the substrate surface, through bond scissioning, and copper oxides on the electrode surface can be formed. Especially, low molecular weight fragments formed by oxidation of substrate surface have polar characteristics and may dissolve in aqueous solution, resulting in reduction of adhesion strength at the interface. The oxides on the electrode surface could be incorporated into plasma film and be attacked by acid in the plating solution penetrating into initial interface through probably existing defects in the plasma film during electrodeposition of copper. But, a hydrogen plasma treatment can minimize the undesirable effects of the oxygen plasma treatment by removing the oxygen of the substrate surface and reducing the surface oxides of copper electrode surface. If the hydrogen plasma removes unstable oxidized carbon and polar groups formed by oxidation, the substrate surface will be more stable in aqueous solution. Therefore, the combination of oxygen and hydrogen plasma treatments can provide clean and favorable surface condition for the better adhesion.

Table I. The Effect of Pre-Treatment on Tensile Lap-Shear Strength.

Preplate Process		Lap-Shear Strength (MPa)
sputtering	wo/c	$3.9 \pm 0.3^{*}$
sputtering	w/c	4.0 ± 0.4
single-graded	wo/c	5.3 ± 0.2
single-graded	w/c	6.7 ± 0.2
double-graded	wo/c	6.1 ± 0.6
double-graded	w/c	8.0 ± 0.7

* 95% confidence limits for five samples are shown.
1. wo/c = without oxygen and hydrogen pre-treatment.
2. w/c = with pre-treatment.
3. If not specified, flow ratio of methane to argon for the graded process scheme is 1.5/1.0.
4. FRP was the substrate used.

Sputter cleaning with a noble gas plasma such as argon, to eliminate such problems, might also be a possibility. However, in this case, metal particles can be deposited during the cleaning process especially when magnetron discharge is employed. In addition, the noble gas treatment may cause bond scissioning of the substrate surface, which will reduce the bond strength, by ion bombardment,[5] too.

Surface morphologies of as-received, oxygen plasma treated, and oxygen-hydrogen plasma treated FRP substrates are shown in Fig. 8. No significant change in morphologies is detected, indicating that there should not be appreciable improvement in bonding due to surface roughening.

No difference in bond strength for samples produced by the sputtering process can be attributed to the combination of etching, cleaning, and bond scissioning effects of high energy argon plasma. The high energy argon plasma will modify the substrate surfaces and produce almost same surface appearance, regardless of the previous pretreatments.

Composition Gradient Effects. The results of tensile lap-shear and chemical composition tests for different process schemes are shown in Tables II and III, respectively. Copper content data of the initial plasma coating layer and those of the failed surfaces after tensile lap-shear tests confirm the fact that the failure occurred at the interface between the substrate surface and the bottom of the plasma polymer

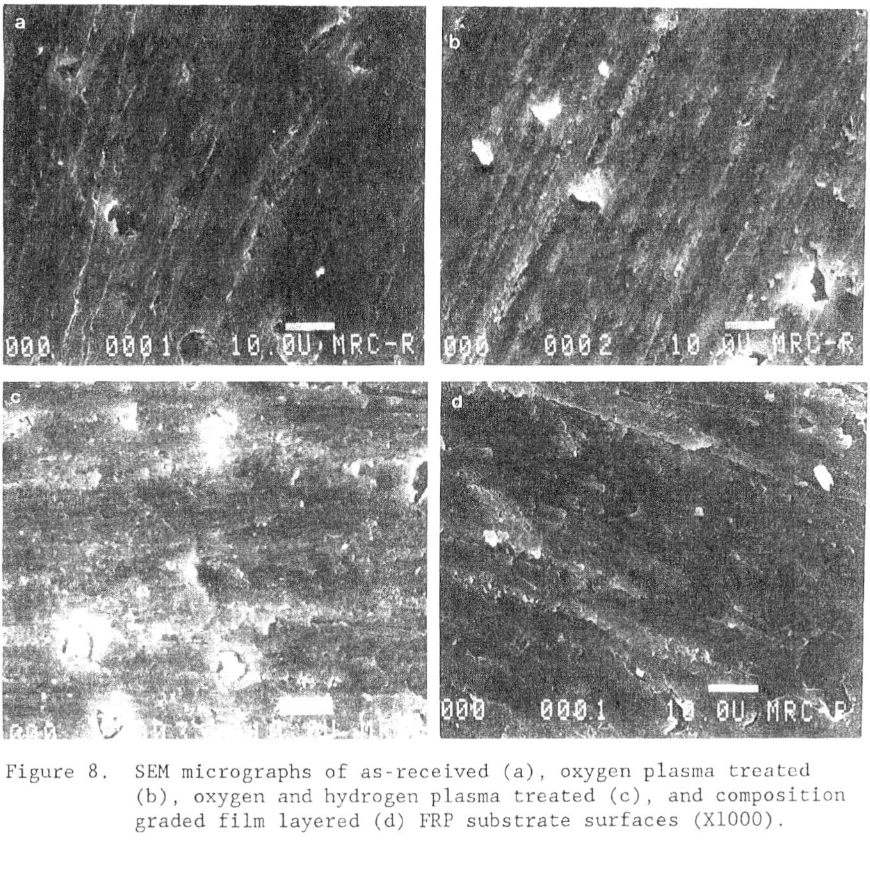

Figure 8. SEM micrographs of as-received (a), oxygen plasma treated (b), oxygen and hydrogen plasma treated (c), and composition graded film layered (d) FRP substrate surfaces (X1000).

Table II. The Effects of Plasma Process Schemes on Tensile Lap-Shear Strength.

Preplate Process	Lap-Shear Strength (MPa)
EVD	$0.7 \pm 0.0^{*}$
sputtering	4.0 ± 0.4
ungraded	5.8 ± 0.5
single-graded	6.7 ± 0.2
single-graded (1.0/1.0)	5.9 ± 0.4
double-graded	8.0 ± 0.7

* 95% confidence limits for five samples are shown.
1. With oxygen and hydrogen pre-treatment except for EVD.
2. If not specified, the flow ratio of methane to argon for the graded process scheme is 1.5/1.0.
3. FRP was the substrate used.

film. These results show the trend that the intermediate layer provided by grading of the metal content throughout the plasma polymer film can improve the strength between the polymer and metal films. The graded metal-containing plasma polymer film can join a polymer and a metal with strong adhesion, and also reduce the stress build-up at the interface, which may come from structural incompatibility between the polymer and metal, by causing a more gradual structural change from "organic" to "metallic".

If the film is deposited by EVD or sputtering only, direct contact and less chemical bonding between the polymer and metal film result. Metallic bonding occurs between the sputter deposited film and an electroplated film. But, the strength of the bonding layer between the substrate and EVD or sputtered copper is mainly dependent on mechanical interlocking or weaker physi- and chemi-sorption interactions which result in low tensile lap-shear strength. The data indicate that sputtering can bring about higher adhesion strength than evaporation deposition, probably due to the high energetic penetration of the sputtered copper onto substrate surface which will increase mechanical interlocking and atomic interfacial mixing, during sputtering by the high energy argon plasma. If a composition ungraded film is deposited as a preplate for electroplating, partial organic to organic bonding between the substrate and the plasma film is achieved and structural change is more or less step-wise. Thus, a little improved adhesion, compared with that of sputtering process, is obtained. Graded films have higher adhesion strengths because very firm organic to organic bonding at the interface between the substrate and initial plasma coating layer exists, and a gradual change of composition and structure over the thickness of film is obtained. If the single-graded process is used, copper can be deposited at the very beginning of the film deposition due to the high sputter yield of copper, and this may prevent establishment of perfect organic to organic bonding. This becomes clearer if adhesion values of the film at different argon to methane ratios are compared. A low methane to argon ratio gives high sputtering of copper, thus decreasing the adhesive strength. The copper content in the initial plasma polymer layer, as indicated in Table III, shows the expected trend.

With the single electrode process, a metal-free substrate-plasma polymer interface can be obtained. However, this results in poorer

Table III. The Chemical Composition of the Initial Plasma Coating Layer and the Failure Surface after Tensile Lap-Shear Testing.

Process	Initial Layer (At%)			Failure Surface (At%)		
	Cu	C	others	Cu	C	others
ungraded (1.5/1.0)*	9.90	88.70	1.40	0.70	96.70	2.40
single-graded (1.5/1.0)	2.70	95.60	1.70	1.90	90.70	7.20
single-graded (1.0/1.0)	6.00	93.10	0.90	1.40	90.90	7.60
	Ti	C	others	Ti	C	others
double-graded	0.10	98.50	1.40	0.05	94.10	5.85

1. The analysis was done on a fresh layer after sputtering off the very thin air-contaminated layer.
2. All the samples were pre-treated with oxygen and hydrogen plasma.

* Represents the flow ratio of methane to argon into reactor.

adhesion of the plasma polymer because low energy input must be supplied in order to avoid sputtering of electrode metal at the initial stage of the plasma operation, whereas adhesion strength of plasma polymer itself increases with energy level.[10] When the double-graded process is used, the metal content can be kept quite low. Also, a high energy level (80 W) can be used to establish very firm organic to organic bonding between the substrate and the initial plasma polymer film using a titanium electrode because it has a low sputter yield.

The Effects of Substrate Type. The results of the tensile lap-shear tests made on samples produced by the double-graded process on different substrates are shown in Table IV. The SEM micrographs of the surface morphology of the different polymer substrates are shown in

Table IV. The Effect of Substrate Type on Tensile Lap-Shear Strength.

Substrate Type	Lap-Shear Strength (MPa)
Nylon 66	3.0 ± 0.4*
Teflon	3.2 ± 0.2
Phenol	4.4 ± 0.3
ULTEM	8.3 ± 0.4
FRP	8.0 ± 0.7

* 95% confidence limits for five samples are shown.

1. Electrocopper was plated onto the composition graded film layered by the double-graded process with oxygen and hydrogen treatment.

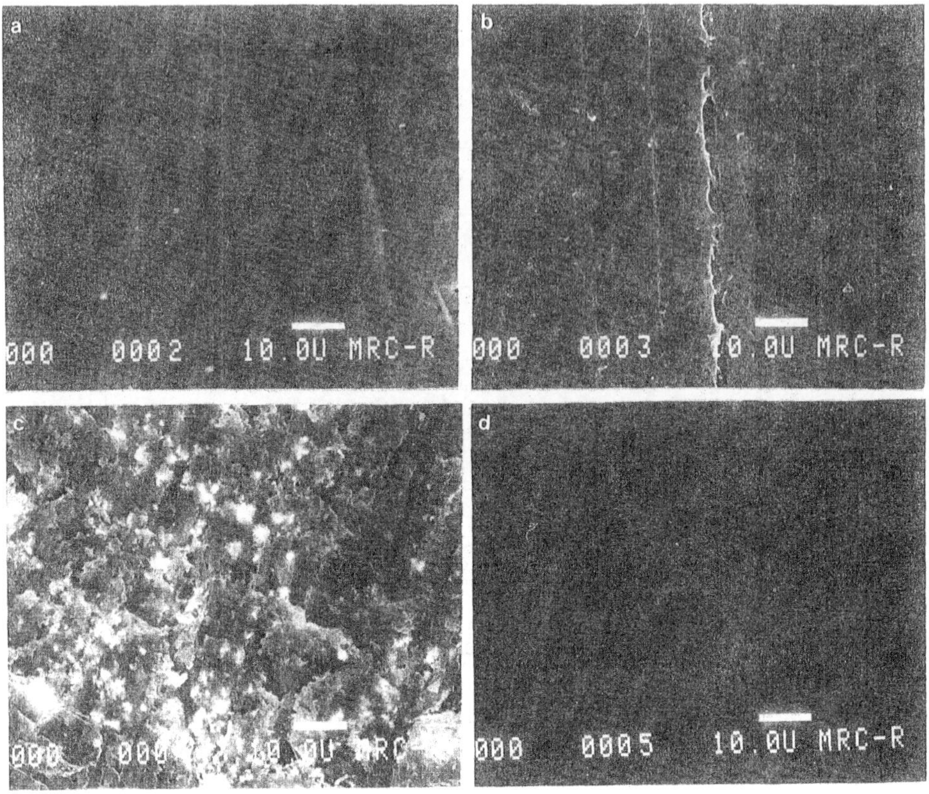

Figure 9. SEM micrographs of as-received Teflon (a), Nylon 66 (b), Phenolic Resin (c), and ULTEM (d) surfaces (X1000).

Fig. 9, and for FRP in Fig. 6-(a). Generally, phenolic resin and FRP seem to have the surface morphological features best suited for improved adhesion provided by mechanical interlocking. However, test results do not correlate the prediction based on surface morphological features, indicating that the adhesion characteristic or chemical affinity of the methane plasma polymer for a substrate is more important and predominant in determining adhesion strength than surface morphological features.

The Effects of Electrolyte Additives. The results of the tensile lap-shear test obtained when different additives are present in the electroplating bath are shown in Table V. The SEM micrographs of electrocopper deposits produced by the double-graded process with pretreatment for different solution chemistries are also shown in Fig. 10.

Shear strength data show that additive-free solution produces a more adhesibe coating while the additives reuce the shear strength of coating. It may be that the intermal stress level is increased in the electrocopper deposit resulting from possible incorporation of additives into the copper lattice, bringing about a reduction of shear strength. Employment of rigorous solution purification seems to be a crucial factor for obtaining sound and well adhering electrocopper deposit, even though additive-free solution can produce relatibely sound and well adhering deposit, as shown in Table V and Fig. 10.

Table V. The Effects of Electrolyte Additives on Tensile Lap-Shear Strength.

Additive	Lap-Shear Strength (MPa)
blank	8.0 ± 0.7*
Cl^- 55 ppm	6.4 ± 0.5
brightener 0.5%	6.1 ± 0.7
brightener 0.5% Cl^- 55 ppm	6.3 ± 0.4

* 95% confidence limits for five samples are shown.

1. Electrocopper was plated onto the composition graded film layered by the double-graded process with oxygen and hydrogen pre-treatment.
2. A commercial brightener was used.
3. FRP was the substrate used.

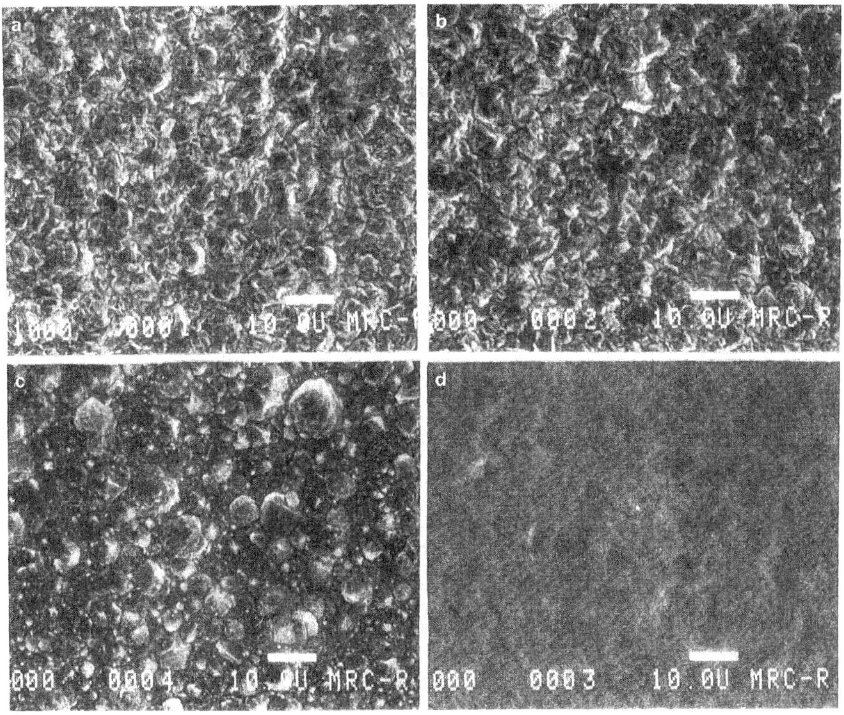

Figure 10. SEM micrographs of electrocopper deposits produced from different solutions: blank (a), Cl^- 55 ppm (b), brightener 0.5% (c), brightener 0.5% and Cl^- 55 ppm (d) (X1000).

Durability Tests

Thermal cyclic was performed on the product coatings according to ASTM G5342-1972, followed by semi-quantitative Scotch tape test to check the adhesion strength. The temperature was cycled five times from 85°C to -40°C, unless failure occurred earlier. Adhesive scotch tape of peel strength 10 N/cm was used for the tape test.

Only the sample which was produced by using the EVD-copper preplate without pretreatment of oxygen and hydrogen failed the durability test. Even though no visible defects appeared during thermal cyclic test, failure occurred by the tape test. No failure occurred with samples produced by employing composition graded film irrespective of whether it was ungraded or graded. Even sputter deposited film passed both tests. The good performance of the sputter deposited sample during both tests can be attributed to inadvertent codeposition of resputtered residual plasma polymer, which had deposited in the reactor. Even though the inside of the reactor was thoroughly cleaned before sputtering, this kind of inadvertent codeposition of residual methane plasma polymer could not be completely avoided because the reactor employed was designed for plasma polymerization rather than sputtering only.

Structure and Stability of Composition Graded Preplate Films

X-ray diffraction analysis, XPS analysis, and SEM and TEM analyses were performed to study structure of composition graded preplate films. In the x-ray diffraction pattern of our composition graded film only the (111) reflection has enough intensity to be recorded. The TEM

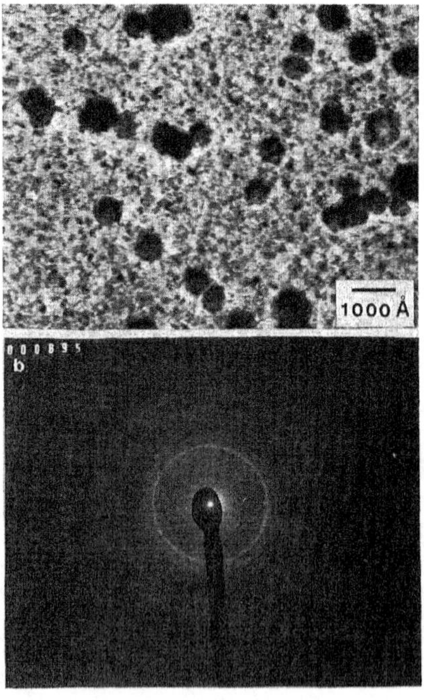

Figure 11. Transmission electron micrograph (a) (X88000) and electron diffraction pattern (b) of the composition graded film layered by double-graded process.

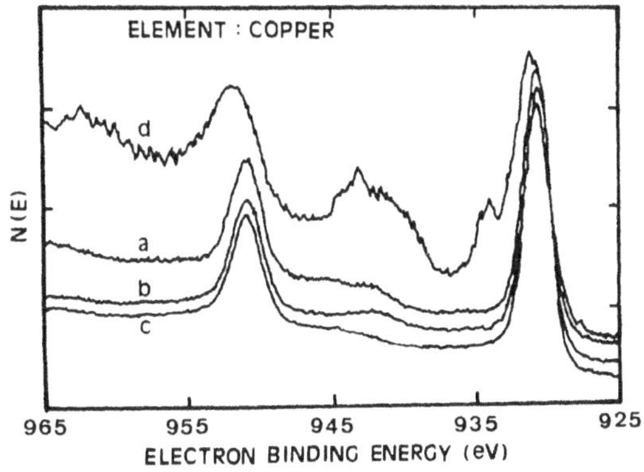

Figure 12. XPS spectra of the outer surface of the composition graded film layered by double-graded process: curve a, surface; curve b, 1 min sputter; curve c, 5 min sputter; curve d, oxidized copper surface.

micrographs (Fig. 11) show the very compact and random distribution of fine copper crystalline particles as a major component with carbon contaminants. The spotty electron diffraction rings indicate the existence of a number of microcrystals in the film. Two distinct particle sizes (approximately, 600 and 100 Å) result from the process procedures. The compactly distributed fine particles come from composition-graded step and the larger particles come from final sputtering step of which energy level of plasma is a lot higher.

The chemical structure of composition graded film was studied mainly by employing a comprehensive XPS examination. A composition graded film deposited by double-graded process onto FRP substrate was used as a sample for XPS measurements. This film was 1500 Å thick and had a electrical resistivity of 3.9 x 10^{-5} ohm.cm. The XPS spectra and chemical composition of the composition graded film as deposited (curve a) and after 1 min. and 5 min. sputtering (curve b and c) are shown in Fig. 12 and Table VI, respectively. The XPS spectra were compared with the data in the literature[12] according to the binding energies. The corresponding binding energies are 932.4 eV (copper), 932.2 eV (Cu_2O) and 933.5 and 943.7 eV (CuO). The figure shows that there is very trace amount of copper oxide (CuO) only at the surface. This probably resulted

Table VI. The Surface Chemical Analysis of Composition Graded Film Layered by the Double-Graded Process after Extended Time (48 hours) Exposure to Atmosphere.

Position	Copper	Carbon	Oxygen
0 min sputter	43.10	35.60	21.20
1 min sputter	93.10	4.10	2.60
5 min sputter	93.50	5.70	0.70

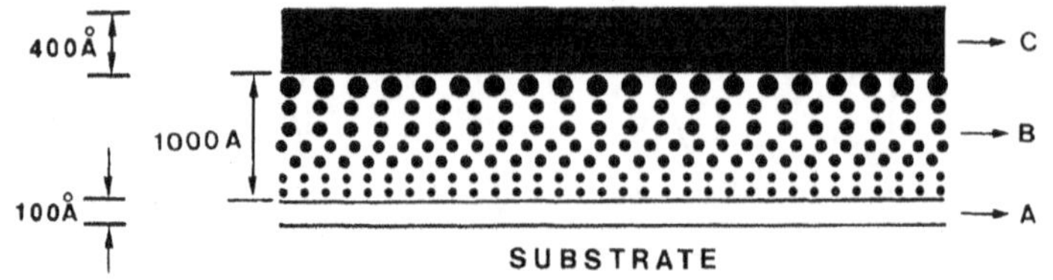

Figure 13. The proposed sectional model of a composition graded film layered by double-graded process. A: Pure CH_4 plasma polymer layer. B: Composition graded layer of CH_4 plasma polymer and copper. C: Sputtered copper layer with carbon contamination.

by reaction with the air due to the atmospheric exposure of sample. No copper oxides are detected in the bulk of film. The negligible amount of copper oxide and the high oxygen content at the surface suggest that most of the oxygen absorbed from the air exist in the other forms such as carbon monoxide or carbon dioxide. The general composition change over the thickness of composition graded film was already shown in Fig. 7. The proposed sectional model of composition graded film layered by double-graded process is shown in Fig. 13.

The stability of composition graded film seems to be good because the adhesion value and chemical analysis data by XPS do not show any significant changes after aging 2 months in a desiccator.

SUMMARY AND CONCLUSION

Composition graded films, ranging chemically from organic to metallic through their thicknesses, generated by a plasma polymerization technique can be used as an intermediate layer for metallizing nonconducting substrate. Strong and durable organic to organic chemical bond is established between the substrate and the initial layer while allowing a metallic bond between the top of the film and desired metallizing material. The gradient is established through gradual change in structure and chemical composition, which results by changing the deposition parameters during film growth. The results of these studies may be summarized as follows.

1) Electroless plating on the polymeric substrate can be performed through vacuum technological activation by the sputter deposition of electrode material with simultaneous plasma polymerization.

2) Composition graded composite film of plasma polymer and metal as a preplate for electroplating can be layered by control of the discharge power and gas flow during the plasma polymerization process.

3) Surface conditioning by plasma pretreatment with oxygen and hydrogen is a crucial factor in obtaining very strong organic to organic bonding at the interface between the substrate and composite plasma polymer film as a preplate for metallization.

4) The strongest and most durable film can be layered by establishing almost 100% organic to organic chemical bonding through the double-graded process with plasma pretreatment.

5) The shear strength of product coating metallized by employing composition graded film is dependent on the extent and strength of chemical bonding between the substrate surface and plasma polymer film, not on those of metallic bonding between the sputtered copper and metallizing material, and is different for various substrates because of difference in the chemical affinity or adhesion characteristics of a certain plasma polymer onto a certain substrate.

6) The composition graded film as a preplate for metallization is stable.

ACKNOWLEDGMENT

The support of this research by the National Science Foundation grant no. MPS-8601205 is gratefully acknowledged.

REFERENCES

1. Robert W. Berry, Peter M. Hall, and Murray T. Harris, "Thin Film Technology", Chap. 3, D. Van Nostrand Company, Princeton, New Jersey, 1968.
2. K. L. Mittal, J. Vac. Sci. Technol., 13, 19 (1976).
3. P. Benjamin and C. Weaver, Proc. Royal Soc., CCLXI, 516-531 (1961).
4. James H. Lindsay and Joseph Lasala, Plating and Surface Finishing, 54 (July, 1985).
5. A. Celerier and J. Machet, Thin Solid Films, 148, 323 (1987).
6. N. Inagaki and H. Yasuda, J. Adhesion, 13, 201 (1982).
7. P. J. Dynes and D. H. Kaelble, J. Macromol. Sci.-Chem, A10(3), 535-557 (1976).
8. D. L. Cho, Y.-S. Yeh, and H. Yasuda, J. Vac. Sci. Technol., submitted.
9. H. K. Yasuda, A. K. Sharma, E. B. Hale, and W. J. James, J. Adhesion, 13, 269 (1982).
10. D. L. Cho, Ph.D. Dissertation, University of Missouri-Rolla, 1986.
11. James Peeling and Gregory Courval, J. Polym. Sci., Polym. Chem. Ed., 22, 419-428 (1984).
12. Yoshihito Osada and Katsuyuki Yamada, Thin Solid Films, 151, 71-86 (1987).

METALLIZATION OF PLASTICS WITH RESISTANCE HEATED SOURCES

Kevin M. Anetsberger

Midwest Tungsten Service, Inc.
7950 South Madison Street
Burr Ridge, Illinois 60521

Since the advent of the use of resistance heated sources for the metallization of plastic substrates in the 1940's, both the sources and the process have become increasingly sophisticated. Evaporative sources have evolved from single strand to multiple strand wires, then to coil and basket forms. Resistance heated crucibles and boats were developed for more difficult evaporative materials. The use of intermetallics made possible the coating of plastic films.

This paper discusses the parameters for metallization of plastics using resistance heated sources and compares this technique to other available plating processes such as: induction heated deposition, ion plating, electron beam deposition, sputtering, and electroplating.

INTRODUCTION

Metallization using resistance heated thermal sources for evaporation is the oldest method of vacuum deposition, yet even today, it remains one of the most pervasive techniques for depositing thin films on a wide variety of materials--especially plastics. Table I presents a partial list of products currently metallized using this technique. The simplicity, low capital equipment cost, versatility, and safety of this process suggest that vacuum metallization using resistance heated sources will continue to be a popular technique for plating plastics for some time.

Michael Faraday performed the first metallization experiments using an electric current to vaporize a wire in a nitrogen atmosphere (Figure 1). Although the nitrogen was inert, its presence severely limited the mean free path, or the distance the vaporized metal could travel without encountering a gas molecule. The metal depositions which resulted from these experiments were generally non-adherent. Pohl and Pringsheim repeated the experiment in 1912 substituting a vacuum for the nitrogen (Figure 2).[1] The absence of atmosphere decreased the temperature needed to vaporize the metal and allowed the vapor to travel farther without being involved in energy robbing collisions with gas molecules. The resulting films were heavier and more adherent.

The need for improved optical equipment during World War II created

Table I. Uses of Metallization.

Decorative	Optical
Auto trim/Medallions	Architectural glass
Casket hardware	Camera Lenses and Filters
Christmas decorations	Compact Discs
Cosmetic caps	Headlamps
Glassware	Jet Plane Canopies
Labels	Mirrors - 1st and 2nd Surface
Light louvers	Road Reflectors
Trophies	Sunglasses
Signage	Telescopes
Electrical	**Physical**
Audio/Video Tape	Solid Lubricants
Capacitors	Tool Hardening
Electrical contacts	
EMI/RFI Shielding	
Floppy Discs	
Integrated Circuits	
Resistors	
Solar Cells	

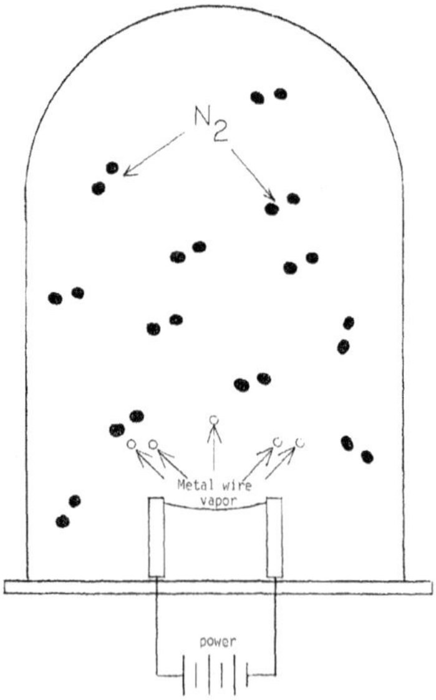

Figure 1. Faraday Deposition Experiment.

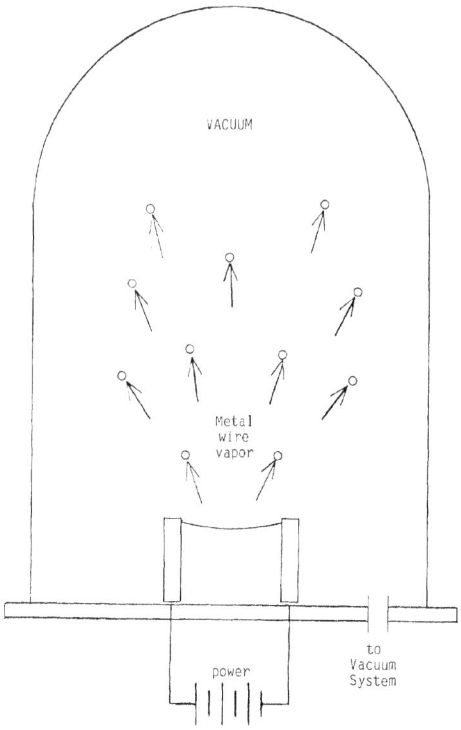

Figure 2. Pohl and Pringsheim deposition experiment.

a need for large scale deposition of anti-reflective coatings of magnesium fluoride onto lenses (Figure 3). To meet these demands, laboratory experiments had to be translated into processes capable of large scale production.[2]

The postwar era saw a dramatic increase in the development of new plastics and plastic products (Table II). This increase in the use of plastics provided an opportunity for vacuum metallizers to enter the commercial sector. Vacuum metallizing plastic materials helped them gain acceptance in the marketplace at a time when they were considered a poor substitute for metal products. This integral relationship helped both the plastic and vacuum metallizing industries to grow and prosper. Since that time, we have seen an increase in the demand for metallized products and in the sophistication of deposition systems.

PARAMETERS IN METALLIZATION BY RESISTANCE HEATED SOURCES

Metallization of plastics using resistance heated sources is a process of evaporation and condensation. As current flows through the source material, the resistance of the source to this current flow generates heat. This heat is used to vaporize the deposition material. The evaporant then radiates out from the resistance heated source in a straight line until it encounters a surface upon which it can cool and condense. Since the vapor radiates out in a straight line only, any surface to be coated must be in direct line of sight with the resistance heated source at some point during the evaporation. The vacuum metallizing system must therefore perform three functions: it must produce a vacuum adequate for deposition, it must provide the energy necessary for vaporization, and it must manipulate the

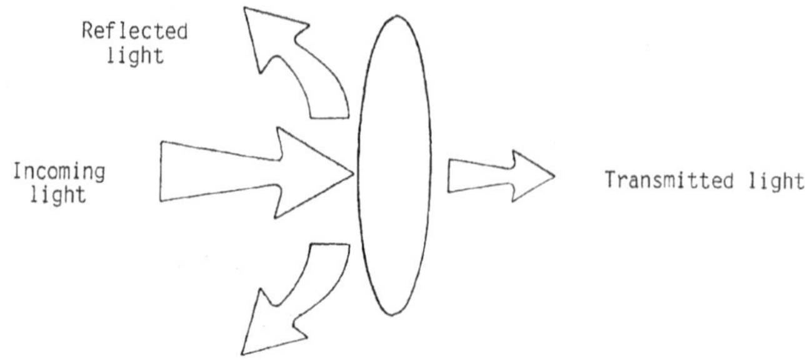

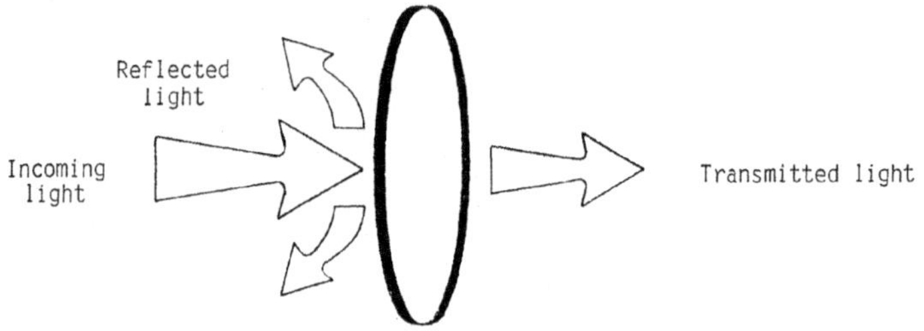

Figure 3. Effects of lens coating with magnesium fluoride.

Table II. Chronology of Plastics (taken from "Industrial Plastics" by Terry Richardson).

Date	Material	Date	Material
1868	Cellulose nitrate	1949	Allylic
1909	Phenol-formaldehyde	1954	Polyurethane or urethane
1909	Cold molded	1956	Acetal
1919	Casein	1957	Polypropylene
1926	Alkyd	1957	Polycarbonate
1926	Analine-formaldehyde	1959	Chlorinated polyether
1927	Cellulose acetate	1962	Phenoxy
1927	Polyvinyl chloride	1962	Polyallomer
1929	Urea-formaldehyde	1964	Ionomer
1935	Ethyl cellulose	1964	Polyphenylene oxide
1936	Acrylic	1964	Polyimide
1936	Polyvinyl acetate	1964	Ethylene-vinyl acetate
1938	Cellulose acetate butyrate	1965	Parylene
1938	Polystyrene or styrene	1965	Polysulfone
1938	Nylon (polyamide)	1965	Polymethylpentene
1938	Polyvinyl acetal	1970	Poly(amide-imide)
1939	Polyvinylidene chloride	1970	Thermoplastic polyester
1939	Melamine-formaldehyde	1972	Thermoplastic polyimides
1942	Polyester	1972	Perfluoroalkoxy
1942	Polyethylene	1972	Polyaryl ether
1943	Fluorocarbon	1973	Polyethersulfone
1943	Silicone	1974	Aromatic polyesters
1945	Cellulose propionate	1974	Polybutylene
1947	Epoxy	1975	Nitrile barrier resins
1948	Acrylonitrile-butadiene-styrene	1976	Polyphenylsulfone

plastic parts in a manner which insures complete and adequate coverage. The vacuum metallizing system performs its functions using three systems: the vacuum system, the electrical system, and the fixturing system.

The vacuum system consists of two components: the vacuum chamber and the vacuum pumps. The vacuum chamber must be able to contain the evaporative sources, the plastic parts to be coated, and their fixtures. The chamber must also be able to maintain a high degree of vacuum for extended periods of time. The chamber is usually a cylindrical steel tank with a diameter of 48 inches or more. Laboratory units may have smaller chambers or glass bell jars to contain the vacuum. Most often, the chamber is mounted horizontally, with a door at one end for loading and unloading the plastic parts. The vacuum pumps are usually located at the opposite end of the chamber. In order to allow for a mean free path long enough for the evaporant to easily reach the plastic parts, a vacuum of 10^{-4} to 10^{-5} torr must be achieved.[3] As a matter of practicality, the pumping system should reach this vacuum level in about twenty minutes. This is the approximate amount of time needed to unload and reload a carriage of plastic parts and to recharge the resistance heated sources with evaporant. Any time longer than this would have workers standing idle as the chamber finished pumping. For metallizing with resistance heated sources, a mechanical vacuum pump is used to bring the chamber to a pressure of about 10^{-3} torr.[4] At this point, and oil diffusion pump is used to bring the chamber to firing pressures.[5]

The electrical energy needed to heat the evaporative sources is provided by a DC power supply. For most metallizers a power supply of between 20 and 55 kVA is installed. The actual power requirements can vary depending on the resistance of the sources, the number of sources used, and the temperature the evaporant vaporizes at. It is important to check with the chamber vendor to insure adequate power is available to evaporate the materials your application requires.

The fixturing system should also be mentioned. A chamber may contain one of three types of fixturing systems. A stationary system, which simply holds the parts in a fixed position during firing, is often used for coating one side only of large planar surfaces, such as 4 by 8 sheets of acrylic. A simple rotary system will move a large number of smaller parts around the centrally located evaporative sources, again coating only the side facing toward center. The most popular system is a rotating planetary system which rotates each rack of parts as the entire carriage is rotating. This planetary motion insures that all sides of the part will face the sources at some point and be coated. Most fixturing systems allow the plastic parts to be lacquered and oven dried on the same racks they are metallized on to reduce material handling.

It is now time to examine the resistance heated source in detail. The resistance heated source relies on the thermal energy generated by a flowing current to generate the heat needed for vaporization of the evaporant (Figure 4). In order to succeed, the resistance heated source must meet several qualifications. First, it must be an electrical conductor with a resistivity which allows it to reach deposition temperatures with a reasonable current flow. Second, the source material must have a melting point which is higher than and a vapor pressure which is lower than the material to be evaporated. This will insure that the source material will not be deposited along with the evaporant. Third, the source should be chemically inert with respect to the evaporant or, at the very least, should form no by-product which would degrade the properties of the film being deposited. The source must also be able to physically contain the evaporant in its solid and/or liquid phases. It should be as economical and easy to handle as possible. Even considering these restrictions, there

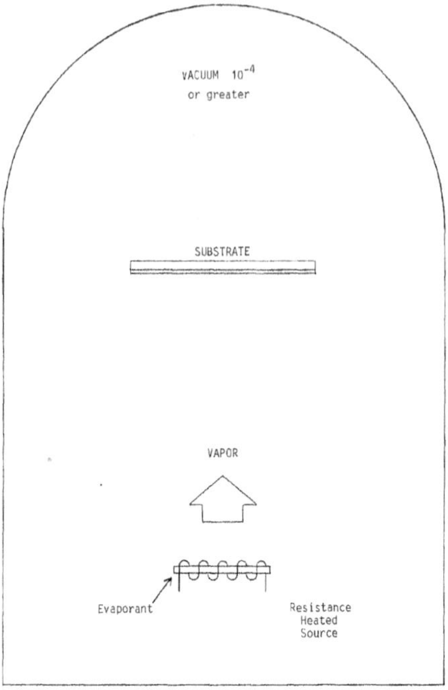

Figure 4. Resistance heated deposition schematic.

remains a wide variety of materials and configurations for the resistance heated source. At present, there are two major source compositions--the refractory metals and the intermetallics, and two major source configurations--coils (filaments) and boats.

The refractory metals, tungsten, molybdenum, and tantalum, are all elemental metals which have high melting points and low vapor pressures. Wherever possible, tungsten is used because it has the highest melting point. Molybdenum and tantalum are used when tungsten would react with the evaporant used or when constructing complex sources. Tungsten is the most brittle of the three metals and is not easily welded or bent through large angles. Tungsten can be formed into rod, wire or sheet stock rather easily, however.

The original tungsten evaporative source was a single strand of wire from which small bits of the evaporant were suspended. Often as the source was heated, the evaporant would simply melt and fall off. To improve the wettability and increase the surface area from which evaporant could be vaporized, several tungsten wires were stranded together. This stranding gave the wire small valleys along which the evaporant could flow. Today, stranded wire represents a majority of the tungsten used for resistance heated evaporative sources. Fabrication of stranded tungsten wire into coil or basket shapes has made it easier to load evaporants and has facilitated the use of evaporants available only in pellet or granular forms (Figure 5).

For evaporants such as powders, which cannot be contained in a coil or basket, a "boat" source is manufactured from refractory sheet stock.

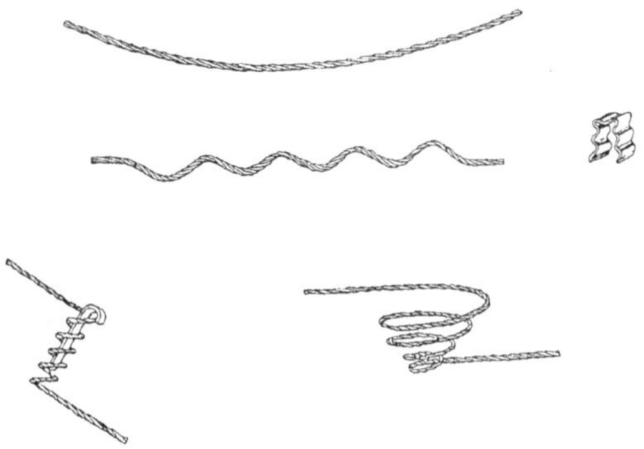

Figure 5. Resistance heated wire sources.

Figure 6. Refractory metal boat source.

Boats are also used when deposition requires heavy charges of evaporant which might fall or drip through a coil or basket. Tungsten is used for simple stamped boat sources (Figure 6). Molybdenum or tantalum are used to manufacture more complicated sources because they are more ductile. Certain evaporants may spatter or spit when heated. To prevent the spatters from disfiguring the plastic substrate, boats can be used which have covers and baffles which prevent a direct line of sight between the solid evaporant and the parts to be coated. For example, silicon monoxide is a hard glass-like material which is often used as a protective top-coating.[7] Since the silicon monoxide spatters upon heating, it is commonly evaporated from a boat like the one shown in Figure 7. Because this source requires both sharp bends and welded joints, tantalum sheet is used for its manufacture.

Although boat sources can solve problems which coils present, boat sources themselves have limitations. Since evaporants must sit inside or upon the boat, the boat itself can block the path of the vapor. In other words, the evaporation will be 180 degrees or less. Filaments, on the other hand, are open structures which allow the evaporant to radiate out in all directions, a full 360 degrees. Boats often are more massive than filaments and, therefore, require more heat to reach evaporation temperatures. Since boats are made from sheet stock, they have large surface areas which radiate heat during deposition. A great deal of this surface area is never in contact with evaporant (for instance, the underside of the boat) so the heat given off by it is simply radiated out as waste heat. Under certain circumstances this heat could be enough to cause warping of plastic parts.

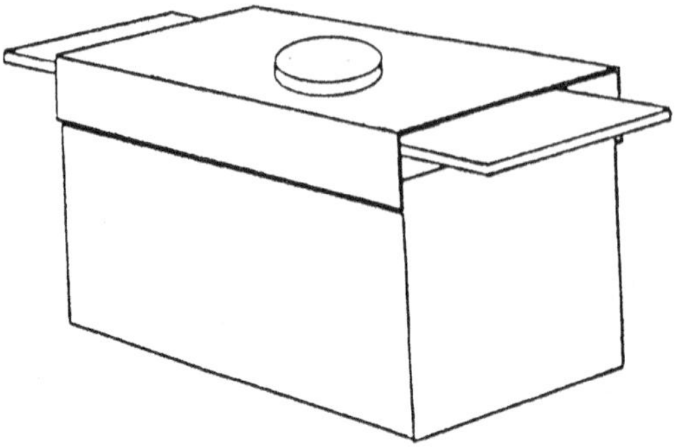

Figure 7. Tantalum source for silicon monoxide evaporation.

In instances where a boat will not hold a sufficient quantity of material, or it will react with the evaporant, a crucible or liner of carbon, aluminum oxide, or ceramic, may be used to contain the evaporant (Figure 8). This crucible is inserted into a filament which will heat the crucible and its contents.

Filaments, boats, and crucibles must be recharged with evaporant after each use. Also, plastic parts must be removed after coating and replaced with new ones. Because of this, these operations are often referred to as batch metallizing, since a "batch" of parts is done all at once, and then preparations are made to coat the next batch.

Metallization can also be done as a semi-continuous operation, extending over long periods of time and depositing large quantities of evaporant. An excellent example would be the vacuum metallization of a roll of plastic film. Clearly, it would be impractical to coat a roll of film several feet at a time in a batch operation. It would be much more desirable to coat the entire roll of film during one vacuum cycle. To do this, evaporant must be replenished throughout the cycle as the film is exposed for coating. Figure 9 shows the details of such a set-up. To coat a large roll of plastic film will often take thirty minutes or longer. Any source we choose must be able to withstand the effects of high heat and contact with the evaporant for this extended amount of time.

For this application, an intermetallic boat source is used. The boat is made from a composition of boron nitride, titanium diboride, and aluminum nitride, mixed to obtain a specific resistivity. The composite is vacuum hot pressed into blocks which are then machined to size. Cavities to hold evaporant may also be milled or ground into the

Figure 8. Crucible.

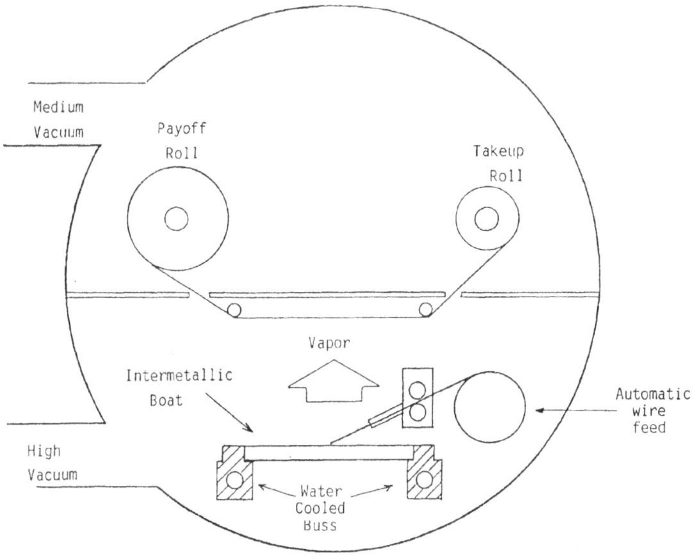

Figure 9. Film coater using intermetallic sources.

intermetallic boat (Figure 10). The boat can be resistance heated for extended periods of time as evaporant is continuously fed to it throughout the vacuum cycle. Intermetallic boats have several special considerations. They are more expensive than other resistance heated sources. They are easily broken if dropped or if exposed to rapid changes in temperature. Because intermetallics expand a large amount upon heating, boat holders must be spring loaded. Intermetallics also generate a large amount of waste heat. This usually requires the boat holders to be water cooled. Intermetallic boats require more current to heat them than other sources. The resistivity must be specified when ordering since several compositions are available.[8]

When choosing an evaporative source, the material to be deposited must be considered also. For resistance heated source metallization the form the evaporant is in, the quantity to be deposited, and its reactivity usually dictate which source should be used. Evaporants made of alloys must be treated with care. Upon heating, these alloys may separate according to vapor pressure into their component metals. Also deposition of tungsten, molybdenum, and tantalum poses a problem since these materials are being used as sources.

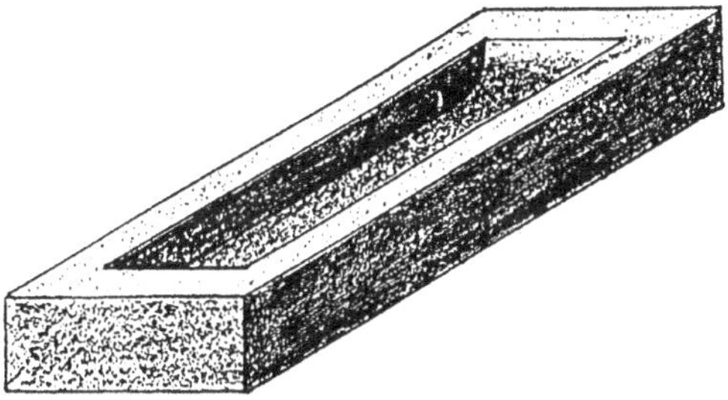

Figure 10. Intermetallic evaporation source.

An evaluation of the plastic to be coated is appropriate also. In many cases, plastics cannot be metallized as molded. Many molders use mold releases to aid in the manufacture of the plastic parts. These mold releases will prevent adhesion of metallization to the plastic and should be avoided at all costs. Some plastics themselves contain internal lubricants or plasticizers which will be drawn to the surface of the plastic under vacuum. These can also cause poor adhesion or discoloration of the metallized film. Plastics in this category include polyolefins, vinyls and butyrates.[9] Plastic surfaces are also often not smooth enough to give a highly reflective surface when metallized.

Often deposition is not made directly onto the plastic itself, but rather onto an organic coating called a basecoat which has been applied to the plastic prior to metallization. This basecoat serves three functions. It seals the plastic surface, it improves adhesion to the metal layer, and it provides brilliance by smoothing out the surface. After metallization, a topcoating is often applied. The topcoat serves to protect the metallized surface from abrasion and corrosion and to provide gloss and depth to the appearance.[10]

The use of these coatings poses two potential problems. The solvents contained in some coatings can attack the plastic substrate. Coating manufacturers will usually provide guidelines for the use of their coatings but it is always wise to be aware of the solvents which may harm a particular plastic part. Table III is a chart which lists commonly metallized plastics according to their vulnerability to solvent attack. Research into water-based topcoats and basecoats may someday eliminate the solvent hazards these coatings now present. After coating, plastic parts are often oven dried. Care should be taken to ensure that the temperature in these ovens never exceeds the heat distortion point of the plastic substrate. Failure to do so could result in warpage or melting of the plastic part. Table IV lists the heat distortion point for commonly metallized thermoplastics. Thermosetting plastics are not affected by bake-out temperatures.

COMPARISON WITH OTHER METHODS

In order to evaluate metallization using resistance heated sources as a potential coating process, it should be compared to other existing technologies. Below several alternative techniques are summarized and compared to vacuum metallizing using resistance heated sources for the plating of plastics.

Thicker, more adherent films, metallized coatings can be obtained using the glow discharge technique to treat the plastic prior to metallization (Figure 11). By electrically exciting a process gas (usually argon) which has been bled into the evacuated chamber, a plasma is created. This plasma will act upon the surface of the plastic to a

Table III. Solubility of Various Plastics.

Aliphatics	Alcohols	Aromatics	Ketones	Esters	Chlorinated	Resistant
Grease Oil	Cellulosics	Acrylics Polycarbonate Styrene	ABS Vinyl		Epoxy Polyethelyne (over 80°) Polypropylene Polysulfone	Nylon Acetals Polyurethane Teflon

NOTE: If a plastic dissolves in a solvent listed in a given column, it will also dissolve in all solvents to the right of it on the chart. Thermosetting plastics are solvent resistant except for alkyds, polyesters, and silicones.

Table IV. Heat Distortion Temperature (°C) for Various Plastics.

ABS	71 - 93
Acetal	90 - 98
Acrylics	54 - 110
Cellulosic	46 - 104
Nylon	82 - 149
Polyethylene	79 - 121
Polypropylene	107 - 149
Polycarbonate	121
Polysulfone	149
Polyurethane	88
Styrene	65 - 77
Teflon	287
Vinyl	60 - 79

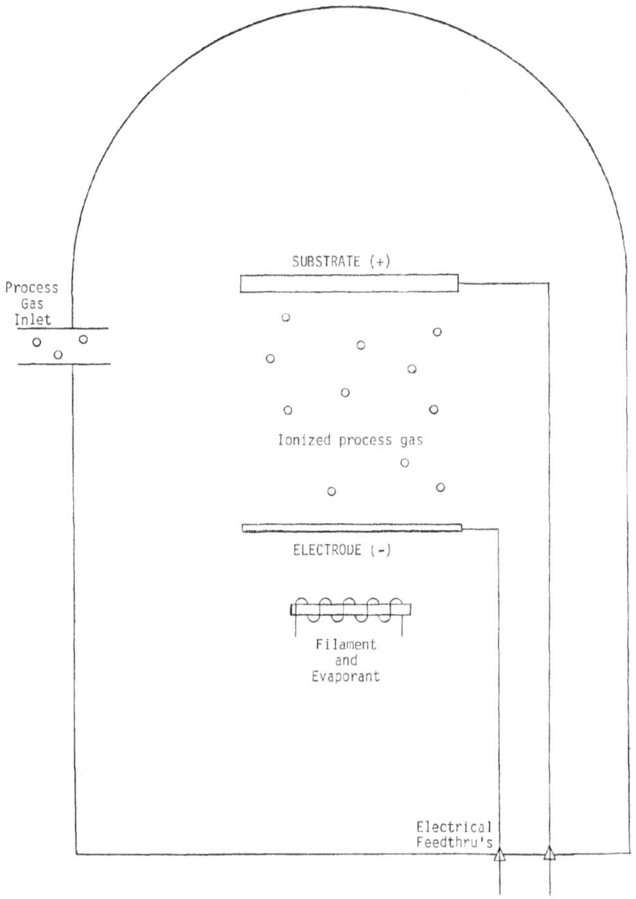

Figure 11. Glow discharge and ion plating schematic.

depth of 1,000 angstroms or less. The plasma action will remove traces of oil and other impurities from the surface of the plastic. The plasma will also create free radicals on the surface which will increase the wettability and surface energy of the plastic surface. All these changes work to improve the adhesion of the metal coating to the substrate. The properties of the part as a whole are unaffected since the changes created are a surface phenomenon only.[11] After glow discharge treatment, the plastic is immediately metallized without exposure to the atmosphere.

A variation of the glow discharge technique is known as ion plating. In ion plating, the plasma is not turned off and the process gas is not removed prior to plating. Instead, the resistance heated sources are used to vaporize the evaporant during the glow discharge process. Since the gas molecules are at such high energy levels, collisions with them by the deposition vapor do not rob the vapor of energy. In fact the opposite is true. The plasma affects deposition in two ways. First, it can redirect the normal straight line path of the evaporant. This can allow the evaporant to reach areas which may not lie directly in the line of sight of the evaporative sources. Secondly, the plasma can impart

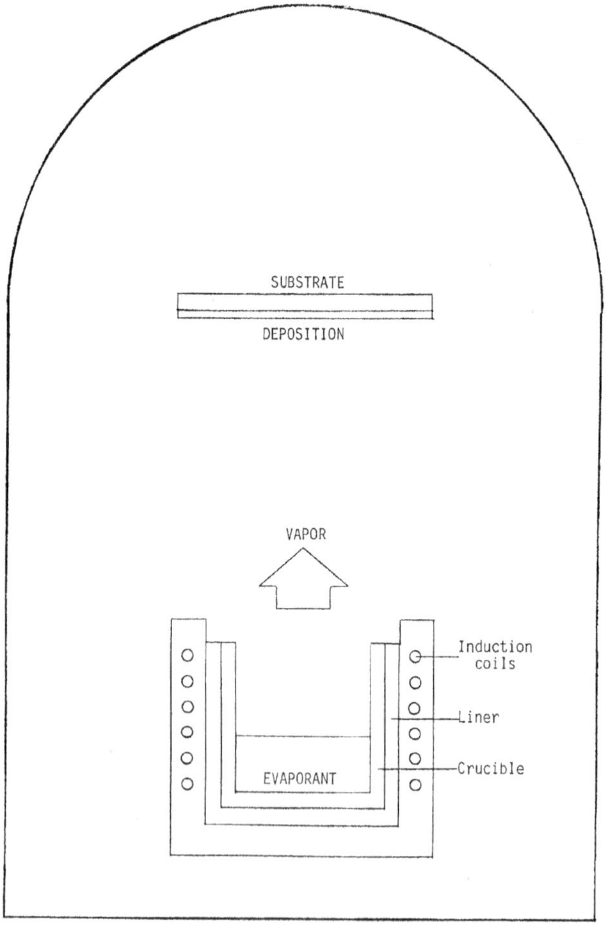

Figure 12. Induction heated deposition process schematic.

additional energy to the vapor during its travel. This additional energy provides improved adhesion between the deposition and the plastic substrate.[12]

Induction Heated Deposition

For induction heated deposition, the evaporant is placed inside a crucible. This crucible may contain conductive bands to aid in heating the evaporant. The crucible is placed inside a primary current coil which generates an electric field. A heat producing electric current is generated by induction either directly in the evaporant or in the crucible heater bands if the evaporant is a non-conductor (Figure 12).[13]

Advantages

- Results are easily reproduced.
- A wide range of currents and voltages can be used.
- Crucible sources are easy to change.

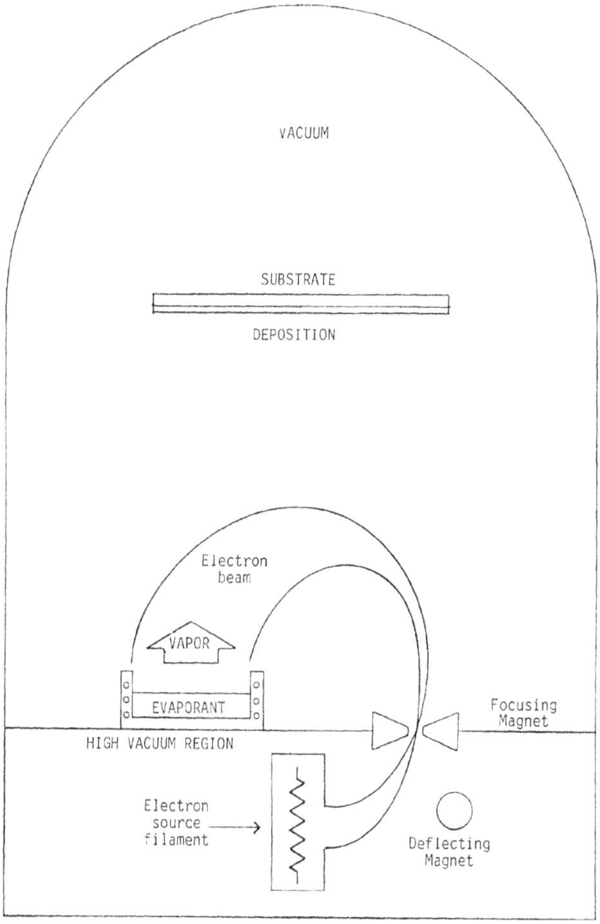

Figure 13. Electron beam deposition schematic.

Disadvantages

- All the evaporant melts at once. Large amounts of excess heat are generated which often require water cooling.
- It is easy to overpower sources and cause spitting of the evaporant.
- Evaporant is directed upward only.

Electron Beam Deposition

In electron beam deposition, a beam of electrons generated by thermionic emission from a tungsten filament is focused on the evaporant. The kinetic energy of the beam is translated into thermal energy upon impact and vaporization occurs (Figure 13).[14]

Advantages

- High energy efficiency.
- High temperatures can be achieved. Electron beams can be used to evaporate the refractory metals.
- Evaporants can be continuously fed.

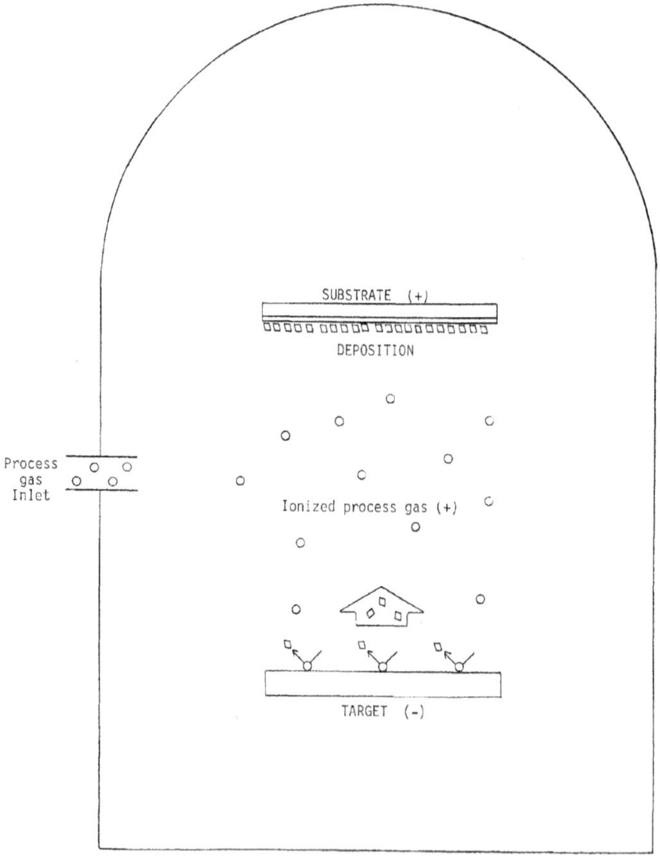

Figure 14. Sputtering deposition process schematic.

Disadvantages

- High currents are needed to generate and control the beam.
- Higher capital equipment costs.
- Lower pressures are required to maintain stable beams.

Sputter Deposition (Sputtering)

Ions from a process gas are accelerated toward a target made of material to be deposited. The ions knock off molecules of the target material which are then deposited on the substrate (Figure 14). Many variations exist.[15]

Advantages

- High impact energy of evaporant yields good adhesion to plastic substrate.
- A wide variety of materials can be deposited including refractory metals, metal alloys, and ceramics.

Disadvantages

- Slow deposition rates.
- High heating of substrate can occur.
- High capital equipment costs.

In addition to other vacuum techniques, metallization using resistance heated sources can also be compared to electroplating on plastics. Electroplating is an electrochemical process. As a non-conductor, plastic must undergo several treatments in order to be plated. Although thirty or more steps may be required, the process can be summarized as follows:

- Cleaning to remove surface contaminants.
- Etching to improve metal to plastic adhesion.
- Electroless deposition chemically reduces metal onto plastic surface to make it conductive.
- Electroplating undercoatings and final finish.[16]

Advantages

- Heavy coatings can be applied.

Disadvantages

- Hazardous chemicals are necessary for deposition.
- Films may contain chemical impurities or gaseous inclusions.
- Chemical attack on plastics is possible during plating.

SUMMARY

To summarize, for many applications metallization of plastics is desirable. A variety of techniques exists to perform the metallization. Metallization using resistance heated sources is an excellent choice for many applications. Low capital equipment costs, high deposition rates, and safety are several reasons why. The techniques are not harmful or degrading to the plastic substrate. Although unsuitable for many high-technology applications because of its inability to deposit many alloys intact, it is still an excellent method for depositing many of the lower melting point metals.

REFERENCES

1. Theodore H. Crane, High vacuum vapor coating, proceedings of Society of Vacuum Coaters technical meeting, 1957, p. 1.

2. Chester Northrup and John Smith, Vacuum deposition of functional coatings, proceedings of Society of Vacuum Coaters technical meeting, 1957, p. 14.

3. John F. O'Hanlon, "A Users Guide to Vacuum Technology," p. 9, John Wiley and Sons, New York, 1980.

4. Ibid, pp. 159-163.

5. Ibid, pp. 189-191.

6. David K. Sugar, Vacuum metallizer maintenance and repair, proceedings of Society of Vacuum Coaters technical meeting, 1972, pp. 11-12.

7. Earl H. Blevis and R. D. Mathis, "Silicon Monoxide--Properties and Evaporation Techniques," pp. 3-6, R.D. Mathis Company, Long Beach, California.

8. John B. Hedge, Continuous vacuum metallizing with intermetallic composite sources, proceedings of Society of Vacuum Coaters technical meeting, 1971, pp. 37-40.

9. Terry A. Richardson, "Industrial Plastics: Theory and Application," pp. 207, 220, 174, South-Western Publishing Co., Cincinnati, Ohio, 1983.

10. Ibid, p. 447.

11. P.W. Rose and E.M. Liston, Plastic Engineering, 41-45 (October 1985).

12. D.L. Chambers, D.C. Carmichael, and T.C. Wan, Deposition of metal and alloy coatings by electron-beam ion plating, Sputtering and Ion Plating, held at Lewis Research Center, pp. 90-91, 1972.

13. Emmett R. Anderson, Induction heated vapor sources, proceedings of Society of Vacuum Coaters technical meeting, 1975, pp. 114-117.

14. H. Harker, Electron beam evaporation, proceedings of Society of Vacuum Coaters technical meeting, 1974, pp. 22-25.

15. J. Frados, Editor, "Plastics Engineering Handbook," pp. 745-749, Van Nostrand Reinhold Company, New York, 1976.

HIGH RATE REACTIVE SPUTTERING ONTO FLEXIBLE POLYMER SHEET

R.P. Howson, A.G. Spencer, K. Oka and R.W. Lewin

Department of Physics
University of Technology
Loughborough, Leicestershire
LE11 3TU, England, UK

The paper considers the use of a d.c. planar magnetron, used with optical emission control of the reactive gases and a controlled magnetic field, to allow plasma to be directed onto the growing film. It is demonstrated that these techniques lead to stable processes allowing the high rate production of compound films, with precisely controlled stoichiometry and structure, onto flexible polymer substrates.

Detailed examples are given for the production of indium oxide and titanium oxide films by the reactive sputtering of the metal in an argon/oxygen atmosphere. Film resistivities of around 4×10^{-6} ohm.m were achieved with indium and indium/tin oxides during high rate production onto flexible P.E.T with films which had a high visible transmittance.

Refractive indices of up to 2.55, measured at 633nm, could be obtained with TiO_2 film made with self biasing above -20V.

INTRODUCTION

Modern polymers are highly desirable materials, having properties which include high strength with low weight and often with high visual transparency. The production of complex materials in thin film form onto polymer foils is complicated by the inability of these materials to tolerate high temperatures. Energy is required by the depositing film in order to produce the desired compound or structure, which cannot be given from substrate temperature. In order to produce stoichiometric compounds or precise alloys the deposition process must be accurately controllable, and energy supplied simultaneously to activate reactions for molecular or structural arrangement.

Vacuum techniques offer a means of creating refractory materials onto polymeric substrates at low temperatures and have the particular advantage that the energy of the depositing species can be controlled through the acceleration of charged particles in gas discharges. This gives substrate cleaning and also preparation of the surface for the growing film, whose properties can also be further modified.

The coating of a flexible polymer wound over a drum in front of a deposition system in a vacuum offers the following advantages which are not seen in other operational arrangements.

(1) High-rate equilibrium processes are possible. Lead-in times which eliminate shuttering disruption allow better cleaning in semi-sealed chambers.

(2) Rapid deposition gives pure deposits as the ratio of arrival rates of depositing species and residual gas impurities is high.

(3) High experimental productivity is obtained; many samples can be produced on one roll in one evacuation. In-situ monitoring facilitates parameter optimization in a short time.

(4) The results are applicable to commercial processes.

It remains however that with the high cost of vacuum equipment the process should have a high rate to give the low costs. This is necessary for products which have to compete with those on more expensive substrates that are not so easily handled but more easily coated. The starting materials have to be of low cost and be available on the scale appropriate for large area coating. A process which meets these requirements is reactive planar magnetron sputtering. A basic system is shown in Figure 1, this allows isolation and bias of the coating drum to give ion plating.

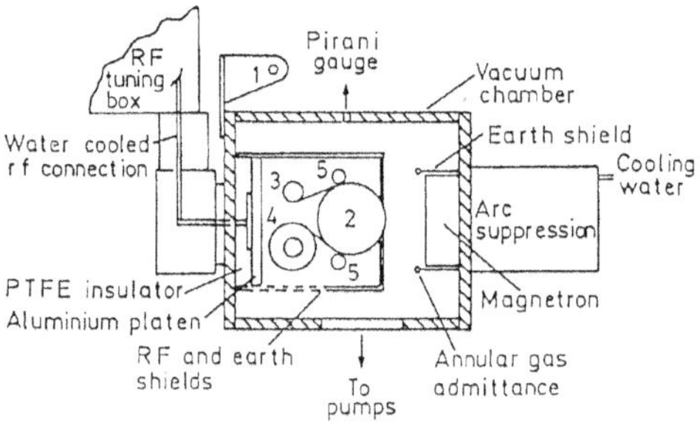

Figure 1. Apparatus for the coating of sheet plastic transferred from roll to roll inside a vacuum using a planar magnetron source and an insulated substrate drum.[1]

REACTIVE PLANAR MAGNETON SPUTTERING

The precision control of rate, combined with a high sputtering-rate capability and operation at a relatively high reactive gas pressure, has led us to the system of d.c. planar magnetron sputtering of the metal or alloy in an atmosphere of argon and oxygen. Ion bombardment of the substrate and growing film surface can lead to better adhesion, film structure and a greater utilisation of the oxygen gas provided for reaction giving the possibility of higher rates.[2,3]

The advantages of planar magnetron sputtering in such a system can be enumerated as follows:

(1) A high rate of sputtering is obtained because of the confinement of the plasma close to the target surface. The limit is determined by a factor involving the thermal conductivity of the target, the efficiency of the water cooling, the melting point of the sputtering material and its sputtering yield.

(2) The heating of the substrate by stray electrons is low. In many cases this is the limiting factor for high-rate conventional sputtering.

(3) Sputtering can take place at a low gas pressure $\gtrsim$ 0.1Pa because of the increased path-length of electrons within the plasma and prevention of their escape. This is achieved by the crossed electric and magnetic fields and the closure of the $\bar{E} \times \bar{B}$ drift path.

When this sputtering source is used in the presence of a reactive gas to create a compound, further advantages ensue:

(1) The rate of deposition can easily and accurately be controlled to allow balanced processes to be maintained. In particular, the high surface-ion current density allows sputtering of the metal from an exposed metal surface even in a reactive gas, with the compound formation taking place on the substrate surface.

(2) The gas pressures required for sputtering and for producing reactive processes are compatible. (Typically of the order of 1Pa).

(3) Alloy metal targets can easily be sputtered to give mixed metal oxides of similar metal ratios.

If an additional field is applied to the substrate surface a discharge can be created which bombards the growing film surface with ions. The advantages of 'ion plating', used in conjunction with sputtering, are:

(1) The reactive gas is activated (ions, metastables, excited species) and the ions driven towards the substrate surface by the ion plating field.

(2) The addition of energy to the growing film surface by reactive and neutral gas atoms results in control of structure and nucleation phenomena without using high-temperature substrates.

(3) The gas pressures needed for the sputtering and ion-plating discharges are compatible. (Also $\cong$ 1Pa).

Problems however remain with this system. Ion bombardment of films on insulating substrates requires r.f. discharges or neutralised ion beams which are difficult and expensive to arrange when required to match the potential deposition rate of a system coating a roll of polymer. The reaction occurring on the substrate also takes place on the sputtering target to form an insulating compound. This leads to arcing, a low sputtering rate, and generally an unstable system.

STABILITY IN REACTIVE SPUTTERING

Reactive sputtering generally uses an atomosphere of a mixture of the reactive gas and argon. The intention being for the argon to sputter the metal target, in the intense discharge at the cathode, before it reacts with the other component of the atmosphere. The sputtered material deposits onto the substrate, which also receives a bombardment of the reactive gas, and the compound is formed. However, the reaction at the cathode is important and can dominate the process and cause instability (Figure 2). Simple solutions include using a reacted target source with d.c., this results in a low rate of deposition, and more importantly a loss of control, and would increase the cost of commercial systems. Low rate d.c. deposition is also possible, here a stable system can be created and the properties of the film controlled to allow optimum films to be obtained.

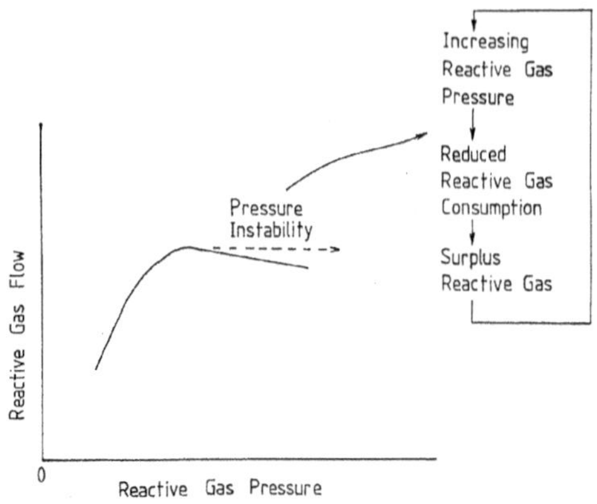

Figure 2. The mechanism that causes instability in reactive sputtering.[4]

In high rate reactive magnetron sputtering, the film deposition results in a substantial pumping rate of the reactive gas. When the depositing film is sub-stoichiometric and metal-rich, the film's consumption of the reactive gas is limited by the metal arrival rate. As the reactive gas pressure is increased, reaction products form on the target (it is 'poisoned') and the metal flux falls, leading to a decreasing consumption of the reactive gas. At pressures where a stoichiometric film is formed the consumption of the reactive gas by the film will be falling. In systems where the rate of pumping by the vacuum pumps is relatively low this can lead to an uncontrollable transition between a metallic and a poisoned target that makes the control of optimum deposition conditions difficult.

For stable production of a compound film we must have at the substrate a constant metal flux, a constant arrival rate of the reactive gas flow (i.e., a constant partial pressure) and a constant utilization of the reactive gas. Unfortunately, these parameters are not independent of each other. The metal flux depends on the state of the target which depends on the reactive gas pressure. Therefore, we must pay particular attention to the reactive gas.

The variation of the reactive gas consumption by the film against the reactive gas pressure is the result of chemical reactions on both the target and the substrate. Ignoring for now the reactive gas removed by the pumps we can consider the amount used up by the growing film.

At the target the surface is constantly being poisoned and sputtered clean. Where the poisoning rate exceeds the cleaning rate, a poisoned surface exists and increasing the reactive gas partial pressure then increases the proportion of the target that is poisoned. In most cases the poisoned surface sputters more slowly than the metallic surface and thus leads to a reduction in the metal flux with increasing reactive gas pressure. This continues until the target is completely poisoned and then the metal flux remains roughly constant. (Figure 3a).

At the substrate, at low reactive gas pressures, the formation of the compound is limited by the arrival rate and utilization of the reactive gas and so a metal-rich film is formed. As the reactive gas pressure is increased the arrival rate of reactive gas increases and the film becomes less metallic; modelled theoretically by Tiedge et al.[13] This continues until the film is saturated with gas (Figure 3b). The consumption of the reactive gas is then limited by the metal arrival rate.

The processes at the target (Figure 3a) and substrate (Figure 3b) combine to give the gas consumption curve (Figure 3c). This has to be superimposed on the pumping effect of the vacuum system.

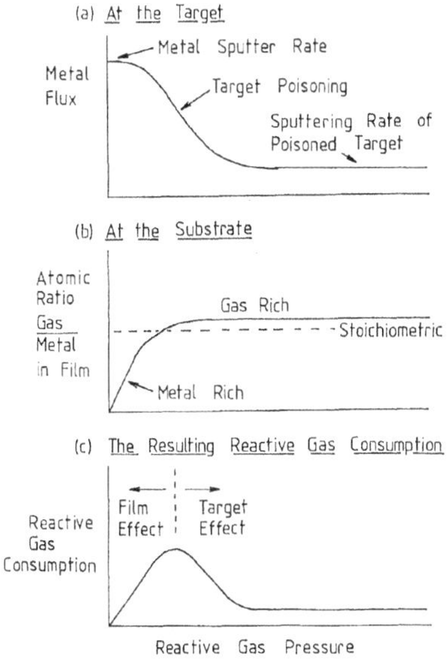

Figure 3. The origin of the consumption of the gas in a reactive sputtering system: a) At the target b) At the substrate and c) The resulting gas consumption.[4]

Figures 4 and 5 illustrate the effect of the pumping rate of the system on the stability of a reactive sputtering process. An increase of pumping speed from Figure 5 to 4 made the process stable. Our object to create a high rate, low cost, technique will be defeated if expensive pumping systems have to be used. Control loops using the partial pressure of the reactive gas and the magnetron potential have been used with limited success[5] as has the use of additional deposition surfaces to shift the operating point to a stable region.[6] A technique promising to allow the stable operation of a high rate reactive process, in a system of limited pumping capacity, is that of controlling the admission of the reactive gas to keep constant the spectral line emission of the metal being sputtered.[7] Such systems have been used to give high rate deposition in equipment having modest pumping capacity.

We have used a continuous high speed piezo-electric valve, or a simple divertor valve to direct the reactive gas flow to the chamber or a dump, to control the total oxygen flow (Figure 6). The signal controlling

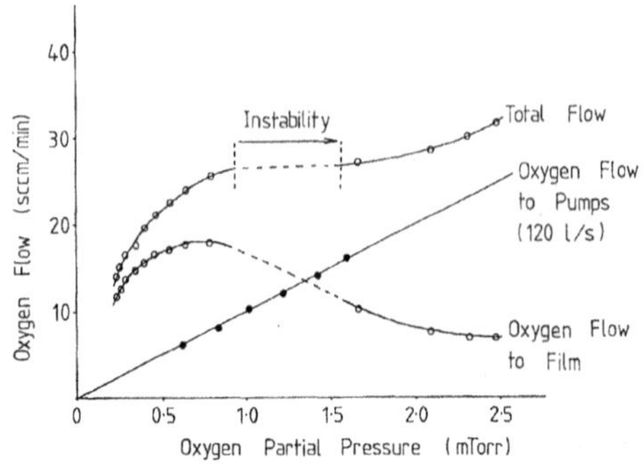

Figure 4. Reactive gas consumption in a stable system depositing indium oxide.[4]

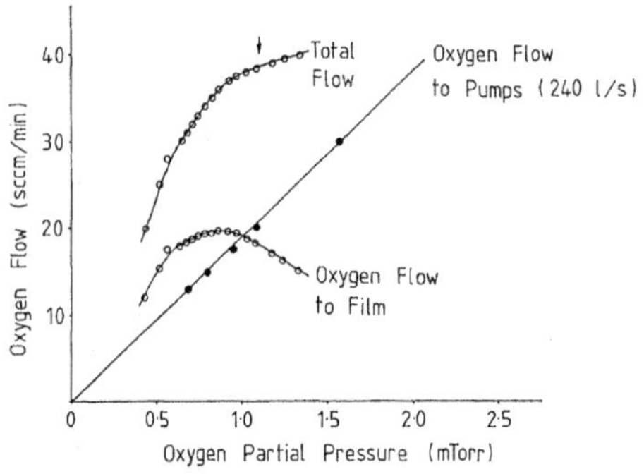

Figure 5. The system of Figure 4 made unstable by reducing the speed of the vacuum pump.[4]

the valve was taken from an optical detector looking at the magnetron plasma through a filter, at a wavelength appropriate for the metal being sputtered. The signal could be normalised to that observed when sputtering the metal in pure argon. For titanium, for example, the operating signal was 40% of the pure metal rate. Both systems allowed control of the reactive systems in regions of partial pressure of oxygen which were otherwise inaccessible due to instability of the process (Figure 7 and 8). With large, high speed systems care has to be taken to limit the capacity of the oxygen manifold to minimize partial pressure oscillations.

ION BOMBARDMENT OF THE SUBSTRATE AND GROWING FILM

As well as having a stable economical system we need to produce films with controlled structure, purity and stoichiometry. One technique which gives us this opportunity is that of ion bombardment of the film during growth. Neutralized ion beam guns and r.f. biasing of the insulating

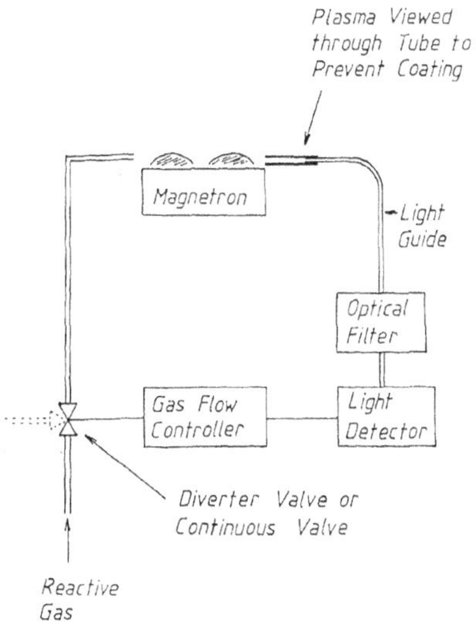

Figure 6. Plasma Emission Monitoring (PEM) control system.

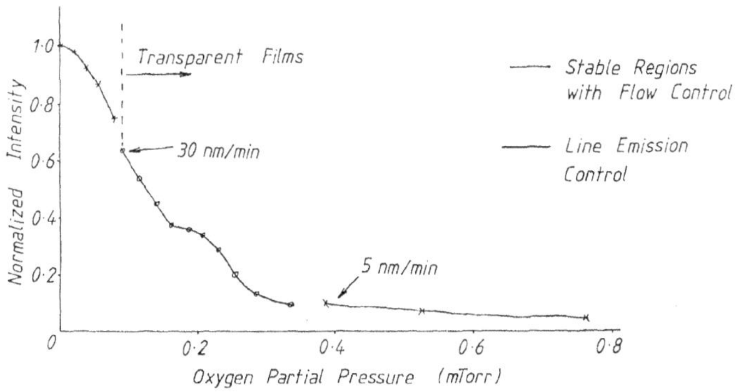

Figure 7. Evidence of the stability attained in the reactive sputtering of Titanium when using PEM control.

substrate have proved successful for low rate systems.[8] A newly applied technique of self-bias arising from immersion of the insulated substrate assembly in a dense plasma offers the potential of a simply applied effective process.[9] This can be achieved very easily with a magnetron source by creating magnetic flux lines along which plasma from the cathode can 'leak' onto the substrate surface. The extent of this 'leakage' can be controlled through manipulation of the magnetic field configuration or, more conveniently, by adjusting the anode potential relative to the system earth. Increased activaton of the reactive process by such means allows a lower partial pressure of the gas to be used for the desired composition and hence operation of the target in a more argon-rich atmosphere which gives a higher rate of sputtering for a given limiting input power.

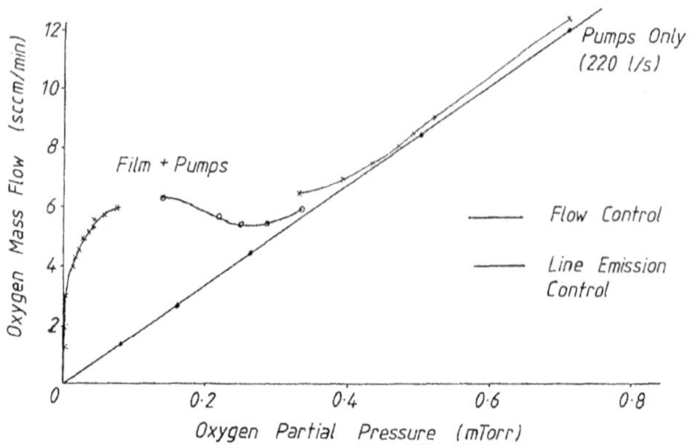

Figure 8. The spectral line emission of titanium when sputtered in a reactive atmosphere in the system of Figure 7.

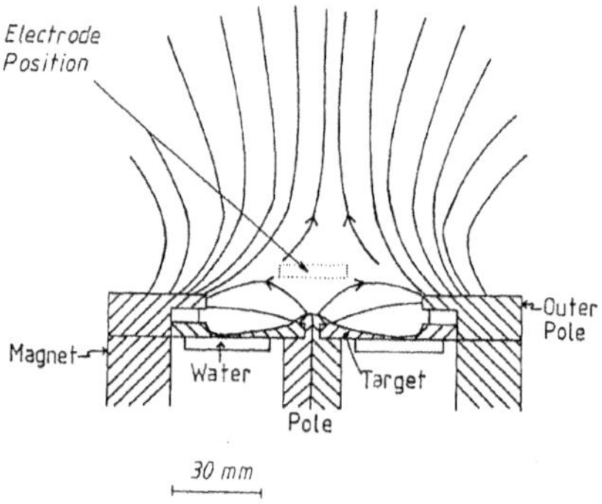

Figure 9. The 'unbalanced' magnetron with anode.

Figure 9 shows schematically an 'unbalanced' magnetron with an anode used to control the spreading of the plasma to the substrate.[10] The floating potential reached by an insulated substrate is shown in Figure 10 together with ion and electron currents flowing at equilibrium. These are extrapolated from I.V. characteristics of a probe. The effect of this bias on the properties of thin films is shown in Figures 11 and 12. For the reactive sputtering of TiO_2, films of higher refractive index were obtained for biases above -20V.[11] In the case of the sputtering of transparent electrically conducting films of indium oxide the influence of the bias was to increase the utilisation of the oxygen giving the potential of higher deposition rates.[12] The properties were essentially unaltered with an electrical resistivity of 6×10^{-6} ohm.m.

CONCLUSIONS

A large area, high rate, vacuum deposition system is available for the

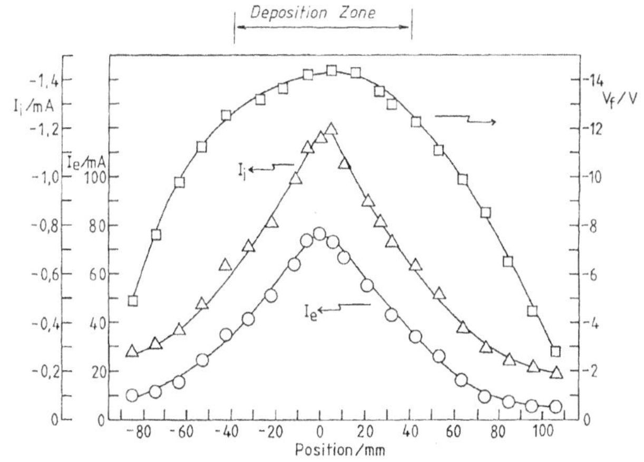

Figure 10. The bias potential (V_f) and electron and ion currents (I_c, I_i) flowing to an insulated substrate immersed in the plasma directed with an 'unbalanced' linear magnetron.

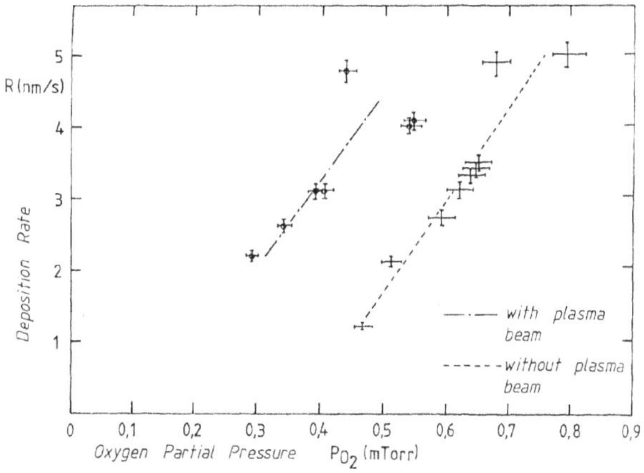

Figure 11. The increase in oxygen utilisation obtained by self-bias ion plating in a system sputtering optimised electrically conducting indium oxide.

production of oxide materials onto flexible polymer substrates. High rate processsing, necessary to achieve the purity and cost of satisfactory coatings, can be achieved with a process using the spectral-line emission, from the metal, being sputtered from a d.c. planar magnetron target, to control the admission of the reactive gas. Ion bombardment of the

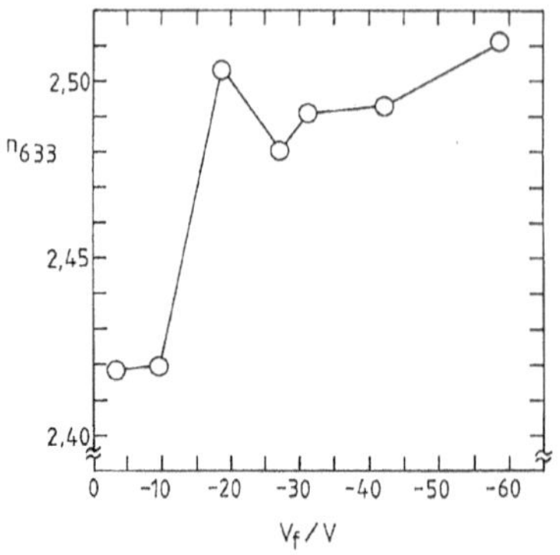

Figure 12. The refractive index at 633nm (n_{633}) of titanium oxide films reactively sputtered onto substrates with a controlled self bias, (V_f).

growing film, to increase utilisation of the reactive gas and to give the depositing film a more desirable structure, can be controlled through immersion of an insulated substrate in plasma provided from the magnetron.

ACKNOWLEDGEMENTS

The authors are pleased to acknowledge the considerable technical assistance they have received from E.M. Stenlake and the financial assistance of Everest Double Glazing, Unitika and Thorn EMI.

REFERENCES

1. R.P. Howson, in "Polymer Surfaces and Interfaces", W.J. Feast and H.S. Munro, editors, Wiley, 1987.
2. J.L. Vossen and W. Kern, editors, "Thin Film Processes" Academic Press, New York, 1978.
3. S. Schiller, U. Heisig and K. Goedicke, in "Proc. Int. Conf. Solid Surfaces", pp. 1545-52, 1977.
4. A.G. Spencer, R.P. Howson and R.W. Lewin, Thin Solid Films, 158, 141-149, 1988.
5. M.I. Ridge, E.M. Stenlake, R.P. Howson and C.A. Bishop, Thins Solid Films, 80, 31, 1981.
6. M. Scherer and P. Wirz, Thin Solid Film, 119, 203, 1984.
7. S. Schiller, U. Heisig, K. Steinfelder, J. Strumpfel, A. Friedrich and R. Fricke, in "Proc. Int. Conf. on Ion Plating and Allied Techniques,"Brighton, 1987, pp. 23-31, CEP Consultants, Edinburgh, 1987.
8. R.P. Howson, K. Suzuki, C.A. Bishop and M.I. Ridge, Vacuum, 34, 291-4, 1984.
9. B. Window and N. Savvides, J. Vac. Sci. Tech., A4, 196, 1986.
10. A.G. Spencer, K. Oka and R.P. Howson, Vacuum, 38, 857-859, 1988.
11. K. Oka, R.P. Howson, R.W. Lewin and A.G. Spencer, Proc. SPIE, 1019, 40-48, 1988.
12. A.G. Spencer and R.P. Howson, Proc. SPIE, 1016.
13. T. Tredge, T.D. Moustakis and J.M. Cebulka, Phys. Rev. B, 23, 5634, 1981.

UV LASER-INDUCED METALLIZATION OF POLYIMIDE FROM ELECTROPLATING SOLUTION

J. Zahavi*, M. Rotel*, D. Katz** and S. Levine***

* Israel Institute of Metals, Technion, Haifa, Israel
** Department of Materials Engineering, Technion Haifa, Israel.
*** I.B.M. Corporation, D/879, P.O. Box 1500 Bromont, Quebec JOE ILD, Canada

The feasibility of direct deposition of gold from an organic gold plating solution (1.0g $HAuCl_4$ in 50ml ethanol) on "Kapton H" polyimide substrates was demonstrated. Specimens immersed in the solution were exposed to UV laser irradiation at 24nsec pulse duration and 193nm wavelength.

Two major processes were evaluated: (A) formation of a dense uniform gold deposit up to one micron in thickness at low laser energy in the range of 100mJ/pulse, and (B) formation of a thin gold deposit followed by bubble-like structures in the range of 200mJ/pulse. An S-type relationship was established between bubble surface coverage and number of laser pulses. Bubble diameter was found to increase linearly with number of pulses.

SEM, EDS, ESCA and AES were used to characterize deposit morphology, structure and composition.

I-V curves were obtained by four-point probe, indicating sheet resistance in the range of 1.45 Ω/□ for the low-energy deposit.

INTRODUCTION

Recently, direct laser metallization of semiconductor substrates immersed in electroplating solution - without use of external current or lithography procedures - has been reported by several investigators[1-4]. No major efforts, however, have been devoted in this context to polymeric substrates.[4] Use of pulsed UV laser irradiation without preliminary surface treatment such as sensitization or activation, and without external power and masking procedures, is the innovation of the present work compared to previous studies.[4,5]

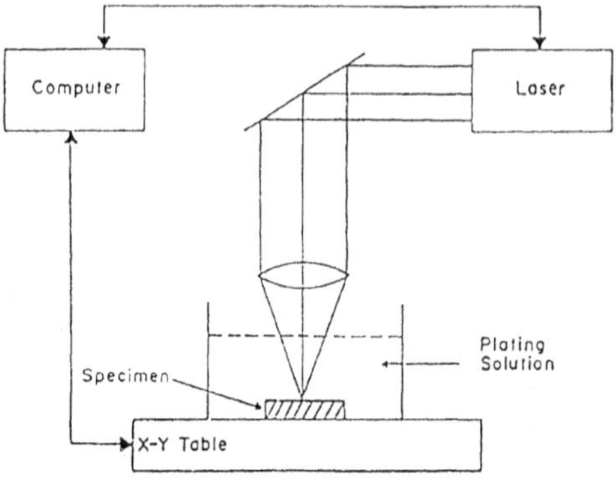

Fig. 1. Experimental setup for direct laser induced plating on polyimide substrate immersed in plating solution.

EXPERIMENTAL

A polyimide "Kapton H" (produced by DuPont, USA) was immersed up to 1mm depth in $HAuCl_4$ (20 g/l) solutions (ethanol, methanol and D.I. water) and irradiated perpendicularly by a UV laser beam at 193 nm, using Lambda Physik excimer laser model 201 MSC (Fig. 1). Organic solutions proved preferable because of their better wetting properties on the polyimide substrate.

Two levels of laser energy, one at 160-200 mJ/pulse and the other at less than 100 mJ/pulse, were used, with laser repetition rates of 2Hz to 30Hz and numbers of pulses from 5 to 2000 for each level. Specimen observation and detection of presence of elemental gold were carried out by SEM+EDS, ESCA and Auger techniques.

SEM RESULTS

1. Gold-Ethanol

Elemental gold deposits produced via UV laser irradiation on polyimide substrate from gold-ethanol solution are shown as bright regions in Figs. 2,3,4. Deposits formed both at the low and high numbers of pulses, but in the first case they were discontinuous (Figs. 2A, 2B). The effect of increase of the number of pulses at constant high energy levels, with repetition rates of 10Hz and 5Hz, is demonstrated in Figs. 2 and 3, respectively. Increase of the pulse number at high energy gives rise to a new phenomenon--formation of bubble-like features. At 10Hz these bubbles formed when the number of pulses exceeded one hundred, and at 5Hz when the number of pulses was about 300. In order to prevent bubble formation, a low energy level of 95 mJ/pulse was adopted. The deposit formed at 5Hz after irradiation with 2000 pulses consisted of a dense layer with a fine morphology overlaid by larger gold particles (Fig. 4).

2. Gold-Methanol

SEM micrographs of the gold deposit produced in the methanol variant are shown in Fig. 5. Single bubbles and clusters were observed at repetition rates of 10Hz and 30Hz at the high energy level of 160 mJ/pulse. These bubbles exhibited elemental gold coverage as was observed also in the ethanol variant. An increase in the number of pulses resulted in increase of surface coverage with a single layer of bubbles (Figs. 2, 3,5).

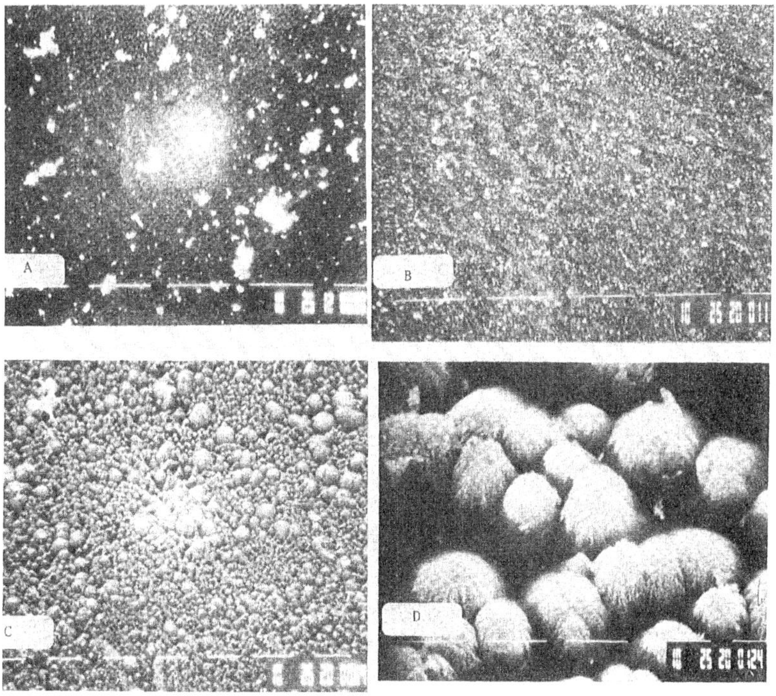

Fig. 2. SEM micrographs of gold deposits on polyimide substrate at different numbers of pulses. (Laser repetition rate (R.R.) 10Hz, laser energy 160 mJ/pulse).
A. After 34 pulses. B. After 54 pulses.
C. After 300 pulses. D. After 600 pulses.

3. Deposit Cross Section

Figs. 6 and 7 show cross sections of the gold deposit at high (173 mJ/pulse) and low (95 mJ/pulse) energy, repectively. The bubble cross section (Fig. 6B) revelaled gold on the bubble surface (point 3), but none inside the bubble (points 1,2).

According to Fig. 7, the gold layer obtained at low energy was about 2μm thick.

ESCA AND AES DATA

1. ESCA

The presence of pure elemental gold was detected by Electron Spectroscopy for Chemical Analysis (ESCA), with typical spectra shown in Fig. 8. Findings were positive under all variants and conditions, without exception.

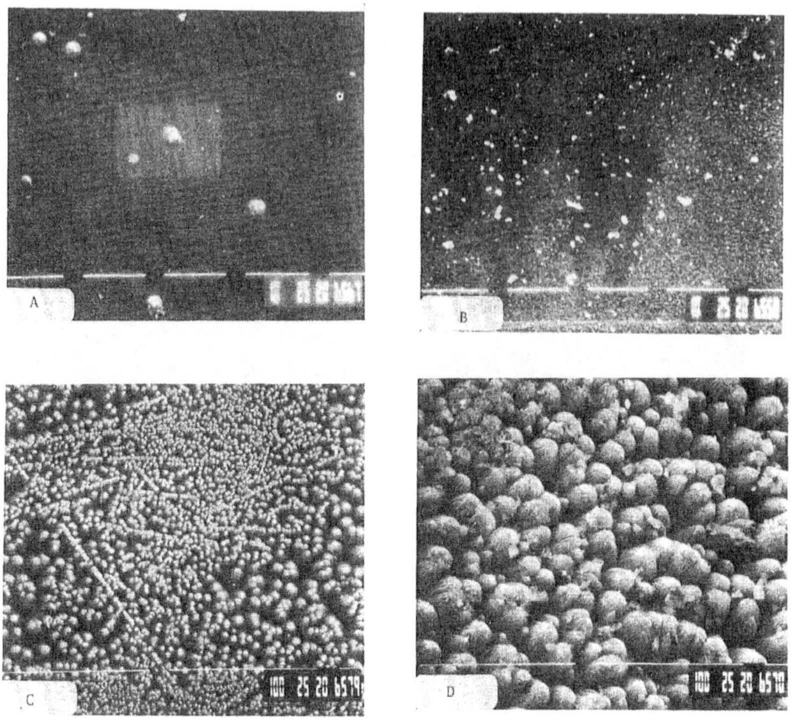

Fig. 3. SEM micrographs of gold deposits on polyimide substrate at different numbers of pulses (laser repetition rate (R.R.) 5Hz, laser energy 173 mJ/pulse).
A. After 100 pulses. B. After 300 pulses.
C. After 600 pulses. D. After 1000 pulses.

2. <u>Auger</u>

Auger analysis of the high-energy deposits revealed presence of carbon together with the elemental gold (Fig. 9). Layer thickness exceeded $1\mu m$ according to the tantalum oxide standard.

Analysis of the bubble-free deposits (95 mJ/pulse) revealed (Fig. 10) a thick layer with a large amount of gold, without carbon or chloride even beyond 5600A. No further depth profile analyses were attempted as the layer apparently exceeded $1-2\mu m$. Fig. 11 shows an AES spectrum of the area from which gold was peeled during irradiation. Overlapping of the gold layer and the polyimide matrix is seen in Fig. 11B.

SHEET RESISTANCE

The sheet resistance of the gold deposits was determined by the four-point probe technique. Calculated values at 95 mJ/pulse were 1.45 and 2.47 $\Omega/\square$, respectively, for 1000 and 2000 pulses.

Both the polyimide and the higher-energy deposits exhibited no conductivity with our instrument.

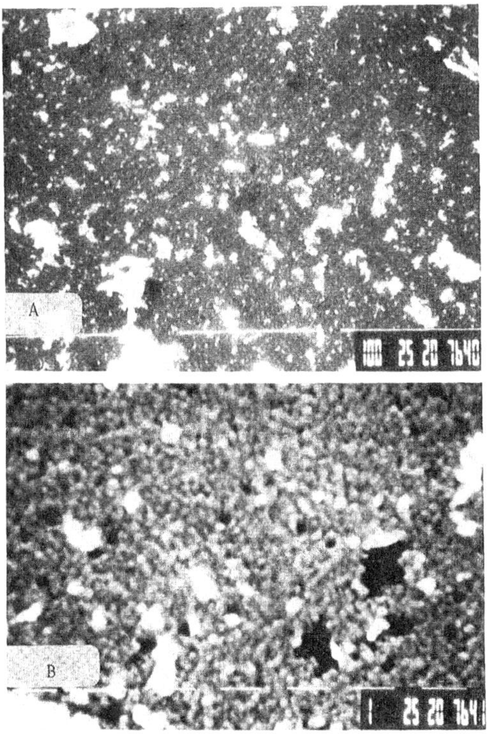

Fig. 4. SEM micrographs of irradiated area after exposure to 2000 pulses at R.R. 5Hz with low laser energy (95 mJ/pulse) in gold-ethanol solution.
A. General view x350. B. Enlargement of A x10,000.

IMAGE ANALYSIS ON BUBBLE FORMATION

Image analysis indicated that bubble formation was associated with nucleation and growth processes. A linear relationship was established between the mean bubble diameter and the number of pulses. The former ranged from about 4.0μm to 24μm in the interval of 100 to 1000 pulses (Fig. 12). The linear bubble growth rate, G, was found to be about 22×10^{-6}[cm sec^{-1}]. A typical pattern of bubble coverage, S, (singles or clusters) versus the number of pulses is shown in Fig. 13, the coverage increasing from about 10% at 200 pulses to almost 100% at 1000 pulses. It should be noted that a single layer of bubbles was produced. The sigmoid in Fig. 13 can be described as follows:

$$S(p) = 1 - \exp(-Kp^{n}) \qquad (1)$$

or, in logarithmic terms

$$\ln(\ln \frac{1}{1-S(p)}) = \ln K + n \ln p \qquad (2)$$

where

$$K = \pi/3[N \cdot G^2]^{n/3} \tag{3}$$

With n found to be close to 3 (see below), eq. (3) becomes:

$$K = \pi/3[N \cdot G^2] \tag{4}$$

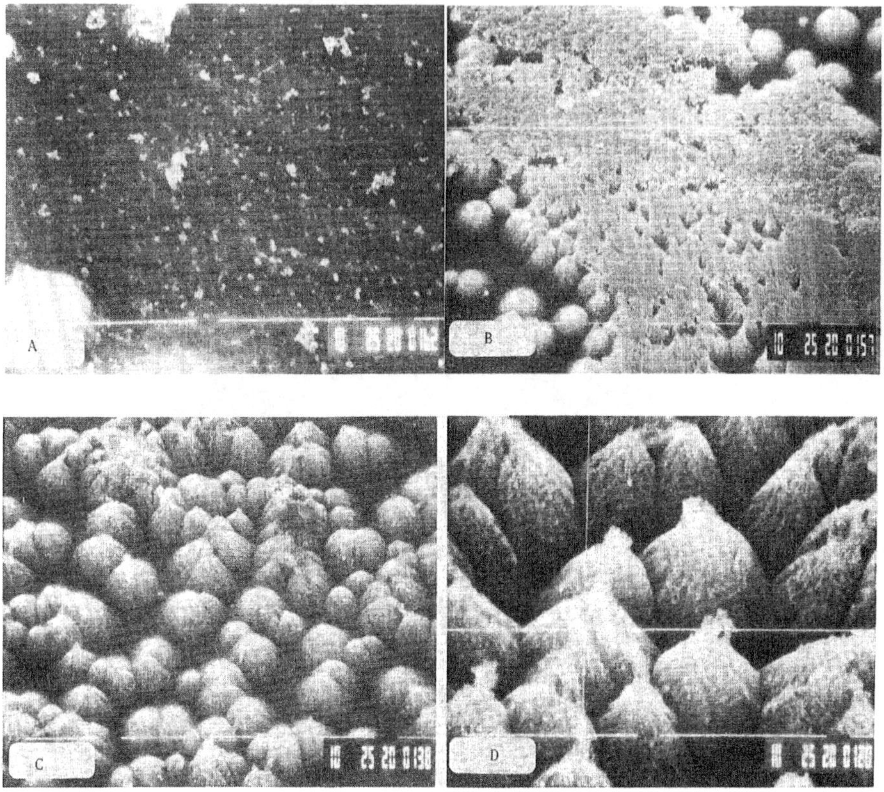

Fig. 5. SEM micrographs of gold deposits obtained from gold-methanol solution at laser energy of 160 mJ/pulse. A. After 300 pulses at R.R. 10Hz. B. After 600 pulses at R.R. 10Hz. C. After 400 pulses at R.R. 30Hz. D. After 640 pulses at R.R. 30Hz.

Notation

S - bubble coverage (formation and growth);
p - number of pulses;
n - constant;
N - bubble nucleation rate $[cm^{-2}sec^{-1}]$;
G - linear bubble growth rate $[cm\ sec^{-1}]$;
K - constant, so long as N and G are constants.

The linearity implied by eqs. (1) and (2) was reasonably confirmed by Fig. 14 from which the values of n and K were found to be 2.8 and 0.21×10^{-4} respectively; thereby equation (4) yielded N as $4.2 \times 10^4 [cm^{-2}sec^{-1}]$.

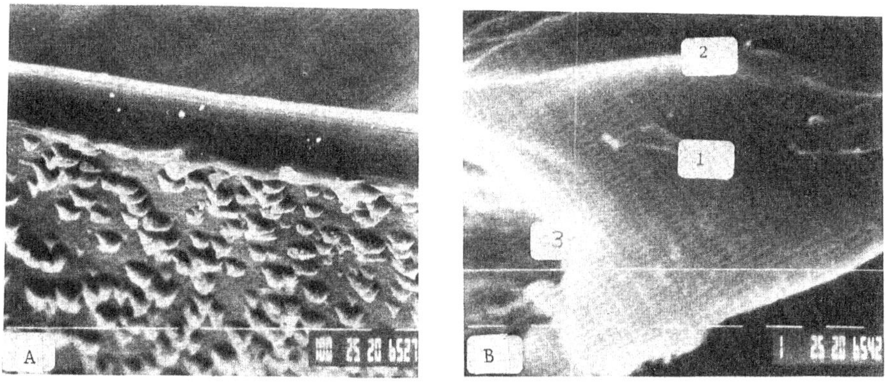

Fig. 6. Cross section (obtained by cutting and viewing at 90°) of gold deposit at laser energy of 173 mJ/pulse, R.R. 10Hz, 400 pulses.
A. Irradiated zone x500.
B. Enlargement of bubble cross section x10,000.

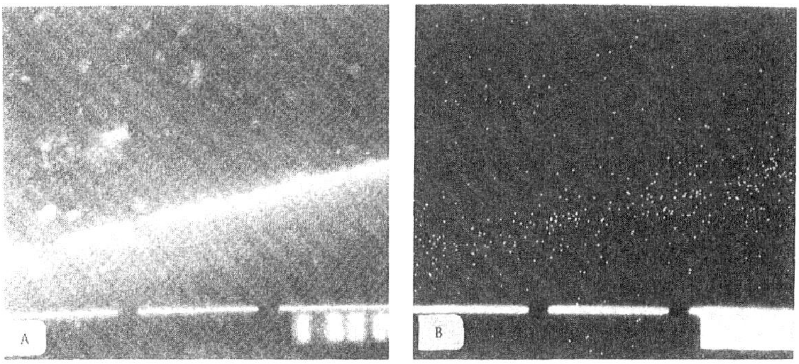

Fig. 7. Cross section of gold layer obtained from gold-ethanol solution at 95 mJ/pulse after 2000 pulses.
A. SEM micrograph of gold layer.
B. X-ray image of cross section.

Finally, the analysis suggested a constant nucleation rate [around 4.2×10^{4} ($cm^{-2} sec^{-1}$)] and a constant linear growth rate [around 21.85×10^{-6} cm sec^{-1}].

DISCUSSION

Recently, several investigators[6-9] have examined the effect of UV laser irradiation on polyimide surfaces in terms of ablation and material removal. Their findings indicate that the effect involved may be associated with photodecomposition (perhaps rupture of chemical bonds) and thermal processes resulting from photon absorption by the substrate. In the light of these studies, as well as of earlier work, possible modes and mechanisms of laser-induced metallization of polymeric substrate immersed in electroplating solution might be advanced as follows.

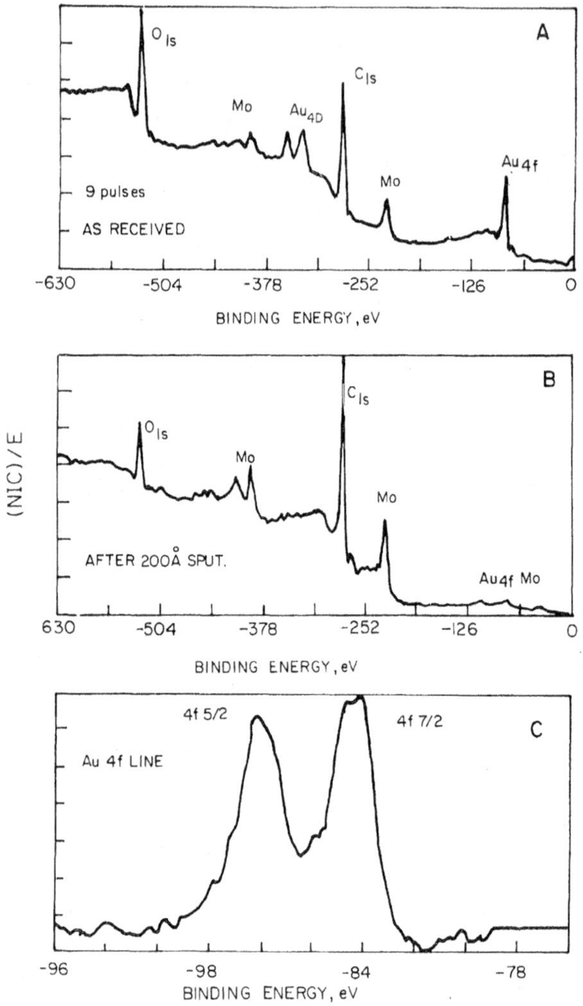

Fig. 8. ESCA spectra of gold deposit obtained from gold-ethanol solution. Laser energy 210 mJ/pulse, 9 pulses, R.R. 10Hz.
A. ESCA surface spectrum showing the presence of gold of surface deposit before sputtering.
B. ESCA spectrum of gold deposit after 200A sputtering.
C. ESCA spectrum showing the presence of pure elemental gold in deposit. (Mo peaks due to sample holder).

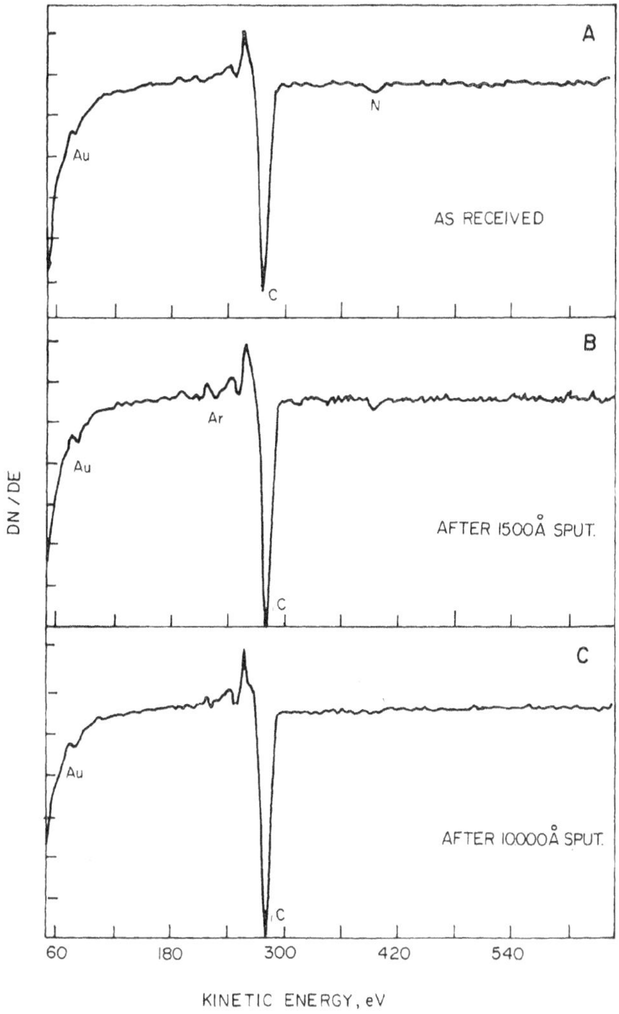

Fig. 9. Auger spectra obtained on bubble in gold deposit.
A. Presence of gold on bubble surface before sputtering.
B. Presence of gold on bubble surface after 1500A sputtering according to tantalum oxide standard.
C. Presence of gold on bubble surface after 10000A sputtering according to tantalum oxide standard.

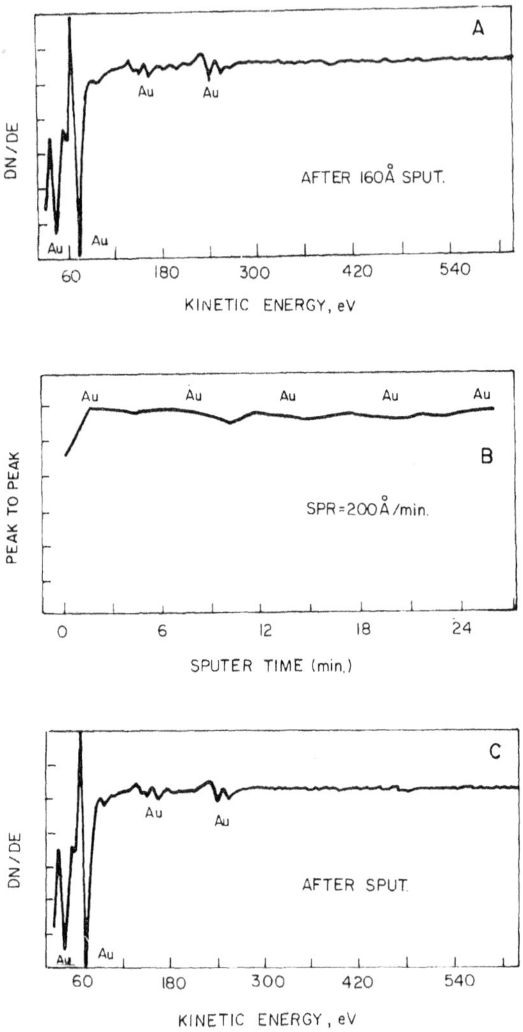

Fig. 10. AES continuous depth profile obtained for gold deposit at low laser energy (95 mJ/pulse), R.R. 5Hz after 2000 pulses (SEM observation is shown in Fig. 4).

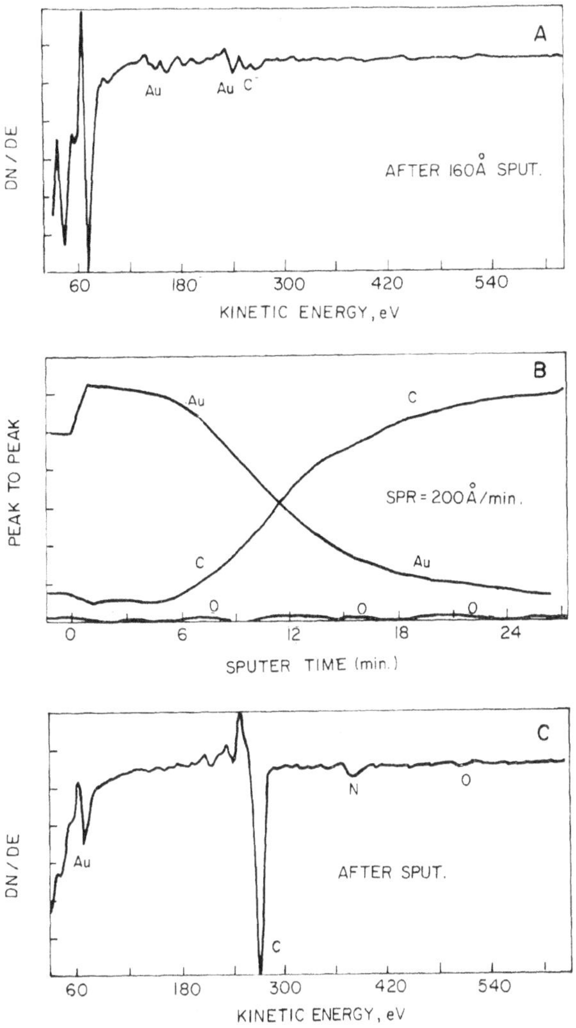

Fig. 11. Overlapping area between gold layer and polyimide matrix obtained at low laser energy (95 mJ/pulse) and R.R. 5Hz. A. Gold deposit. B. Depth profile to 5400A. C. Gold in polyimide matrix after 5400A.

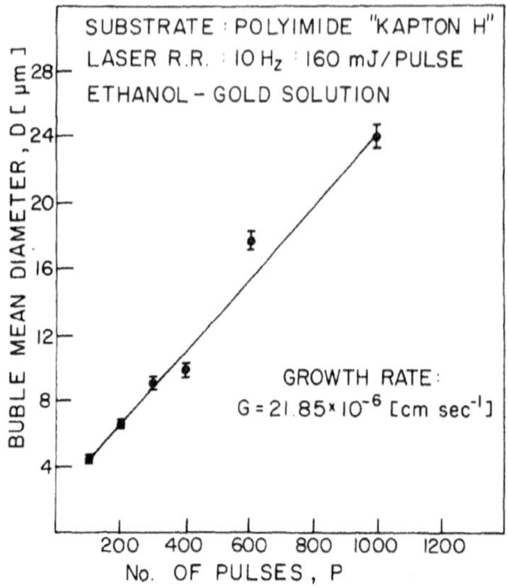

Fig. 12. Mean bubble diameter vs number of pulses. Spot data representing the mean of 20 image-analysis measurements.

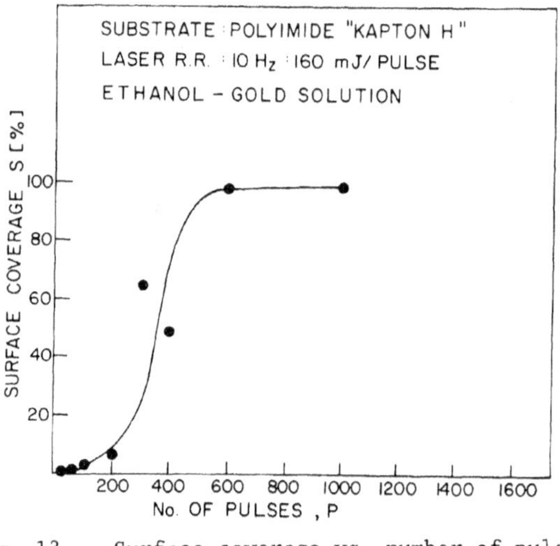

Fig. 13. Surface coverage vs. number of pulses.

The temperature increase in the polyimide is high at energy density of 10^6 watt/cm^2 even when the number of pulses is very low (below 50). Under these conditions surface temperature could exceed 300°C[6-8], probably due to the adsorptive capacity of the film[6] ($B=10^5 cm^{-4}$) which confines the penetration depth to around 10^3A. This local heating induces pyrolysis of the polyimide, whereby its surface acquires semiconductor-type properties[9-11], with the attendant increase in the number and mobility of charge carriers.[9] Further irradiation gives rise to photoemission electrons which reach the plating solution and are trapped by metal ions (cathodic reaction), as a result of which metal is deposited in the irradiated zones--as was in fact observed by several investigators.[12,13] The cation reduction to zerovalent atoms at the polyimide-electrolyte interface was also demonstrated[14,15] confirming that the polyimide serves both as substrate and reducing agent.

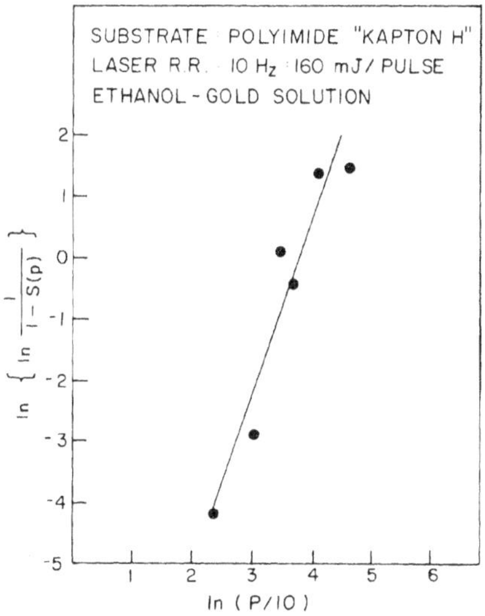

Fig. 14. Linear dependence between ln(1/1-s(p)) and ln(p/10).

Further investigation is called for with a view to better understand these new processes of laser-induced direct metallization of polyimide and other polymeric substrates from electroplating solutions without external power and without the need for pre- or post-treatment such as activation or sensitization.

CONCLUSION

We have demonstrated the feasibility of direct deposition of gold from a plating solution on polyimide substrate under open-circuit conditions. The deposited metal is pure and free of contamination by the electrolyte. The deposition takes place by photochemical and thermal processes under excimer laser irradiation.

REFERENCES

1. R.F. Karlicek and V.M. Donnelly, J. Appl. Phys. 58(2), 1084-1090 (1980).
2. Y.C. Kiang, J.R. Moulic and J. Zahavi, IBM Technical Disclosure Bulletin, 26, No. 1, 327 (June 1983).
3. J. Zahavi, S. Tamir and M. Halliwell, Mat. Research Soc. Symp. Proceeding, 7, 387-393 (1986).
4. J. Zahavi, S. Tamir and M. Halliwell, Plating Surface Finishing, 47-64 (1986).
5. J. Zahavi, S. Tamir and M. Halliwell, Laser gold deposition on polyimide, Proceedings Interfinish 1984, 11th World Congress on Metal Finishing, Jerusalem, Israel, 497, (October 21-26, 1984). Published by International Ltd., 10 Rothschild Blvd., 65121 Tel-Aviv, Israel.

6. V. Srinivasan, M.A. Smrtic and S.V. Babu, J. Appl. Phys. 59(11), 3861-3867 (1986).
7. R. Srinivasan and B. Braren, J. Polym. Sci., 22, 2601-2609 (1984).
8. P.E. Dyer and J. Sidhu, J. Appl. Phys. 57 (4), 1420, (1985).
9. H.B. Brom, Y. Tomkiewicz, A. Aviram, A. Broers and B. Sunners, Solid State Commun. 35, 135 (1980).
10. A. Burger, E. Fitzer, M. Heym, and B. Termiesch, Carbon, 13 (3), 149-157 (1973).
11. D. Bruck, Polymer (London) 6, 319 (1965).
12. A. Felske and J.W. Plieth, J. Opt. Soc. Amer. B3, 815 (1986).
13. D.V. Podlesnik, H.H. Gilgen, A.E. Willmer and R.M. Osgood, J. Opt. Soc. Amer., B3, No. 5, (May 1986).
14. S. Mazur and S. Reich, J. Phys. Chem., 90, 1365 (1986).
15. S. Mazur, P.S. Lugg and C. Yarnitzky, J. Electrochem. Soc., 134 (2), 346 (1987).

SIMPLE METAL-ON-PLASTICS PATTERNING BY PULSED ELECTRON BEAM IN SOFT VACUUM

J. Krishnaswamy, M. Eyolfson, L. Li, G. J. Collins

Department of Electrical Engineering
Colorado State University, Fort Collins, CO 80523

H. Hiraoka*

IBM Almaden Research Center, 650 Harry Road
San Jose, CA 95120-6099

M. A. Caolo

Hewlett-Packard Inc.
3404 East Harmony Road, Fort Collins, CO 80525

Direct patterning of polyimide images with a built-on-metal layer can be accomplished by electron beam exposures of doped polyamic acid films, followed by wet development in NMP and subsequent heating and curing to polyimide films. During curing at 300°C the metal salts converted to metal and migrated to the top surface of polyimide films. Additional electron beam exposures of images assisted metal film formation. Auger electron and Rutherfold backscattering spectroscopic analyses were carried out to confirm metal migration to the surface. Silver nitrate, gold mercaptide and cupric chloride were used as dopants in polyamic acid films.

* *To whom all correspondence should be addressed.*

INTRODUCTION

Metal incorporation into polymers is of great interest in many technical areas. Interactions between metals and polyimide have been extensively studied with the idea of improving mechanical[1] and adhesive properties[2]. Enhanced surface reflectivities have also been reported for polyimides doped with copper[3] and silver salts[4]. In the case of palladium salt incorporation into polyimide, it has resulted in a ten-fold increase in surface conductivity[5]. These report s mainly discuss the chemical state of the elements present in the film, their relative concentration, and the effect of subsequent thermal curing on metal migration, using a variety of techniques including SEM, AES, XPS, thermomechanical and thermogravimetric methods. Successful metallization by such a method suggests useful applications such as direct seed-layer image fabrication with good adhesion to polyimide film, and others. We would like to report here our study of direct patterning of polyimide images incorporating a built-on-metal layer. This is achieved by combining our direct imaging method of polyimide films by pulsed electron beam exposures in soft vacuum and polyamic acid films incorporating metal salts. We have already reported the direct patterning of polyamic acid using pulsed electron beams in soft vacuum[6,7].

EXPERIMENTAL

Electron Beam Image Fabrication

With the apparatus already described[7], a beam of 28 keV pulsed electrons produced in soft vacuum in air at 30 mTorr impinge upon the metal doped polyamic acid film after traversing an electron transparent stencil mask. The electron interaction with the metal doped polyamic acid results in crosslinking and the resulting reduced solubility leads to patterning of the relief image following conventional wet development in 1-methyl-2-pyrrolidinone (NMP). The experimental procedure is summarized in Fig. 1.

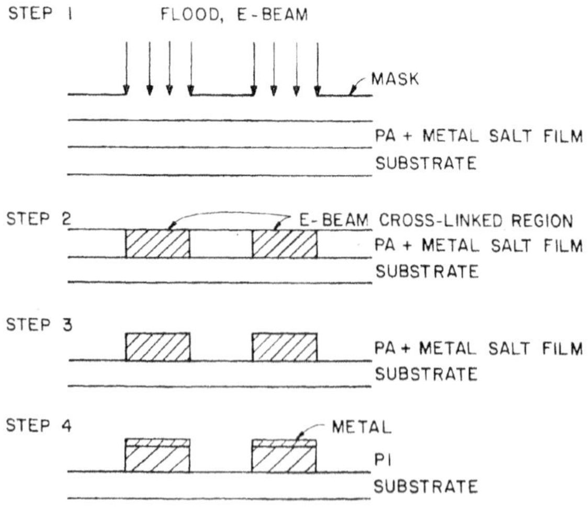

Figure 1. Experimental steps for metal-on-polymer images.

The AES and EDX analyses were carried out at Hewlett-Packard Co., Fort Collins, the SIMS analysis at CSU and the Rutherford Backscattering Spectroscopy (RBS) measurements at Charles Evans and Associates, Redwood City, CA. AES analysis was carried out on Perkin-Elmer Phi-600 instrument with a LaB_6 cathode. A probe current of 0.01 μA at 10 kV was used with a sputter current of 1 mA at 2 kV for the depth profiling corresponding to an average sputter rate of ~25 nm/min. The SIMS depth profiling was carried out on a Perkin-Elmer model 6300 SIMS instrument using oxygen with a sputter current of 82 nA at 7 kV. The RBS measurements of a $AgNO_3$-doped polyamic acid film and of the cured film were carried out with 2.275 MeV He^+ ion beam at a 160° normal detector angle and at a 100° grazing detector angle. Computer generated theoretical spectra were obtained by iterative fitting to the experimental spectra in order to determine the composition of each sample.

Silver Nitrate Doped Polyamic Acid Films

Two grams of silver nitrate was dissolved in 5 ml of NMP and homogeneously mixed with about 20 ml of polyamic acid, PI2555, a Du Pont product. The $AgNO_3$-doped polyamic acid was spin coated onto a silicon wafer with a 6000 Å silicon dioxide film at 2000 rpm. The resulting film was soft cured at 50°C for 30 minutes, giving a final thickness of 0.6 μm. The film was then exposed in a proximity mode to 50-100 pulses of 28 keV electrons produced in 30 mTorr air, with a total dose of 5 x 10^{-4} to 1 x 10^{-3} C/cm^2. The image was

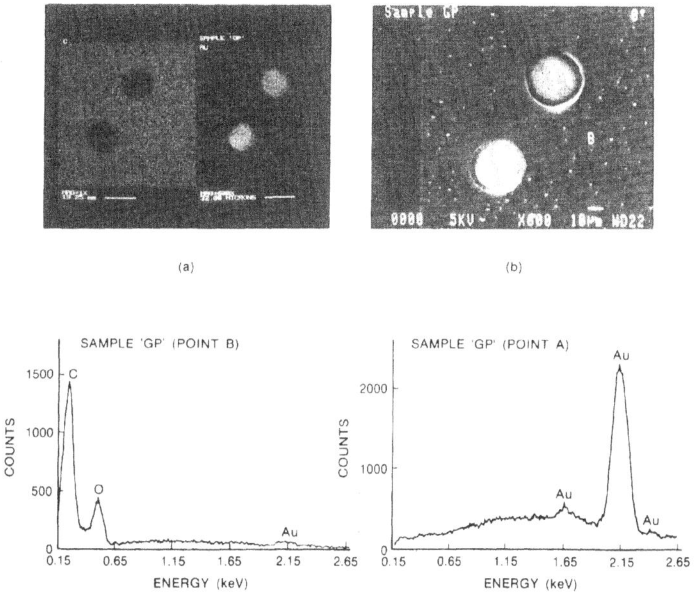

Figure 2. Gold segregation in gold mercaptide (Engelhard's gold solution #6854) + polyamic acid composite. (a) EDX elemental map of segregated regions; (b) SEM micrograph of the segregated region; (c) EDX analysis of gold depleted region; (d) EDX analysis of gold rich region.

completely developed in NMP after about 1 min. The pattern was heated to 300°C for 1 hr. The exposed surface turns shiny with a foggy metallic luster.

Gold Mercaptide Doped Polyamic Acid Films

Ten ml of Engelhard's gold compound (#6854) was mixed with 30 ml of NMP and the mixture was then slowly added to polyamic acid, PI2555. For this ratio a homogeneous mixture is formed. However, the stability of the mixture to gelation is very sensitive to the mix ratio. An adherent film can be made by spin coating this mixture onto a silicon surface. When this film is cured and then heated to about 350°C for 2 hours, segregation of gold takes place in the form of circular, highly reflective isolated surfaces on the film. The degree of segregation seems to depend upon the cure cycle. Fig. 2(b) shows the SEM surface micrograph of such sample in which two gold-rich regions are shown. Fig. 2(d) shows the EDX analysis of the gold-rich region. These gold rich regions are mostly circular with a diameter of 30-50 μm but do vary depending on the cure cycle.

Copper Doped Polyamic Acid Films

One gram of cupric chloride was dissolved in 10 ml of NMP and to this was added 25 ml of Du Pont PI2555 polyamic acid solution. The resulting solution was spin coated onto a silicon wafer at 4000 rpm to obtain a thin film of the

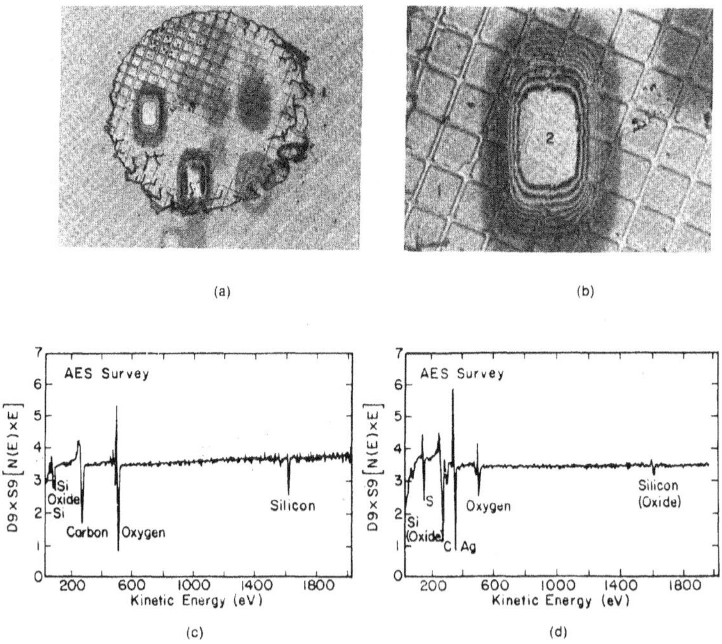

Figure 3. Patterning of silver nitrate-doped polyamic acid with pulsed electron beams. (a) Photomicrograph of patterned surface of silver nitrate-doped polyamic acid; (b) details around a diagnostic "Dig" made with a SIMS machine; (c) Auger surface survey at point [2]; (d) Auger surface survey at point [1].

polyamic acid doped with copper. When the film was heated to around 300°C in air for 30 min, it turned brown with a metallic luster.

RESULTS AND DISCUSSION

Silver On Polyimide Imaged Film

Fig. 3(a) shows the patterned surface of silver nitrate doped polyamic acid. The two large rectangular areas show sputter crater made with a SIMS instrument. The dig depth is around 2 μm and the total film thickness is about 4 μm. The SIMS profiling indicated a constant silver signal to a certain depth followed by a monotonic decrease to the substrate. Fig. 3(b) shows details around one of the digs. The figure shows that although a relief image of the SEM grid mask is still visible, the development of the image is not completed. The area eroded [2] by the ion beam and a control area [1] on the surface were then analyzed by AES. The areas [1] and [2] are shown in Fig. 3(b). Fig. 3(c) shows the AES survey made on surface [2] which is depleted of silver, while Fig. 3(d) shows the AES survey on area [1] which clearly shows enhanced silver signal.

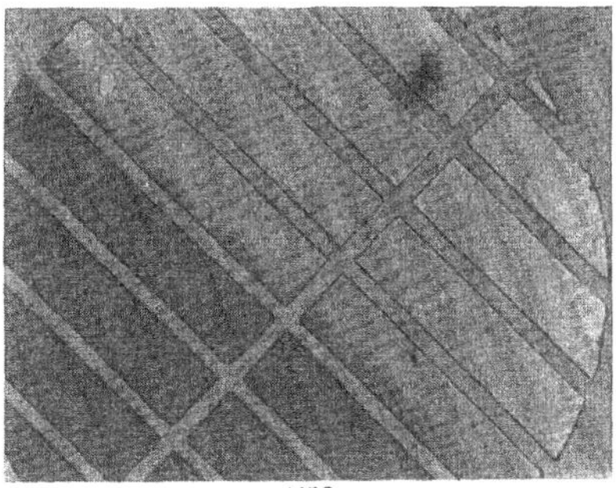

Figure 4. Completely developed images of $AgNO_3$-doped polyamic acid and effect of electron beam hardening: imagewise exposure at 28 keV, 50 pulses, electron beam hardening with 150 pulses only in dark region, curing at 300°C.

Fig. 4 shows a completely developed polyimide image. The bright region of Fig. 4 was obtained just after curing at 300°C. The darker region of Fig. 4 has been subjected to an additional process of electron beam hardening prior to 300°C heating. Clearly the darker area has more metallization than the bright area. The distribution of each element in the film shown in Fig. 4 as a function of Ar sputter time (and so depth from the surface) before heating, after 300°C curing, and after electron beam hardening and 300°C curing are shown in Fig. 5A, 5B and 5C, respectively.

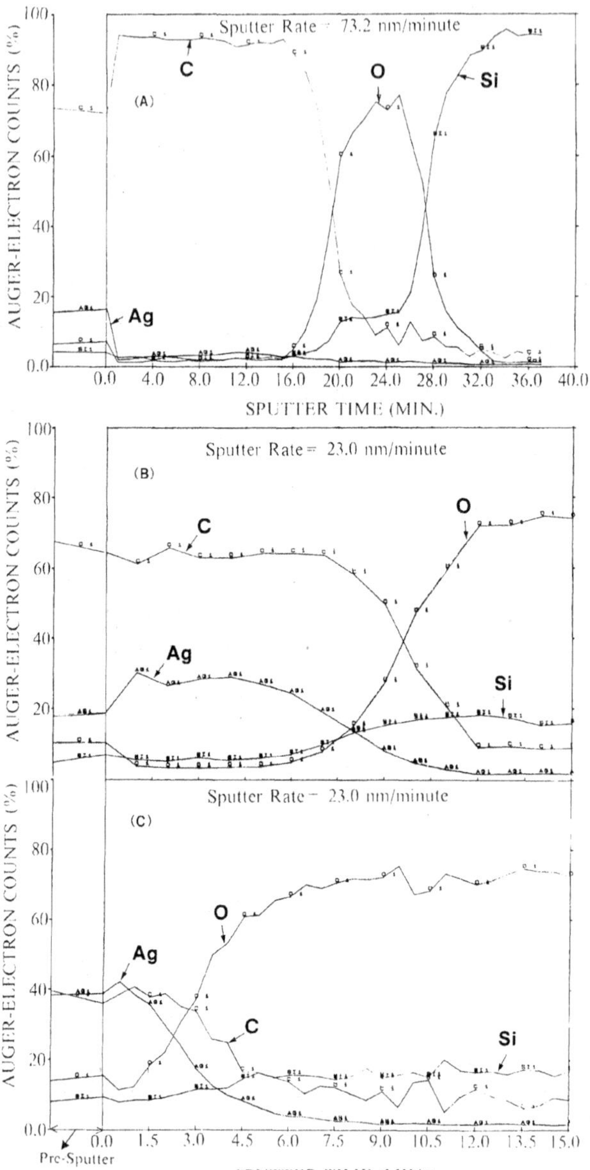

Figure 5. Elemental distributions as a function of depth from the surface. (A) before curing, sputter time from 0 to 40 min with sputter rate of 73.2 nm/min, (B) after curing at 300°C, sputter time from 0 to 15 min with 23.0 nm/min rate, and (C) after e-beam hardening and curing, sputter time from 0 to 15 min with 23.0 nm/min rate.

The result clearly demonstrates silver migration to the surface, particularly after the electron beam hardening. The electron beam exposure may assist silver reduction from the cationic state to the neutral state. The resultant polyimide film becomes electrically conducting after the curing. RBS measurements on a $AgNO_3$-doped polyamic acid film and on its cured film at 260°C in air also support the major silver migration to the polyimide film surface , as shown in Fig. 6. The polyimide film has a shiny silver surface after the curing and has an electrically conducting surface.

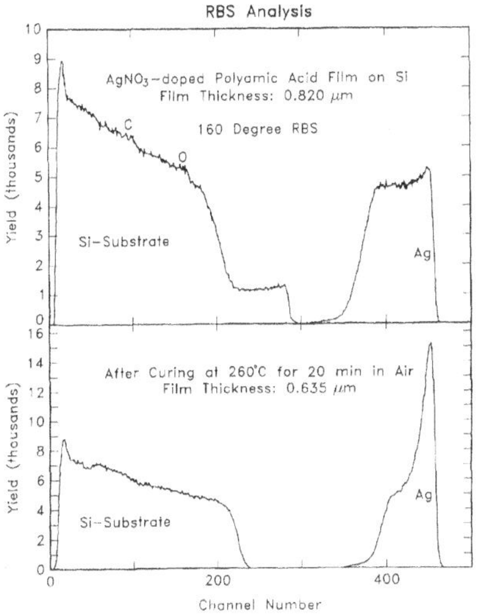

Figure 6. Rutherford backscattering spectroscopy measurements using 2.275 MeV He(4) ion; (a) 0.82 μm thick $AgNO_3$ -doped polyamic acid film spin-coated on a silicon wafer; (b) RBS data for the film after heating at 260°C for 20 min in air.

Gold On Polyimide Images

Because the segregation of the gold mercaptide-doped polyamic acid film into 30 to 50 μm diameter gold globules was evident from the experiment described in the experimental section, a smaller feature size was chosen for patterning with the soft vacuum electron beams. A thin film of about 1 μm of Engelhard gold containing polyamic acid was spin coated onto a silicon wafer and exposed to about 150 pulses of 28 keV electron beams at 30 mTorr through a micromesh screen (a product of Buckbee Mears, St. Paul, MN) with approximately circular holes of 3 to 5 μm in diameter. The exposed film was developed in NMP for 2 minutes. The developed pattern can be seen in the optical micrographs of 100 and 500 times magnifications in Fig. 7(a) and 7(b), respectively. The developed pattern turns golden yellow when heated to 325°C for 2 hours. Fig. 7(c) shows the EDX analysis of one of the 3 to 5 μm diameter crosslinked regions and Fig. 7(d) shows the EDX analysis off the pattern. These EDX data and the unmistakable gold color of the dotted patterns clearly indicate gold on the polyimide images. However, it will be necessary to carry out further surface analytical studies to determine the depthwise distribution of gold in the film.

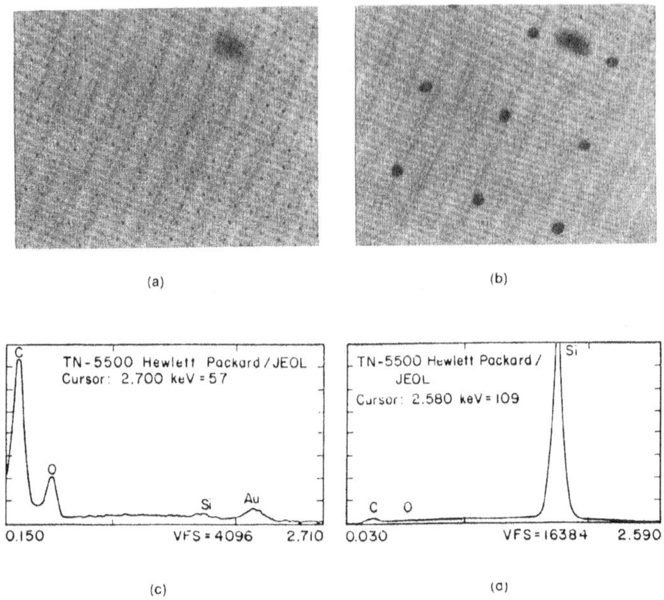

Figure 7. Patterning of gold mercaptide + polyamic acid composite. (a) 3μm patterns from a micromesh screen before heating, x 100; (b) same as in (a) at x 500; (c) EDX analysis on pattern heated at 350°C for 30 min; (d) EDX analysis at off-pattern position.

Copper On Polyimide Images

The soft cured, spun films of cupric chloride-doped polyamic acid were exposed to about 200 pulses of electrons produced in 30 mTorr at 28 keV. Faint relief images could be seen immediately after the exposure. The electron exposed regions cross-linked. The unexposed regions could be removed either by AZ-351 or NMP. Fig. 8(a) shows the photomicrograph of the crosslinked regions which is still not completely developed. Fig. 8(b) shows the AES survey off the pattern after heating. Fig. 8(c) shows the AES survey on the unheated pattern while Fig. 8(d) shows the AES survey on the heated pattern. There is increased oxygen on the air-heated surface. On the unheated, patterned surface the copper content is too low to be seen by the AES. However, copper clearly migrated to the surface, as shown by the copper AES signal of the heated film while no copper signals were seen in the unheated film and in the film prior to complete image development. In order to avoid copper oxidation in air, however, the curing must be carried out in vacuum or in absence of oxygen. These AES studies alone are insufficient to clearly define the chemical state of copper, i.e., whether it is in its elemental form or in an oxidized state or in the form of a copper-oxygen-carbon complex[8]. Also, it is known from earlier studies[9,10,11] that copper-polyimide interphase forms initially a polyamic acid-copper complex which during the imidization cure decomposes to form cuprous oxide.

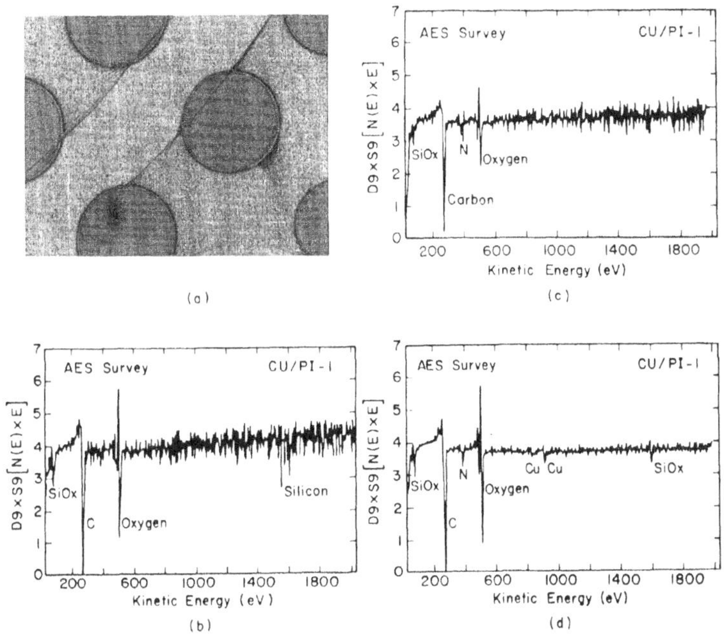

Figure 8. Patterned Cu-doped polyamic acid. (a) Pattern magnification in x100; (b) AES survey at off-pattern position; (c) AES survey on pattern position, unheated; (d) AES survey on pattern position, heated to 325°C for 30 min.

CONCLUSIONS

The following conclusions can be drawn from these results. (1) Metal salt-doped polyamic acid can be successfully patterned using soft vacuum electron beams (50 to 100 pulses, 30 mTorr, 5×10^{-4} to 1×10^{-3} C/cm^2). (2) Heating the patterned surface of a metal salt-doped polyamic acid renders the surface preferentially rich in metal, as evidenced by the enhanced electrical conductivity on the silver surface. These preliminary studies need further confirmation by surface analytical techniques and volume to surface electrical conductivity ratio measurements. (3) Creation of metal-on-polymer image in some cases, therefore, can be carried out in a fewer number of steps. The method looks most hopeful for silver and palladium. (4) Another advantage of the present process is the natural good adhesion of the metal film to the polyimide film.

ACKNOWLEDGMENTS

We sincerely thank Chuck Messer, Marlene Henderson, and Terry Potts of Hewlett Packard, Fort Collins, for their help. This work was supported by AEC, the National Science Foundation, and IBM.

REFERENCES

1. L.T. Taylor, and A.K. St.Clair, J. Appl. Polym. Sci., 28, 2393 (1983).
2. L.T. Taylor, and A.K. St.Clair, U.S. Patent 284,461 (1981).
3. S.A. Ezzell, T.A. Furtsch, E. Khor, and L.T. Taylor, J. Polym. Sci. Polym. Chem. Ed., 21, 865 (1983).
4. A. Auerbach, J. Electrochem. Soc., 131, 937 (1984).
5. T.L. Wohlford, J.S. Schaaf, L.T. Taylor, T.A. Furtsch, E. Khor, and A.K. St.Clair, in "Conductive Polymers," R.B. Seymour, Editor, Plenum Press, New York, p.7, 1981.
6. J. Krishnaswamy. L. Li, G.J. Collins, H. Hiraoka, and M.A. Caolo, paper presented at the Fall Meeting, Materials Research Soc., Boston, 1987.
7. J. Krishnaswamy, L. Li, G. J. Collins, H. Hiraoka, and M.A. Caolo, SPIE Vol. 773, "Electron-Beam, X-ray and Ion-Beam Lithographies VI," p.159 (1987).
8. J.M. Burkstrand, J. Vac. Sci. Technol., 16, 363 (1979).
9. Y.-H. Kim, G.F. Walker, J. Kim, and J. Park, J. Adhesion Sci. Technol., 1, 339 (1987).
10. Y.-H. Kim, J. Kim, G.F. Walker, C. Feger, and S.P. Kowalczyk, J. Adhesion Sci. Technol., 2, 95 (1988).

OPTICAL PERFORMANCE AND DURABILITY OF SILVERED POLYMER MIRRORS

Gary Jorgensen and Paul Schissel

Solar Energy Research Institute
1617 Cole Boulevard
Golden, CO 80401

Metallized polymeric reflector materials for solar energy applications offer significant system weight reduction and cost savings. Such materials must maintain high optical performance over extended service lifetimes. Durability studies of candidate silvered polymer mirrors have been carried out as a function of real-time outdoor weathering and accelerated exposure. Major progress has been made toward achieving a long-term goal of demonstrating a specular reflectance of 90% that is maintained for at least five years.

INTRODUCTION

Use of solar energy to generate electric power and process heat frequently requires large mirrors to provide high levels of concentrated sunlight. Hundreds of millions of square meters of collection area could ultimately be needed to allow substantial displacement of conventional fuel usage. The weight and fragility of glass/metal mirrors impose considerable structural support constraints that, coupled with the material costs, are economically infeasible. Metallized polymeric reflector materials offer significant system weight reduction and cost savings, while affording greater flexibility in terms of system design. Such materials must maintain high optical performance over extended service lifetimes.

Solar thermal technology typically concentrates solar flux via tracking mirrors onto a receiver where energy is collected as heat for conversion to electricity or for use as process heat. The two primary solar thermal technologies are central receiver and distributed receiver designs. Present central receiver systems use fields of heliostats to focus sunlight onto a single receiver cavity. Such fields may be comprised of thousands of heliostats, each typically having 75-100 m^2 of reflecting area mounted on a two-axis drive. In distributed systems, each collector module is paired with its own receiver. Point-focus collectors may be up to 17 m in diameter and use parabolic dish reflectors that track in two directions to concentrate sunlight onto a receiver. Parabolic troughs are line-focus, single-axis tracking collectors that concentrate sunlight onto receiver tubes positioned along their focal lines.

On the basis of optical performance, silver is the reflector material of choice. The hemispherical reflectance of a freshly deposited silver film weighted over the solar spectrum (250-2500 nm) is greater than 97%. A transparent front surface protective layer is required to shield the silver from abrasion, soiling, and corrosion.

Typically, an acrylic polymer with UV absorbers (to inhibit UV-photon activated degradation) is used. The solar-weighted hemispherical reflectance of such an unweathered silvered acrylic material exceeds 92%.

To attain high system energy conversion efficiencies, concentrated sunlight must be used. This means that hemispherical reflectance is not a complete measure of optical performance; candidate mirror materials must exhibit very good specular reflectance as well. Depending upon the solar technology application, the system requirement is 90% reflectance into a full cone angle of 4-8 mrad. Such performance has been demonstrated for unweathered materials; the present thrust is to maintain this level over a minimum five-year lifetime.

The durability of candidate metallized polymer reflector materials is a critical issue. Resistance to optical degradation caused by UV radiation, elevated temperature, atmospheric pollutants, abrasion, and soil retention must be demonstrated. Adhesion at the silvered/polymer interface is another potential problem exacerbated by absorption of moisture. Several approaches have been undertaken to mitigate these effects.

EXPERIMENTAL APPROACH

The current standard construction of a metallized polymer reflector for solar applications consists of a polymer film, a highly-reflective metal layer, and an adhesive, as fabricated by the 3M Company. A substrate material capable of carrying structural loads is also required (Figure 1). Polymethylmethacrylate (PMMA) is

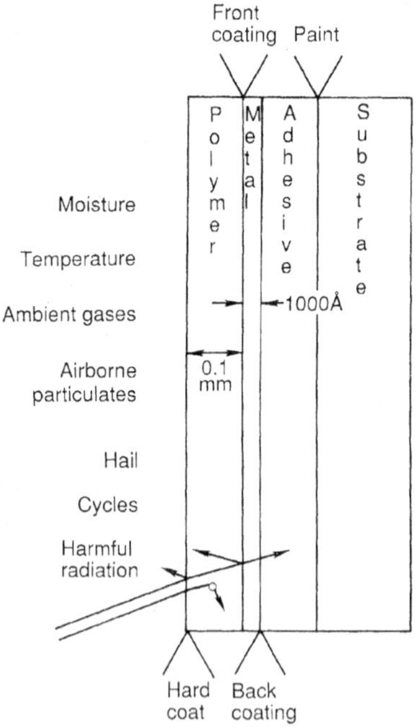

Figure 1. Metallized polymer reflector construction.

the most promising polymeric film to date due to its high transmittance and excellent inherent weatherability. A number of additives are typically incorporated into the bulk polymer to improve the durability of the overall construction.[1] These include UV absorbers to prevent UV radiation from reaching the metal/polymer interface, and anti-oxidants to inhibit oxidation of the polymer.

The metal reflector material of choice is silver. Typically, the silver is thermally evaporated onto the polymer film. The nominal thickness of the silver layer is 100 nm.

The adhesive layer must maintain a good bond between the metallized polymer and the substrate without introducing adverse loss in optical performance. Metals (especially aluminum and stainless steel), polymers, and composites have been considered as substrate material candidates. Both sheet stock and thin membrane substrates are possible.[2] The mechanical properties of the substrate must be able to provide stability against wind and gravity loads and to avert the introduction of significant slope errors due to creep. The surface finish of the substrate must be fine enough to minimize additional specularity loss.

Optical Measurements

Two types of optical measurements are routinely used to quantify degradation effects. The first is specular reflectance at selected wavelengths and acceptance angles. A specular reflectometer, which uses variable circular source and collection apertures, allows absolute specular reflectance to be determined as a function of acceptance angle.[3] Typically, measurements are made at 4-, 8-, and 12-mrad full cone angles. Bandpass filters with central response peaks at 400 and 650 nm are routinely used to monitor optical performance. Measurements at 400 nm are particularly sensitive to the optical performance of silvered polymer mirrors, which tend to degrade in this spectral region.

Hemispherical reflectance throughout the solar spectrum is also regularly measured. Spectral hemispherical reflectance measurements are carried out using dual beam UV-VIS-NIR spectrometers equipped with integrating sphere attachments. Two commercial instruments have been used, namely, a Perkin-Elmer model 340 and the newer model Lambda-9. Data are collected over 250-2500 nm (Figure 2) to allow solar-weighted (air mass 1.5) hemispherical reflectance to be calculated.[4,5] Typically, the response upon weathering of specular reflectance at a particular wavelength has been found to be closely associated with hemispherical reflectance at the same wavelength. That is, the usual effect of weathering is a general loss in reflectance rather than specularity. Thus, hemispherical reflectance measurements provide an excellent basis for comparative durability studies.

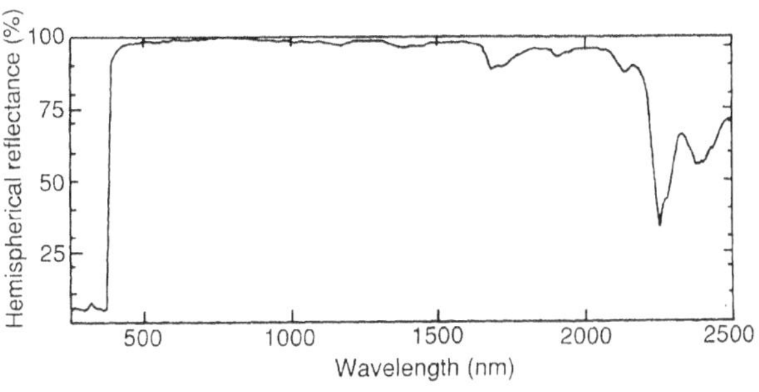

Figure 2. Spectral hemispherical reflectance of typical unweathered standard silvered-polymer mirror.

Weathering

Several weathering methods are used to allow optical durability to be characterized. Following initial (unweathered) optical measurements, samples are subjected to either real-time outdoor exposure or to a controlled, accelerated weathering environment. Ambient weather conditions are a function of geographic location, so outdoor exposure has been utilized for a variety of candidate sites throughout the country. Ultraviolet photons are known contributors to optical degradation of reflector materials. Consequently, a knowledge of the spectral distribution of incident light upon test samples is critical. Figure 3 gives the typical (air mass 1.5) insolation level of the solar spectrum.

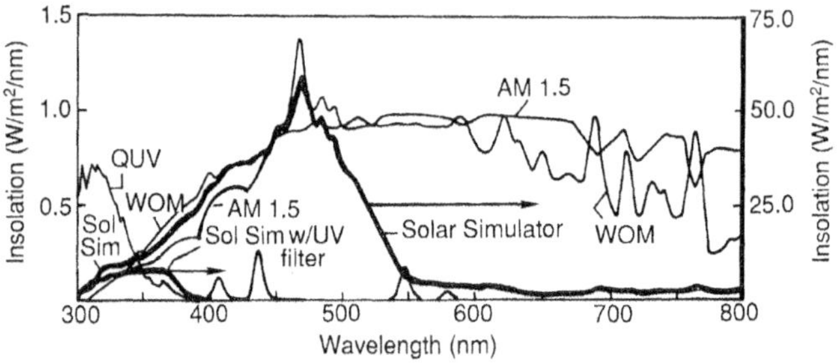

Figure 3. QUV, WOM, and solar simulator (50X with and without UV bandpass filter) vs. air mass 1.5 solar spectrum.

The purpose of accelerated exposure is to be able to predict lifetimes on a time scale appreciably less than the desired or anticipated real-time service lifetime. To accomplish this, the level of probable harmful factors is increased in the hope that the observed durability on an abbreviated time scale correlates with degradation effects experienced during normal service conditions. Three devices have been used to accelerate exposure. These are Weather-Ometers (WOM) (Atlas Electric Devices Co., Chicago, IL), a QUV apparatus (Q-Panel Co., Cleveland, OH), and solar simulators (Oriel Corp., Stratford, CA). The spectral distribution that exposed test samples experience for each of these devices is shown in Figure 3. Controls of other environmental factors are described below for the solar simulator and are given by Schissel, Goggin, and Shinton[6] for the WOM and QUV.

The Weather-Ometer provides a fairly close match to an air mass 1.5 solar spectrum and so does not really accelerate light exposure. Abbreviated results can be obtained, however, because this machine can be run 24 hours each day as opposed to 8 hours per day real-time exposure. General agreement is found between WOM and outdoor exposure results.

QUV exposure is intended to accelerate exposed samples in the UV part of the spectrum. However, because substantial radiation from the QUV exists at wavelengths shorter than the terrestrial solar spectrum, damage can be induced in samples very rapidly that might never be experienced outdoors.

The solar simulator uses the same light source as the WOM, namely, a xenon arc lamp. A dichroic filter is typically used to reduce the heat load of exposed samples. Thus, the spectral distribution is again similar to that of outdoor light between 300-500 nm. By concentrating the filtered beam, acceleration factors of 50X can be obtained in this wavelength region, as shown in Figure 3.

EXPERIMENTAL RESULTS AND DISCUSSION

Metallized reflector materials during the early 1980s were designed for indoor lighting applications. They were thrust into outdoor service environments by the demand for alternative mirrors. Only a low amount of UV stabilizer was typically contained in the bulk polymer film. Compared to silver, aluminized polymers exhibited greater durability in terms of corrosion-induced optical degradation. However, the reflectance of unweathered aluminized polymers is considerably lower than silvered polymers. The optical constants of aluminum are such that when they are coupled with a polymeric film (having a typical index of refraction of 1.4-1.5) the solar-weighted hemispherical reflectance drops from 92% (for front surface aluminum) to roughly 86%. A comparable value for silvered polymers is 95% (versus 98% for the unprotected metal). The spectral hemispherical reflectance of a typical unweathered, standard silvered polymer mirror is shown in Figure 2. Excellent optical performance is exhibited. To maintain such performance during service, however, required vastly improved durability of silvered polymers.

Experimental data strongly suggested that optical durability of these early silvered polymers was a strong function of exposure to UV radiation.[3] One obvious way to mitigate this effect was to increase the amount of UV absorbers added to the bulk polymer film. Early reflectors contained roughly 0.02 wt % of UV stabilizers. This constituent was increased to 1.0 wt % of the polymer film. Such an additive is designed to substantially attenuate terrestrial solar radiation below 385 nm. In blocking these suspected harmful low wavelength photons from reaching the silvered/polymer interface, roughly 2.5% of the solar resource is sacrificed.[5] Coincidentally, this energy band would be forfeited by the substantial decrease in reflectance of silver (due to plasmon absorption[7]) in this wavelength region.

The increased UV absorber content of more recent silvered polymer reflector materials has dramatically improved the durability of these mirrors.[3] The stability of the current best "standard" silvered polymers is very good. Real-time outdoor weathering results indicate that such materials can plausibly resist corrosion for five years in certain geographical locations (such as Golden, CO, and Minneapolis, MN).

For other sites, further improvement in durability is needed. In Arizona, California, and Florida, for example, a factor of three or four improvement is required to achieve the durability goal. The regional dependence of durability cannot be explained in terms of radiation levels; the insolation in Phoenix, AZ, is only 10% greater than in Golden. This suggests that insolation is not the controlling factor for the degradation rate in Phoenix. Note that this is in contrast to accelerated weathering devices such as the QUV and WOM where degradation is strongly coupled to insolation. A possible explanation may be the elevated temperatures experienced in the accelerated tests; temperatures may be high enough that the rate of corrosion is insensitive to changes in temperature but susceptible to changes in levels of light. Outdoors, the rate of corrosion may not have been saturated by thermal effects and so may be more closely linked to changes in ambient temperature than changes in insolation.

Temperature sensitivity may be related to permeation through the polymer film. Although optical degradation is greater in Phoenix than in Golden, preliminary data suggest that the concentration levels of some pollutants are lower in Phoenix compared with those in Golden. However, the lower concentration of pollutants (or O_2, H_2O) could be more accessible to the silvered/polymer interface because of the higher ambient temperatures in Phoenix.

Several experiments were performed to try to isolate the various suspected deleterious effects and their possible synergistic interactions and to further quantify rate-limiting factors in optical degradation of metallized polymers. A vacuum furnace was used with a quartz window to allow UV exposure of standard silvered polymer mirrors (mounted on painted aluminum substrates) under controlled temperature and external environmental conditions.

The light source was a 1000-W xenon arc lamp solar simulator. The spectral bandwidth was limited to between 305 nm (by a 305 nm longpass filter) and roughly 550 nm (by a dichroic filter). This radiation was further filtered by a UV bandpass filter selected to match the UV absorber characteristics of the polymer film (i.e., a Schott UG-11 filter with a peak of 75% transmittance at 330 nm and a 90 nm FWHM). Spectral traces of these distributions are shown in Figure 3.

In one experiment, a total of six samples experienced a constant purge of ultrapure nitrogen for exposure times up to 500 hours. A second experiment used a compressed air purge for 250 hours. In each case, three samples were maintained at 30°C by a water-cooled substrate, and three samples were maintained at 80°C by a thermocouple-controlled pad heater. One each of the high and low temperature samples was shielded from the solar simulator, one each of the high and low temperature samples was exposed to the light source described above (including the bandpass filter), and one each of the high and low temperature samples was exposed to the light source described above without benefit of the UV bandpass filter.

Spectral hemispherical reflectance measurements of each set of six samples were compared with data taken prior to exposure. Results after 238 hours exposure for nitrogen purge and 250 hours exposure for compressed air purge are shown as difference curves $[R_{2\pi}(\lambda,t=0) - R_{2\pi}(\lambda,t=238,250 \text{ hours})]$ in Figures 4 and 5, respectively.

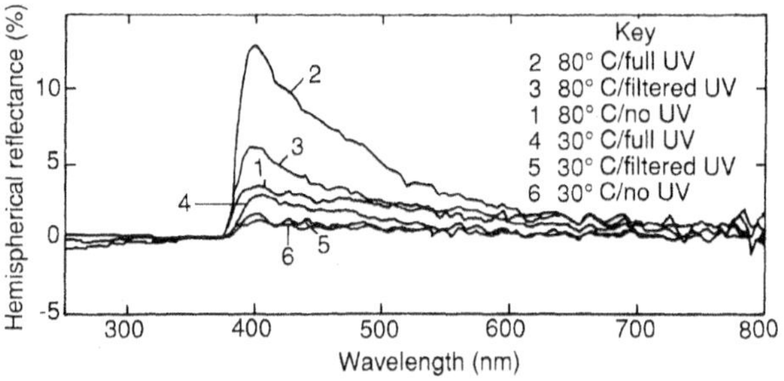

Figure 4. Difference curves (0-238 h) for solar simulator/vacuum furnace, nitrogen purge.

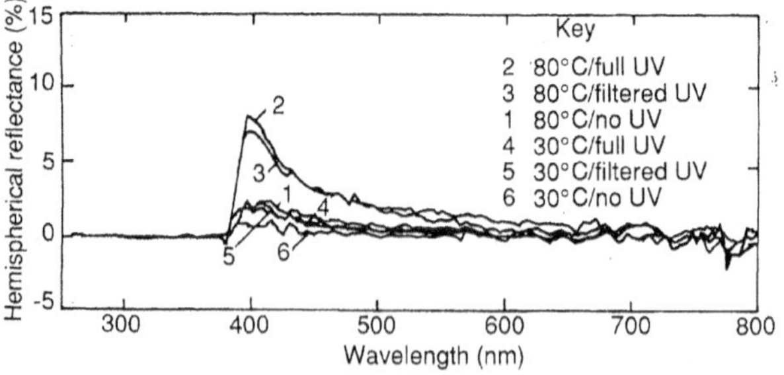

Figure 5. Difference curves (0-250 h) for solar simulator/vacuum furnace, compressed air purge.

Table I. Results of Analysis of Variance of Solar Simulator Experiments; Xs represent effects that are statistically significant at a 95% confidence level.

Radiation	Time of Exposure (hr)	Factors			
		Atmosphere	Light	Temperature	Light + Temperature
Filtered	100	X	X	X	X
Full	100		X	X	X
Filtered	250			X	
Full	250		X	X	

In the case of the nitrogen environment, Figure 4 indicates that samples subjected to the elevated temperature (80°C) degrade most severely. At that temperature and in the spectral band of 400-500 nm, samples exposed to "full UV" exhibit roughly twice the degradation experienced by samples that see "filtered UV" light. The filtered light, in turn, roughly doubles the optical degradation that occurs at 80°C for the shielded (no UV) sample.

As can be seen in Figure 5, exposure in air for 250 hours apparently causes less severe degradation than that experienced in a pure nitrogen environment. Although initially surprising, there is some evidence in the literature for greater sensitization of silver compounds to light in a nitrogen atmosphere relative to an environment that contains oxygen.[8] Compared with the nitrogen purge data, the relative amount of peak degradation (around 400 nm) is similar in magnitude although somewhat different in spectral characteristics. Additionally, the elevated temperature/filtered UV exposure is almost as harsh as the elevated temperature/full UV condition in the presence of air. This was not the case for a nitrogen purge environment where the full UV condition was markedly more severe than the filtered UV exposure of the 80°C samples.

An analysis of variance was performed on the results of these experiments. For each type of light exposure (full vs. filtered, corresponding to the two curves labeled as "Solar Simulator" and "Sol Sim w/UV filter" in Figure 3), the analysis was run for data after 100 hours and 250 hours of exposure. In each case, the three factors and their levels of influence were (1) atmosphere (nitrogen vs. air), (2) light (none vs. full or filtered), and (3) temperature (30°C vs. 80°C). Results are shown in Table I; effects that are significant at a 95% confidence level are indicated by Xs.

Other interaction terms (light + atmosphere and temperature + atmosphere) were not statistically significant for any of the cases listed in Table I. As can be seen, after 100 hours, an interaction of light and temperature is statistically significant for both short-wavelength UV light (filtered case) and for UV + visible (full case). The fact that this factor does not show up after 250 hours of exposure is indicative of an induction period effect; after some initial degradation is caused by a synergism between light and temperature, the importance of this cross term decreases. Exposure to UV + visible (light only) remains important in the degradation process after 250 hours. In all cases, temperature is seen to be a significant factor. There is some indication that atmosphere plays an initial role for shorter wavelength exposure.

Other concerns with the standard silvered polymer reflector include potential leaching of stabilizers out of the bulk polymer film during weathering,[1] photoreaction of light with absorber species, and transmittance of UV radiation through

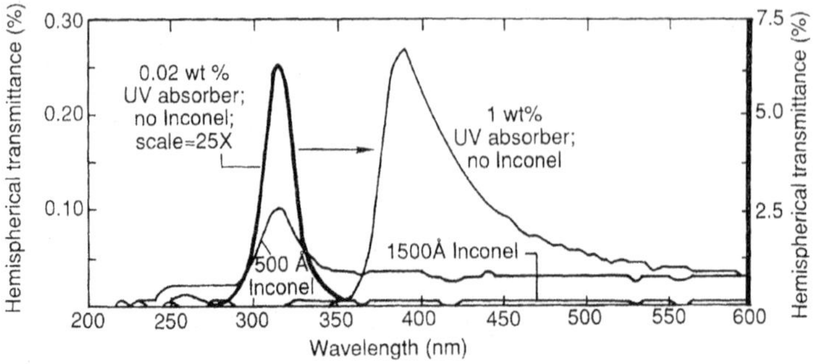

Figure 6. Hemispherical reflectance transmittance of various silvered-polymer reflectors.

the mirror. Figure 6 shows the measured hemispherical transmittance for various silvered polymer reflectors. The standard material, labeled "1 wt% UV absorber, no Inconel," exhibits a peak transmittance of about 0.3% just below 400 nm. The peak transmittance of silver occurs near 320 nm, as can be seen by the curves for older reflectors having only 0.02 wt % UV absorbers in the polymer film. Note that for such a low concentration of UV absorbers, transmittance through the silvered polymer exceeds 6%. The peak for the standard mirror has been "shifted" to higher wavelengths due to the effective cutoff at 385 nm caused by the abundant presence of UV absorbers in the polymer film.

Conceptually, the durability of the standard construction can be improved in a number of ways, as denoted in Figure 1. Enhancements can be made at each interface either separately, or in combination with each other or with further improvements within the bulk polymer film. Some options may address several concerns (for example, treatment of the silvered/polymer interface to reduce corrosion and increase adhesion).

Front Surface Abrasion Resistant Hard Coatings

The addition of a hard coating at the outer surface of the polymer film has the potential to improve durability in several ways. If such a layer were an abrasion-resistant coating (ARC), a more aggressive cleaning regimen could be prescribed to combat the loss of optical performance by soiling. Some measure of protection could also be provided against scattering losses due to crazing of the polymer surface caused by wind-borne particulates and hail. The dense nature of such hard coatings may serve as a barrier against potentially harmful atmospheric pollutants and moisture.

Two types of ARCs have been investigated, namely, temperature-cured and UV-cured coatings. The approach has been to apply such coatings to standard silvered polymer mirror materials and to measure specular reflectance at 650 nm as a function of abrasion. Cyclic abrasion of samples is carried out according to Ref. 9 using a Tabor Abraser model 505 (Teledyne Taber, N. Tonawanda, NY). Improved abrasion resistance has been demonstrated for both types of ARCs. Good weatherability of temperature-cured ARCs has also been demonstrated, and some data suggest that these coatings may reduce the loss of reflectance and improve the cleanability of such reflectors.

Bulk Polymer

Another way to improve the durability of silvered polymer mirrors is to modify the constituents of the base PMMA resin. Stabilizers added to the polymer film have been shown to improve the weatherability of such reflectors.[1] Alteration of the amount and/or type of UV absorbers could conceivably enhance performance. The elimination from the bulk polymer of trace quantities of chemicals believed to be harmful (such as chlorine and especially sulfur) is another way of potentially increasing durability.

To evaluate the effectiveness of increasing the degree of UV absorptance and decreasing the permeability to ambient air (O_2, H_2O, pollutants), a series of screening experiments were run. Standard silvered polymer mirrors were further shielded by unsilvered polymer films and glass filters having longpass cutoff wavelengths (at which transmittance is 50%) of 305, 375, 395, and 495 nm. Some filters were mounted in such a way that light was filtered but ambient air was not excluded; others were sealed along the edge with tape to provide increased barriers to both light and atmospheric effects. Samples were exposed to both WOM- and QUV-accelerated weathering and compared with results for standard silvered polymer mirrors.

After three months exposure, optical measurements indicate that additional UV screening does in fact improve durability. Photons above 495 nm cause essentially no degradation; photons between 395 and 495 appear to cause some degradation but cannot be extensively filtered because significant solar radiation would be sacrificed. Photons between 305 and 395 nm cause damage in spite of the UV filtering between 280-385 nm of the standard polymer film. Results also reveal that exclusion of ambient air is less important than restricting incident photons, at least for accelerated testing.

To further quantify the effect of externally screening photons, another round of filter experiments was performed. This time, a finer bracketing of longpass cutoff wavelengths was used (305, 360, and 395 nm) in an attempt to more accurately identify which bands of photons are most harmful. Results are presented in Table II for hemispherical reflectance measurements (at 400 nm and weighted over the solar spectrum) of standard silvered polymer mirrors that were shielded by various longpass cutoff filters during WOM exposure. No effort was made to exclude the ambient atmosphere; filters were "spaced off" the reflector surface rather than mounted with edges sealed. Data in Table II represent averages for three replicates of each sample/filter type.

Table II. Hemispherical Reflectance at 400 nm ($R_{2\pi}$[400 nm]) and Solar-Weighted Hemispherical Reflectance ($R_{2\pi}$[250-2500 nm]) of Standard Silvered Polymer Mirrors Shielded by Longpass Cutoff Filters During WOM Exposure.

Filter Wavelength (nm)	WOM Exposure (mo)	$R_{2\pi}$(400 nm) (%)	$R_{2\pi}$(250-2500 nm) (%)
305	0	90.1	92.4
	1	85.7	91.2
	2	84.0	90.7
	3	81.2	88.6
	3½	79.1	87.1
360	0	90.2	92.4
	1	86.3	91.5
	2	84.4	90.0
	3	81.9	89.1
	3½	79.9	87.6
395	0	89.9	92.3
	1	87.8	91.9
	2	85.9	91.3
	3	84.6	91.0
	3½	83.6	90.7

As with the earlier filter experiments, some optical degradation occurred even for the 395-nm filter samples. However, the decrease in performance is significantly lower than that of the 305-nm and 360-nm filter samples. Little difference can be seen between the 305-nm and the 360-nm filter results, indicating that photons in that spectral band are adequately screened by the UV absorbers within the polymer film. Comparison of the 395-nm vs. 305-nm and 360-nm filter data suggests that photons between 360 and 395 nm are damaging to optical durability. Since the cutoff wavelength of the UV absorbers in the polymer film is roughly 385 nm, photons between 385-395 nm are particularly suspect.

Based upon results of experiments using external filters to impede the influence of light and atmospheric constituents on silvered polymer materials, alternative filtering within the bulk polymer film was explored. The amount and type of added UV absorbers were varied in an attempt to control the level and spectral content of light reaching the silvered/polymer interface. Increasing the amount of absorber from 0.02 wt % to 1.0 wt % had previously produced substantial improvement in durability in terms of corrosion and failure of the adhesive layer.[10] Consequently, increasing the amount of the standard (type A) absorber additive from 1.0 to 1.5 wt % was expected to result in further improvement. Such an addition of more stabilizer not only provides a greater reserve of absorbing sites within the bulk polymer, but it also pushes the cutoff wavelength (at which the reflectance of the mirror drops to 50%) from 386 nm to 393 nm.

Another approach is to use a different type of UV absorber material (so-called type C), which exhibits a different spectral absorptance profile. The cutoff wavelength for 1.5 wt % of type C is 401 nm. Reflector samples with 1.0 wt % type A, 1.5 wt % type A, and 1.5 wt % type C absorbers (all with the standard PMMA base resin) were subjected to three months WOM exposure. Spectral hemispherical reflectance was measured and compared with initial (unexposed) data via spectral difference curves. The results for the three samples are virtually identical (Figure 7). As can be seen, no improvement occurs by increasing the amount of type A absorber from 1.0 to 1.5 wt %; the spectral shifts are due to the difference in cutoff wavelength. Furthermore, the alternative type C absorber is also no more effective than the standard 1.0 wt % type A.

These results appear to contradict those of the external filter experiments in predicting the effect of photons between 385-400 nm upon optical durability. The discrepancy may be due to the difference in transmittance properties of the glass filters compared with UV stabilized PMMA films. In general, the polymer film achieves a maximum transmittance of 92% between 400-1700 nm, and the 395-nm glass filter reaches 87% above 475 nm. In addition, the cutoff is much sharper for

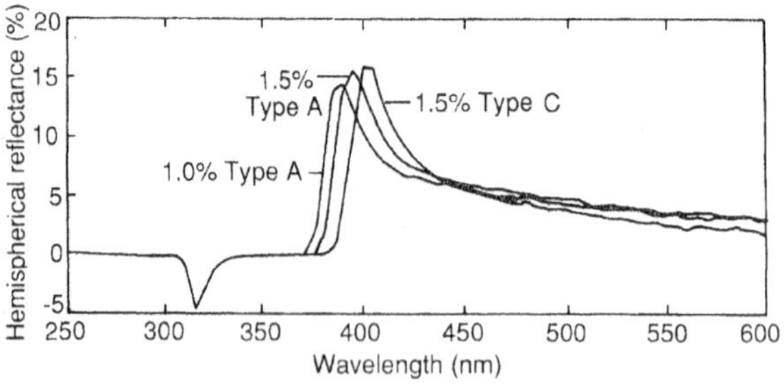

Figure 7. **Spectral hemispherical reflectance difference curves (initial to 3 months WOM exposure) for standard 1.0% type A (386 nm, 50% cutoff) 1.5% type A (393 nm, 50% cutoff) and 1.5% type C (401 nm cutoff) UV absorbers.**

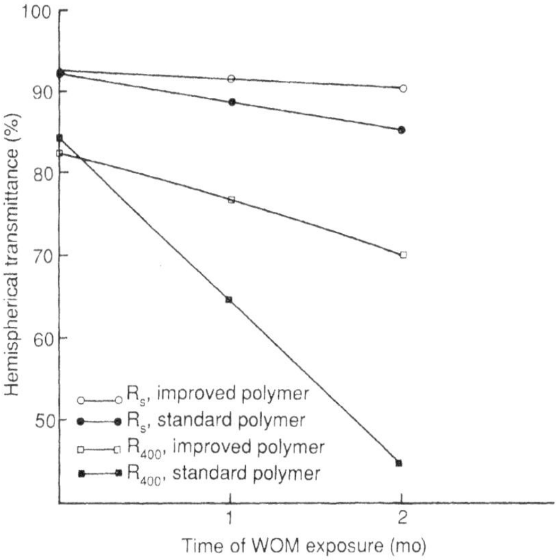

Figure 8. Solar-weighted hemispherical reflectance R_s and hemispherical reflectance at 400 nm R_{400} for standard vs. improved polymer formulation (reflector-mounted on aluminum substrate) as a function of WOM exposure.

the polymer film, going from 5% to 95% of maximum over a 25 nm interval compared to a 75 nm roll-off for the filter. Consequently, the polymer films all transmit considerably more radiation between their respective nominal cutoff wavelengths and 400 nm than the 395-nm glass filter.

Another factor which may contribute to the disparity is uncertainty associated with the chemical composition of the additives used with the base polymer resin. For example, the type C absorber material contains chlorine, which may contribute to degradation and possibly obscure some otherwise beneficial effects of shifting the cutoff to higher wavelengths. Sulfur containing compounds are also suspected of participating in optical degradation.

One attempt to control the trace constituents of the base PMMA resin has been an improved formulation which contains sulfur-free antioxidants. Films made from this resin have been silvered and mounted on an aluminum substrate (to accelerate degradation as described below) for comparison with the standard silvered polymer reflector (also mounted on aluminum). Results of optical measurements after two months WOM exposure are shown in Figure 8. Significant improvement in both hemispherical reflectance at 400 nm and in solar weighted hemispherical reflectance is clearly evident.

Silvered/Polymer Interface Coating and Metal Back Coating

Another approach to preventing degradation is to deposit a passivation layer at the silvered/polymer interface. Such a layer may take the form of thin inorganic or organic coatings. By creating a barrier at the interface, protection against both externally and internally generated (i.e., within the polymer film) reactants can be provided. At the same time, such a coating could also promote adhesion.

Evidence also exists that metallic back coatings can improve durability of silvered polymer mirrors. As was indicated in Figure 6, a metallic back coating can virtually eliminate the UV transmittance of an unbacked silvered polymer reflector. This would avert the possibility of harmful effects associated with such radiation reaching and interacting with the rear adhesive layer.

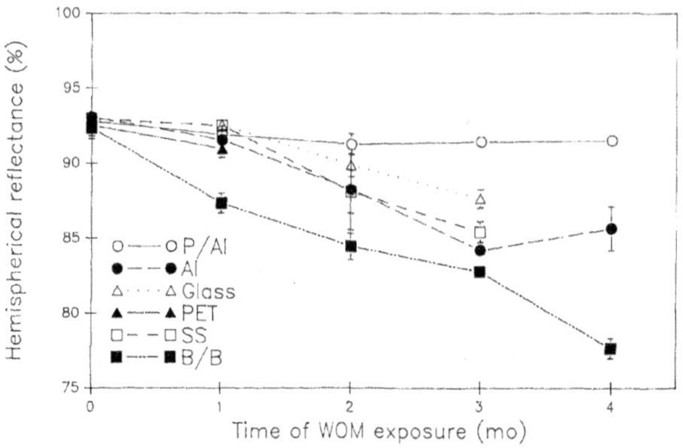

Figure 9. Solar-weighted hemispherical reflectance as a function of substrate type.

One concern of metallic back layers is their effect upon specularity caused by mechanical mismatches during the deposition process or during thermal cycling while in service. Earlier work with metal back coatings[10] found that the specularity of silvered polymer mirrors with aluminum backings decreased rapidly with weathering. Inconel was more promising in terms of specularity but seemed to accelerate general corrosion during weathering.

Substrate Coating

Optical durability of silvered polymer mirrors during weathering has been found to be strongly dependent upon the type of substrate to which the reflector is mounted. Mirror samples were mounted to a variety of substrates and subjected to accelerated weathering in the QUV and WOM. Substrates included painted aluminum, bare aluminum, 416 stainless steel, glass, and polyester. A "back-to-back" configuration was also used whereby the adhesive layers of two silvered polymer samples were directly applied to each other.

Results after four months WOM exposure are shown in Figure 9. Clearly the greatest resistance to degradation is provided by an aluminum substrate with a baked white polyester-based paint (P/Al). By two months, all other substrates significantly degraded. The back-to-back configuration (B/B) was clearly the worst; bare aluminum (Al), glass, polyester (PET), and 416 stainless steel (SS) were all intermediately poor. Error bars in Figure 9 represent 1 standard deviation associated with a number of samples.

Little difference was detected between polished vs. unpolished aluminum or stainless steel. Bare 416 stainless steel was found to degrade in unique ways; corrosion was found even when samples were shielded from the light source, suggesting sensitivity to temperature/atmospheric effects that are not photon driven.

Results from outdoor exposure are in general agreement with the WOM experience; painted aluminum is presently the best substrate. Samples exposed in the QUV did not show a sensitivity to substrate; samples mounted on all substrate types exhibited rapid optical degradation. This suggests that either longer wavelength photons give rise to the substrate variability experienced outdoors and in the WOM, or else the shorter wavelengths found only with the QUV cause such severe degradation that it masks any potential benefit of one substrate compared with another. Experiments are under way to investigate other paint/substrate combinations.

CONCLUSIONS

Durability studies of candidate silvered polymer mirrors have been carried out as a function of real-time outdoor weathering and accelerated exposure. Major progress has been made toward achieving a long-term goal of demonstrating a specular reflectance of 90% that is maintained for at least five years. Real time outdoor weathering suggests that this goal has been achieved for some geographic locations (for example, Minnesota, Colorado, and New Mexico). For other sites (for example, Arizona, California, and Florida) a 3-4X factor increase in service lifetime is needed. A variety of approaches have been identified to achieve further improvement of the optical durability of silvered polymer mirrors. Many of these strategies have shown high initial promise.

A better understanding of the effect of substrate materials on durability is needed. Other issues such as cleanability and adhesion of the silvered/polymer interface must be addressed in greater detail.

ACKNOWLEDGMENTS

This work was sponsored by the U.S. Department of Energy under contract DE-AC02-83CH10093. Numerous test samples were provided by B. A. Benson of the 3M Company. Additional sample preparation, weathering, optical characterization, and useful discussions were provided by members of the Materials Research Branch of the Solar Energy Research Institute.

REFERENCES

1. H.H. Neidlinger, and P. Schissel, Photostabilization studies of silver/ polymethylmethacrylate films, SPIE Proc. 823, 181 (1987).

2. C. Kutscher, A. Lewandowski, R. Wood, T. Wendelin, and G. Jorgensen, "Membrane Dish Research Task FY 1987 Progress Report," Solar Energy Research Institute, Golden, CO (1988).

3. P. Schissel, and H.H. Neidlinger, "Polymer Reflectors Research During FY 1986," SERI/PR-255-3057, Solar Energy Research Institute, Golden, CO (1987).

4. ASTM Standard E903-82, Solar absorptance, reflectance, and transmittance of materials using integrating spheres, American Society for Testing and Materials, Philadelphia, PA.

5. ASTM Standard E891-82, Terrestrial direct normal solar spectral irradiance tables for air mass 1.5, American Society for Testing and Materials, Philadelphia, PA.

6. P. Schissel, R.M. Goggin, and Y.D. Shinton, Durability of silvered polymer mirrors, SPIE Proc. 823, 190 (1987).

7. P. Schissel, H.H. Neidlinger, and A.W. Czanderna, "Identification of Chemical and Physical Phenomena Affecting the Durability of Silvered Polymer (PMMA) Mirrors," SERI/PR-255-2493, Solar Energy Research Institute, Golden, CO (1985).

8. T.H. James, Some effects of environment on latent image formation by light, Photo. Sci. and Eng., 14, No. 1, 84 (1970).

9. ASTM Standard D1044-82, Resistance of transparent plastics to surface abrasion, American Society for Testing and Materials, Philadelphia, PA.

10. P. Schissel, H.H. Neidlinger, and A.W. Czanderna, "Polymer Reflectors Research During FY 1985," SERI/PR-255-2835, Solar Energy Research Institute, Golden, CO (1986).

DEPOSITION OF ADHERENT, THICK COPPER COATINGS ON GLASS

W. C. Cowden, T. G. Beat, T. A. Wash and J. W. Dini

Lawrence Livermore National Laboratory
P. O. Box 808, L-332
Livermore, CA 94550

Glass parts were successfully coated with copper as thick as 1 mm using the combined technologies of vacuum deposition and electroplating. The key elements to the process were: preparation of the glass substrates, deposition of a low stress titanium binder layer by physical vapor deposition followed by a 6 μm thick layer of copper without breaking vacuum, and then electroplating 1 mm of copper in acid sulfate solution. An extremely important part of this work was the deposition of copper with as close to zero stress as possible. Data were generated on the influence of plating current density on stress and mechanical properties.

INTRODUCTION

Thick (1 mm), adherent copper was needed on low coefficient of expansion glass to serve as a heat sink for a laser calorimeter. Another requirement was for 0.4 mm of copper on glass for an x-ray spectrometer system and lastly, diamond machinable coatings approximately 125 μm thick were sought for possible improvements in the optics of systems currently fabricated from metal. To meet these requirements, vacuum deposition and electroplating were combined. An initial adherent, thin layer was deposited by magnetron sputtering or electron beam evaporation and then thick copper was applied by electroplating. This combined usage of these two different coating processes is but one example of applications wherein these technologies can complement one another[1,2].

The conventional technology used by platers for metallizing nonconductors via stannous chloride/palladium chloride activation followed by electroless deposition prior to final electroplating is not acceptable for glass when thick coatings are required. With this process, deposits thicker than around 12.5 μm easily separate from a glass substrate.

The generally accepted criterion for adhesion between an oxide substrate such as glass and a metal film is that the metal must be oxygen active to react chemically with the oxide surface, forming an interfacial reaction zone[3]. Table I gives the heat of formation of various metal oxides. The higher the negative heat of formation, the higher the affinity for oxygen. Deposition of one of these "glue" or "binder" layers by physical vapor deposition (PVD) followed by a thin layer of copper without breaking vacuum produces an adherent base for thick electroplating.

ULTRA LOW EXPANSION GLASS

Our first efforts were with ultra low expansion glass (ULE). The ULE substrates (2.5 cm x 15 cm x 2.5 cm thick) were prepared by first fine grinding the surfaces to be coated with 3 µm particles of aluminum oxide followed by an etch in a six percent solution of nitric acid at 100°C. The glass was then placed in the vacuum chamber and preheated to 250°C for a minimum of eight hours. Next the substrates were coated with 300 Å of a chromium binder using electron beam evaporation. Titanium has also been used successfully as a binder layer. Without breaking the vacuum between the coating steps, the binder layer of chromium was overcoated with 4000 to 5000 Å of copper. The samples were then electroplated with copper to final thickness of approximately 125 µm, which was thick enough to allow for precision machining and polishing.

Table I. Heat of Formation of Various Metal Oxides.

Oxide	Heat of formation, kcal/mol
Nb_2O_5	-463
V_2O_3	-290
Cr_2O_3	-270
TiO_2	-218
WO_3	-200
MoO_3	-180

Values taken from "Handbook of Chemistry and Physics", Chemical Rubber Co., Cleveland, OH (1971).

LASER CALORIMETER APPLICATION

Neutral density (common window glass) glass substrates (15 cm by 15 cm) coated with 1 mm of copper on one side were needed for a laser calorimeter. The final assembly will contain nine layers, with the first eight consisting of PVD films and the last layer the thick electroplated copper deposit (Figure 1). The work reported herein was directed at our efforts at producing the thick, adherent copper deposit. Efforts at fabricating the entire assembly are proceeding now that we have demonstrated success with the thick copper film. Since this is the most recent example of glass parts that have been coated, complete details on the cleaning steps prior to coating, substrate in-situ treatment, PVD coating and electroplating are included.

Substrate Preparation

Initial Cleaning

A. Substrate was immersed in 2% liquid laboratory detergent cleaner and scrubbed with gauze pad on both sides. It was then rinsed in flowing distilled water.

B. While still wet, the substrate was submerged in the vapor from a boiling ethyl alcohol solution for ten minutes. It was then removed from the solution and placed in a vented hood for 30 minutes to dry.

C. Prior to loading in the vacuum chamber, the final cleaning step consisted of a lens paper drag across the surface finish with ethyl alcohol.

Substrate In-Situ Treatment

Substrates were preheated to 250°C for a minimum of eight hours prior to the start of the vapor deposition cycle. This temperature was maintained throughout the PVD cycle. Inside the vacuum chamber, the substrates were mounted on a rotatable box shaped housing that provided indexing capabilities to each sputter source. Substrate heating was achieved using quartz lamps located inside the rotatable stainless steel box. Two dc magnetron sputter sources were mounted 90° off axis from the box (see Figure 2).

Physical Vapor Deposition

A. The binder layer for these substrates was titanium deposited by magnetron sputtering. Deposition rate was 0.7 Å/sec to final thickness of 1000 Å.

B. The substrate holder/heater was then exposed to a copper magnetron sputter source and copper was deposited at a rate of 30 Å/sec to a final thickness of 6 µm.

C. The substrate was allowed to cool to ambient temperature prior to venting the vacuum chamber. It was then placed in a desiccator containing an argon atmosphere and transferred to the electroplating area.

Electroplating

A. The part was removed from the desiccator and placed in a plating fixture designed to hold it and provide electrical contact and shielding during plating.

B. The fixture containing the part was placed in the copper plating solution for 15 to 30 seconds before current was applied.

C. After completion of plating, the part was rinsed and dried and removed from the fixture. Figure 3 shows two panels plated with 1 mm of copper; the smaller panel was machined by single point diamond turning after plating.

Copper plating was done at current densitites that resulted in zero of slightly compressive stress in the deposits. This stress was measured by using a commercial device* which relies on the rigid strip principle. During plating, opposite sides of a two legged strip are plated and the resulting deposit causes the strip to spread apart. Figure 4 shows a test strip and the device used to measure deflection.

---

*Electrochemical Co., Inc., York, PA, Model 683 EC

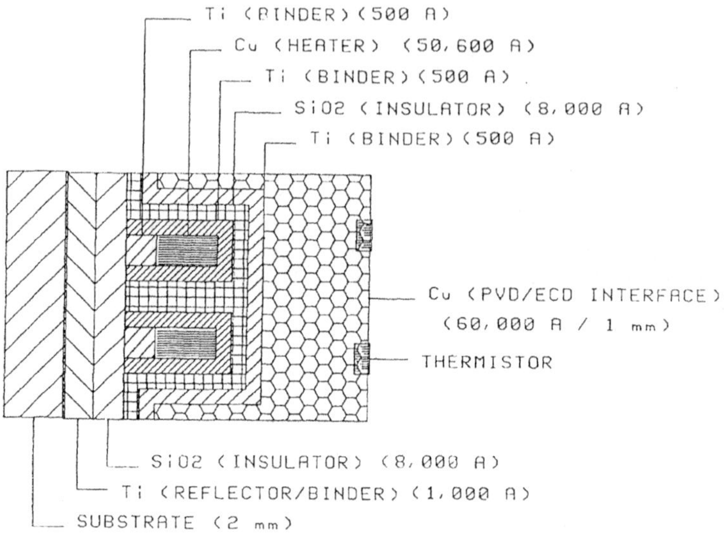

Figure 1. Cross Section of NOVA Calorimeter Assembly.

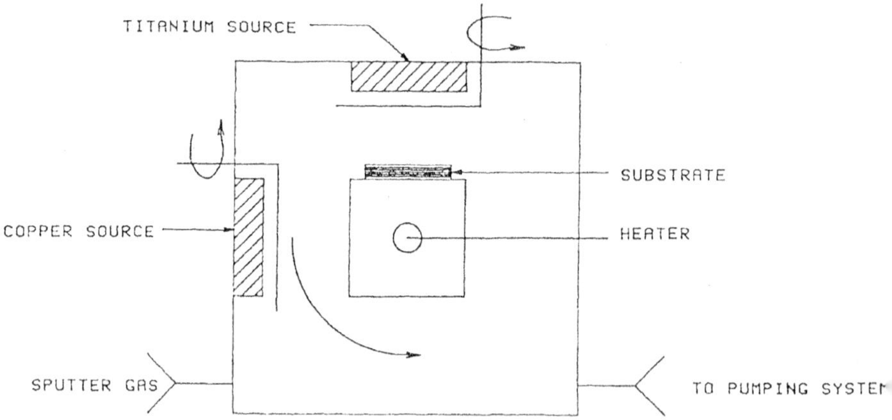

Figure 2. PVD Coating System Schematic.

Figure 3. Glass Panels Plated with 1 mm of Copper.

Figure 4. Device Used to Measure Electrodeposit Stress.

Two copper plating solutions were used in this work. One was a proprietary acid sulfate solution and the other was a low copper-high acid solution containing no brighteners. Stress in the deposits produced in these solutions is shown in Figures 5 and 6 as a function of plating current density. During plating of glass substrates, the proprietary solution was operated at a current density of 161 A/m^2 and the solution containing no additives was operated at a current density of 108 A/m^2.

Mechanical property data for the electroplated deposits are presented in Tables II and III. These data clearly show that the deposits produced in the solution containing no additives were considerably softer and less stronger than those produced in the solution containing the proprietary additive. These results were not unexpected and agree well with other published data for deposits produced under similar conditions[4,5]. Since the deposit produced in the no additive solution was so soft and yielding, it was our preferred deposit for the larger glass substrates (15 by 15 cm square and 22 cm diameter parts).

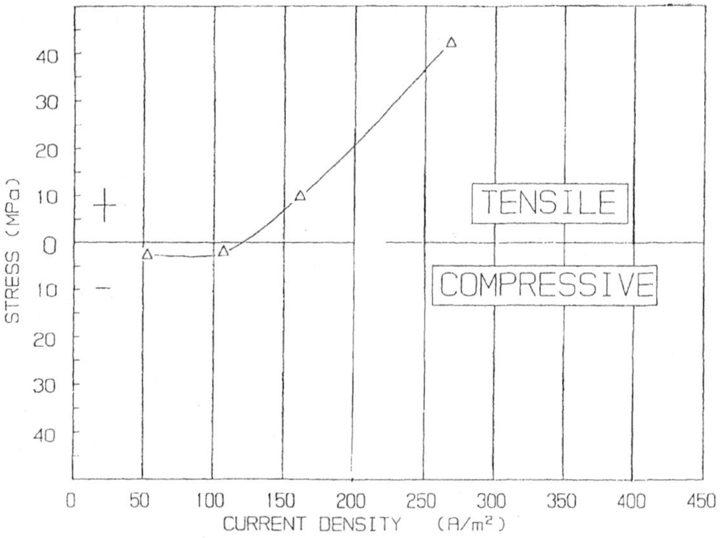

Figure 5. Deposit Stress as a Function of Current Density for a Low Copper/Sulfuric Acid Solution Containing No Brightener.

OPTIC FOR SPACE PROGRAM APPLICATION

A 22 cm diameter optic for a space program application which was fabricated of Pyrex glass required a 0.4 mm thick copper deposit on one side in a region encompassing the center 15 cm diameter of the part. The same procedure discussed above for the laser calorimeter was used for this part.

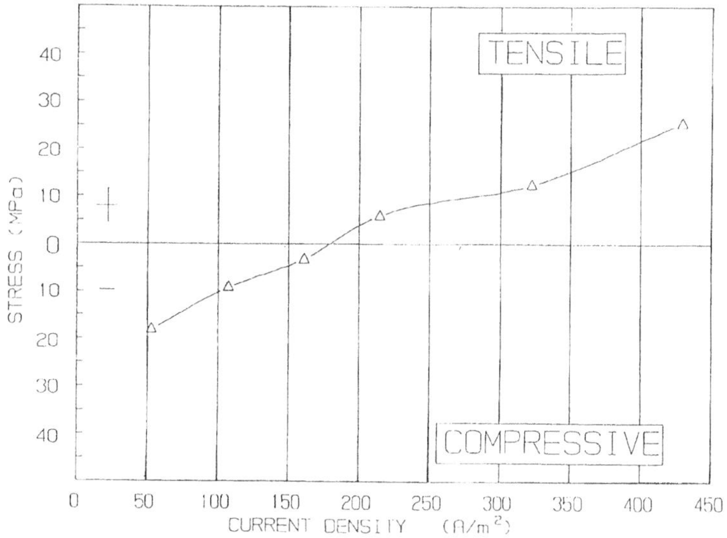

Figure 6. Deposit Stress as a Function of Current Density for a High Copper/Low Sulfuric Acid Solution Containing a Proprietary Brightener.

Table II. Mechanical Property Data for Low Copper High Acid No Brightener Solution[a][b]

Current Density (A/m^2)	Yield Strength (MPa)	Tensile Strength (MPa)	Elongation[d] (%)	Hardness ($V_{100}g$)
54	28	95	4.7	51
108[c]	28	95	5.3	61
161	37	119	9.5	57
269	181	344	10.7	133

a The solution contained 36 g/l $CuSO_4.5H_2O$, 173 g/l H_2SO_4 and 50 mg/l chloride. Operating temperature was 24°C.

b Thickness of all deposits was 0.75 mm.

c Condition used for plating on glass substrates.

d Gage length was 2.5 cm.

Table III. Mechanical Property Data for High Copper Low Sulfuric Acid Proprietary Brightener Solution[a][b]

Current Density (A/m^2)	Yield Strength (MPa)	Tensile Strength (MPa)	Elongation[d] (%)	Hardness (V_{100g})
54	--	252	1.5	142
161[c]	479	696	14.7	125
387	--	673	15.7	115

a The solution contained 200 g/l $CuSO_4.5H_2O$, 68 g/l H_2SO_4, 50 mg/l chloride, and 0.06 to 0.225% by volume, UBAC No. 1 additive (OMI International Corp., Warren, MI). Operating temperature was 24°C.

b Thickness of all deposits was 0.75 mm.

c Condition used for plating on glass substrates.

d Gage length was 2.5 cm.

After electroplating, the part was diamond turned to a final thickness of 0.25 mm and then coated with a thin gold film and a layer of SiO_2. The PVD process used to apply these latter coatings required heating at 250°C for four hours and this caused no damage to the glass-copper interface.

SUMMARY

A variety of glass substrates were successfully coated with adherent copper as thick as 1 mm. These parts were capable of withstanding single point diamond machining of the copper to remove as much as 0.5 mm of the deposit to provide an optical surface. They were also able to withstand heating to 250°C for four hours without degradation of the metal-glass bond. The key steps in the process included preparation of the glass substrates, deposition of a low stress binder layer of titanium by magnetron sputtering followed by a 6 µm thick layer of copper without breaking vacuum, and then electroplating in an acid copper sulfate solution. Copper deposition was done under conditions which resulted in as close to zero stress as possible. Data showing the influence of plating current density on stress in the deposit were generated to help define optimum conditions to use during coating. Although two different copper plating formulations were used effectively, mechanical property data showed that deposit produced in a solution containing no proprietary additives was noticeably softer. For this reason, this deposit was favored for larger substrates (20 cm diameter parts).

ACKNOWLEDGEMENT

All work was performed under the auspices of the U.S. Department of Energy by the Lawrence Livermore National Laboratory under Contract No. W-7405-ENG-48.

REFERENCES

1. J. W. Dini, Plating Surface Finish, 72, 48 (July 1985)

2. J. W. Dini, Synergism of electroplating and vacuum processes, Proceedings First International SAMPE Metals Conference (August 1987)

3. D. M. Mattox, Thin Solid Films, 18, 173 (1973)

4. V. A. Lamb and D. R. Valentine, Plating, 52, 1289 (1965)

5. V. A. Lamb and D. R. Valentine, Plating, 53, 86 (1966)

DEPENDENCE OF STRESS ON DEPOSITION CONDITIONS FOR SPUTTERED COPPER FILMS ONTO FLEXIBLE POLYIMIDE SUBSTRATES

A. Entenberg, V. Lindberg, L. Fendrock, Sang-ki Hong
T. S. Chen

Department of Physics and
Center for Materials Science and Engineering
Rochester Institute of Technology
Rochester, New York 14623

R. S. Horwath

Systems Technology Division
IBM Corporation
Endicott, New York 13760

We have investigated the dependence of internal stress on argon pressure and deposition rate for 0.25 μm copper films sputtered onto a 1-mil (25 μm) - thick polyimide substrate. A dc planar magnetron was used to deposit the copper onto a flexible substrate (Kapton) which was held flat by top and bottom edges. We observed two types of stress: tension, in which the film is trying to contract on the substrate and compression, in which the film is trying to expand on the substrate. The stress is primarily of an "intrinsic" nature, relating to the interfacial mismatch between the natural atomic structures of the growing film and the substrate surface. Using a formula due to Stoney, stress was estimated directly from the resulting radius of curvature of the relaxed film and substrate. At a fixed deposition rate of 2.0 Å/s, there is a stress transition from compression to tension at a pressure of about 2.5 mTorr. At a fixed pressure of 5.0 mTorr, there is a stress transition from tension to compression at a deposition rate of about 4.8 Å/s. Scanning electron micrographs (SEM) show columnar grains with a void network for films under tension and smooth, tightly packed surfaces for films under compression. The origin of the tensile morphology is a "shadowing" mechanism operative at higher argon pressures; at lower pressures, an "atomic peening" mechanism is responsible for the compressive morphology. The morphology data are consistent with the Movchan-Demchishin zone model diagram as extended by Thornton for the growth of sputtered films. Film resistivity and reflectivity were closest to their bulk values for compressive films deposited at low argon pressure and high deposition rate. At a deposition rate of 2 Å/s, the critical pressures at which resistivity starts to rise and reflectance begins to decrease are very close to the aforementioned stress transition pressure. In general, the data are consistent with the dependencies of physical properties on pressure and deposition rate observed for other metals. We are currently examining the effects of film thickness and residual gas pressure on stress. The ultimate goal of this research is to correlate stress with the adhesion between film and substrate.

INTRODUCTION

The control of sputtered thin film properties such as morphology, stress, and resistivity is of great importance to the semiconductor packaging industry. The dependence of these properties on working gas pressure and substrate temperature has been studied for various metals.[1-8] The influences of deposition rate, substrate bias, and substrate orientation have also been studied.[9,10]

As part of an ongoing study of the adhesion between metals and polymers, we have measured the internal stress of sputtered copper films on a polyimide substrate for (1) a range of argon pressures between 1 and 35 mTorr (at a fixed deposition rate of 2 Å/sec) and (2) a range of deposition rates between 1 and 5 Å/sec (at a fixed pressure of 3 mTorr).

A dc planar magnetron was used to deposit a 0.25 µm film onto a 1-mil (25 µm) thick polyimide substrate held flat by its top and bottom edges. The stress was estimated directly from the resulting radius of curvature of the relaxed film and substrate. In addition to stress, resistivity was also measured as a function of argon pressure; a scanning electron microscope (SEM) was used to study the dependence of film morphology on pressure and deposition rate.

Intrinsic and thermal components make up internal film stress. The intrinsic stress comes from the mismatch between the atomic structure of the substrate and the "natural" structure of the growing film. The thermal stress is a result of the mismatch of the thermal expansion coefficients of the film and the substrate as the sample cools from deposition temperature to room temperature. Since the temperature of our substrate did not exceed 85°C during deposition, we have estimated the thermal stress to be less than 8% of the maximum total internal stress.

The two stress states are: compression, in which the film is trying to expand on the substrate, and tension, in which the film is trying to contract. At sufficiently low pressures, a mechanism, known as "atomic peening", is believed to cause compression in films.[2,4] At higher working pressures, peening becomes inoperative. A "shadowing" mechanism causes isolated columnar structures and an extensive void network.[2] Tensile stress in sputtered films may come from this columnar morphology. The film is believed to contract (i.e., go into tension) when the isolated columns try to close the voids between themselves.

Whatever the cause of internal stress in sputtered films, there is, for a given set of system parameters (i.e., magnetron geometry, deposition rate, film thickness), an abrupt transition from compression to tension at a critical pressure which depends on the atomic mass of the material.[4,5,7] Above this pressure, tensile stress quickly attains a maximum value and then gradually decreases to zero. The relaxation of stress beyond some maximum tensile stress point has been reported for chromium sputtered with an argon plasma[5] and molybdenum sputtered using xenon[6]. We also see this for copper on polyimide. Stress relaxation is probably due to cracks which develop in the film. The number of cracks increases for higher argon pressures, as can be seen using the scanning electron microscope.

The resistivity of films sputtered at different argon pressures can also be understood in terms of film morphology, as we will show below.

EXPERIMENTAL SETUP

A dc planar magnetron (US-Gun Inc) was used to deposit films. The copper targets (2-inch diameter by 3/8-inch thick) were 99.9% pure. The substrate was a 3 x 3-inch sheet of 25 μm thick Kapton polyimide (Type H) manufactured by Dupont.[15] Following storage in a desiccator, the substrate was degreased for 10 minutes in an ultrasonic bath of Freon-12 and then quickly rinsed in deionized water. The Kapton sheet, suspended by two horizontal straps at top and bottom edges,was positioned directly in front of the copper target at a distance of 20 cm as shown in Figure 1.

A 50 liter vacuum chamber was evacuated with a liquid-nitrogen-trapped diffusion pump and rotary pump to a base pressure of 2×10^{-5} Torr. Research grade argon (99.997% pure) was introduced through a mass flow controller at rates of 5 to 15 sccm. After the plasma was ignited, the target was presputtered for 2 minutes while a shutter covered the substrate. The discharge voltage, current, and power were in the ranges of 400 to 600 volts, 0.2 to 0.9 amperes, and 100 to 450 watts respectively depending on the operating pressure (1-35 mTorr) and deposition rate (1-6 Å/sec).

The deposition rate and film thickness were read on an Inficon quartz-crystal rate deposition monitor which was calibrated with a Dek-tak surface profilometer. Copper was sputtered to form a film which was nominally 2500 Å thick with an uncertainty of ± 10%.

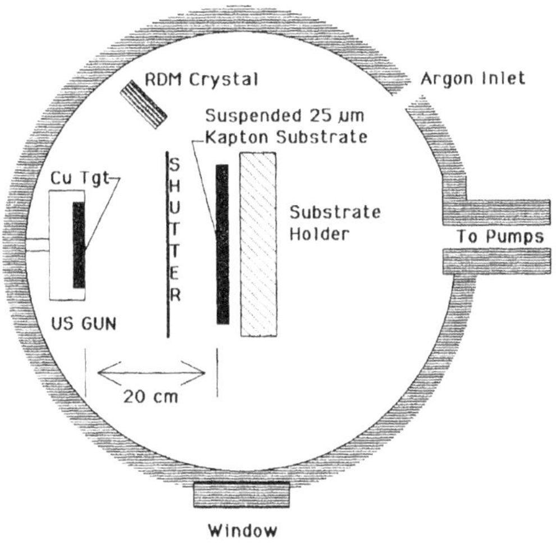

Fig. 1a. Geometry of the sputtering chamber.

During deposition, the maximum substrate temperature, as measured by a Tempilabel temperature-indicator fastened behind the Kapton sheet, was about 85°C.

After the run, the coated substrate was removed and various physical properties were measured. Once released from the substrate holder, the Kapton curled according to the internal stress in the sputtered copper. In most cases the film curled along a horizontal axis parallel to the metal retaining strips of the substrate holder. A number of the samples curled along one of the diagonals of the rectangular substrate. No attempt was made to force the films to curl along a specific axis.

The internal film stress, intrinsic and thermal, was estimated by direct measurement of the radius of curvature, R, of the metallized substrate. Compressive stress produced a convex surface curvature (negative R); tensile stress produced a concave curvature (positive R). The following expression, due to Stoney, was used to calculate stress:

$$\sigma = \frac{E_s\, t_s^{\,2}}{6\, R\, t_f}$$

where R is the radius of curvature, t_f is the film thickness (0.25 μm), E_s is the Young's modulus (3 GPa) of the substrate, and t_s is the substrate thickness (25 μm).[11,12] This expression for film stress should only be regarded as an estimate of the true stress value. Moreover, a systematic correction factor needs to be applied to account for Poisson's ratio and other second order effects. Such a correction would alter only the magnitude of the peak stresses, but would not affect the position of the stress transition pressure or the general shape of the curve shown in Figure 2.

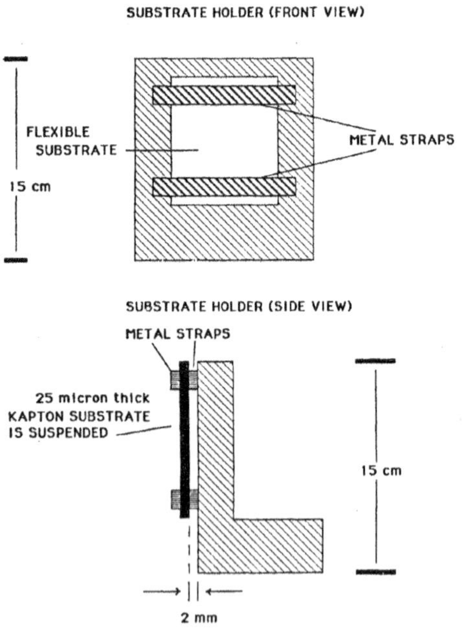

Fig. 1b. Substrate holder for flexible polyimide substrates.

The surface morphology of the films was studied with a scanning electron microscope (SEM). Special additional runs, using a glass coverslip as substrate, were taken in order to obtain fracture cross sections of the copper films for various argon pressures.

The sheet resistance was measured at the center of the film using a four point inline probe with a point spacing of 1.5 mm. Using the known film thickness of 2500 Å, the resistivity was computed.

RESULTS AND DISCUSSION

The dependence of stress on argon pressure, for a fixed deposition rate of 2 Å/sec, is shown in Figure 2. The data indicate that the stress behavior of sputtered copper films is similar to that reported for other metals in the literature.[5] Below 2.5 mTorr, the film is compressive; at 1.7 mTorr, the stress is -0.22 GPa which corresponds to a radius of curvature of -0.67 cm. There is a transition from compressive to tensile stress at a critical pressure of 2.5 mTorr.A peak tensile stress of 0.30 GPa occurs at 3.3 mTorr for a measured radius of curvature of +0.48 cm. At 8 mTorr, the stress is at 20% of its peak value and gradually falls to zero for higher pressures. As shown in Figure 2, the stress estimates are repeatable; the slight spread in the data at 8.3 mT represents five separate trials.

The thermal stress component of the total internal stress is roughly constant over the pressure range and is estimated to be -0.02 GPa. The following expression was used to calculate the thermal stress:[13]

$$\sigma_{th} = E_f (\alpha_f - \alpha_s) (T_s - T_a)$$

where E_f is Young's modulus (about 120 GPa) of the film, α_f and α_s are the average of thermal expansion-coefficients (1.65×10^{-5} K^{-1} and 2×10^{-5} K^{-1} respectively) for the film and substrate, T_s is the average substrate temperature during deposition (75°C) and T_a is the temperature (23°C) at which the thermal stress is measured. The bulk values used

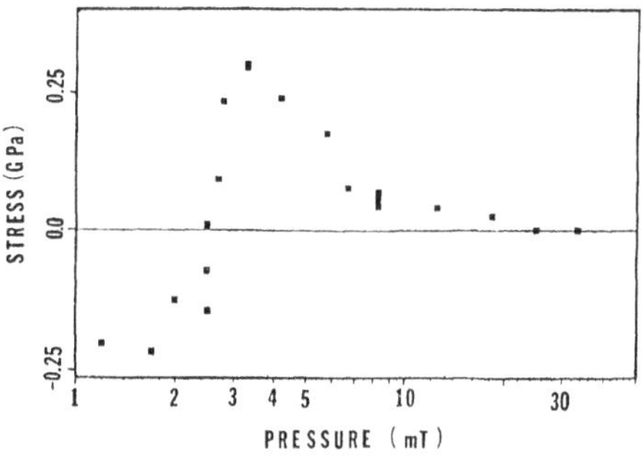

Fig 2. Internal stress vs argon working pressure for 0.25 μm Cu films deposited at 2 Å/s onto flexible polyimide sheet using a dc planar magnetron. Negative stress corresponds to compression; positive stress indicates tension. Statistical uncertainties in measured quantities on this and other graphs are less than 10%.

for E_f, and α_f, were taken from reference 14; the value for α_s was taken from reference 12. The thermal stress is compressive and its magnitude is small (less than 8%) compared to the peak tensile stress. The thermal stress contributes most significantly near the critical stress transition pressure at which the total internal stress is near zero; at this pressure, the compressive thermal stress is balanced by the tensile intrinsic stress.

It is worth noting that the peak tensile stress of 0.30 GPa is close to the 0.35 GPa tensile strength of bulk copper.[14] It is not unusual for metal films to obtain tensile strengths significantly larger than the bulk values.[15] We attribute the cohesiveness of this highly tensile film to the adhesion bond between the copper film and the flexible polyimide substrate. When the same thickness of copper is sputtered under identical conditions onto a glass substrate at a pressure of about 3.3 mTorr, the maximum tensile stress point, the resulting film tears itself apart into small shreds which slide off the glass. At pressures corresponding to lower tensile stress, the copper film has good cohesiveness and adheres well to the glass.

The argon stress transition pressure has been shown to depend monotonically on the atomic mass of the sputtered material.[5,9] Moreover, the critical pressure is also influenced by source geometry[5] (i.e., planar magnetron versus cylindrical post magnetron), by source to substrate distance, deposition rate, and residual gas pressure. Our measured critical pressure of 2.5 mTorr for copper (atomic mass 63.5) is in good agreement with the "cylindrical-post" curve of transition pressures found in Reference 5; straightforward interpolation of this data leads to an expected value of 3.5 mTorr for copper. Interpolation of the "planar" data (also Reference 2) leads to a transition pressure which is lower by a factor of ten. At this time, we are unable to explain our agreement with the cylindrical post rather than with the planar data. A detailed comparison between the systems of Ref. 5 and ours is required.

The morphology of the copper films at various pressures was studied with a scanning electron microscope (SEM). Surface characteristics and fracture cross sections (using Kapton and glass microscope coverslips) were examined. Our observations are consistent with the zone model of Movchan and Demchishin (M&D) as extended by Thornton and with those reported in the literature for copper and other sputtered metal films.[2,10] The substrate temperature range during deposition is from 25°C to 85°C which corresponds to an absolute temperature ratio of (T/T_m) = 0.22 to 0.26, where T_m is the melting point (1358K) of copper.

At lower pressure (less than 5 mTorr), close-packed fibers are seen in the fracture cross-section while the surface appears to be smooth and slightly granular (see Figure 3a.). These characteristics resemble the structures described for the transition zone T of the modified M&D model (TMD).[2,10] At higher pressure (greater than 10 mTorr), the surface appears crazed and brittle under the SEM; distinctly separate columnar grains (vertical one-dimensional growths) are found in the fracture cross-section (see Figures 3b and 6a.). These columnar structures and the intervening void network most nearly correspond to Zone 1 of the TMD model. This dependence of film morphology on argon pressure agrees with previous studies of the structure of sputtered copper films.[16]

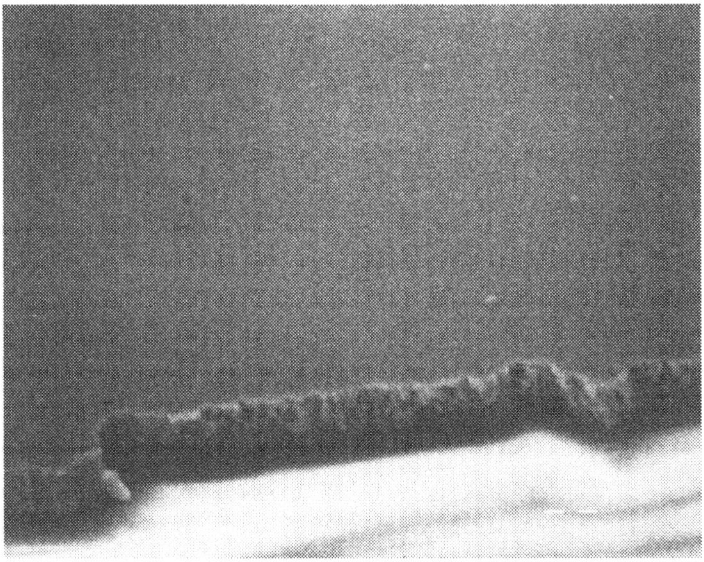

Fig 3a. Scanning electron microscope photograph showing close-packed fibers in a Cu film deposited at low argon working pressure (2 mT). On each of the SEM photos, the long white line at the bottom represents 1 μm.

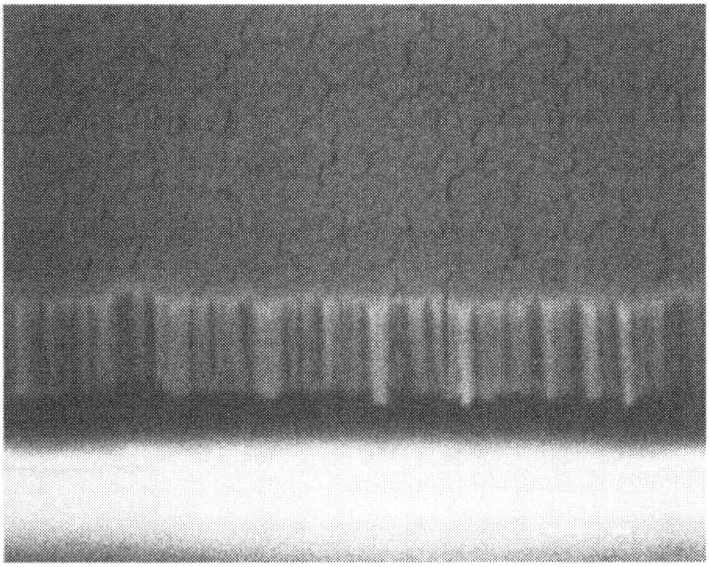

Fig 3b. Scanning electron microscope photograph showing columnar grains in a Cu film deposited at high argon working pressure (10 mT).

The dependence of film resistivity on argon pressure is shown in Figure 4. Below 3.5 mTorr, the resistivity curve is relatively flat with a value of 3 times that for bulk copper (1.7×10^{-8} Ωm); we attribute this result mainly to impurities which are being incorporated into the film. (When the standard base pressure of 2×10^{-5} Torr is lowered to 3×10^{-6} Torr, the plateau region goes from three times bulk to about two times bulk resistivity.) Above 3.5 mTorr, the resistivity rises; at 25 mTorr, the resistivity is about 30 times its bulk value. This rise in resistivity with pressure can be attributed to an accompanying enhancement of the void network in the sputtered film.

The dependence of stress on deposition rate, for a fixed pressure of 3.0 mTorr, is shown in Figure 5. Starting at a relatively low deposition rate (1.0 Å/sec), the stress rises to a tensile maximum (2.0 Å/sec) and then undergoes a transition into compression at about 4.8 Å/sec. This behavior has been observed before with sputtered molybdenum.[17]

The effect of deposition rate on film morphology may be seen by a comparison of Figures 6a and 6b. As mentioned earlier, Figure 6a is an SEM of a 7500 Å tensile film on glass made at a pressure of 10 mTorr and a deposition rate of 2 Å/sec. For the SEM in Figure 6b, the thickness and pressure are similar, except that the deposition rate is twice as high at 4 Å/sec. The columnar structures of Figure 6 are distinct both in the film's cross-section and at the top surface. In Figure 6b, the "columns" seem to fade and are no longer distinct; the top surface of the film shows some cracks, but no evidence of columns.

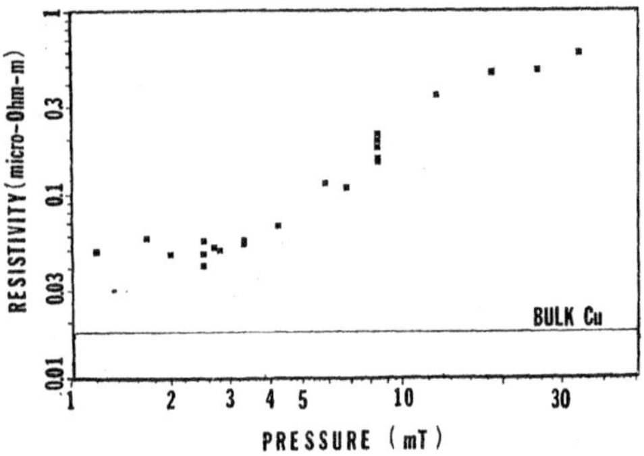

Fig 4. Electrical resistivity vs argon working pressure for 0.25 µm Cu films deposited at 2 Å/s onto flexible polyimide sheet.

The transition from tension into compression at a deposition rate of 4.8 Å/sec may be explained by the increased flux of energy to the growing surface of the film. The increased flow of energy provides surface adatoms with enough mobility to counteract the shadowing mechanism which produces a Zone 1 film with columnar structures and voids. Deposition rate is somewhat analogous to substrate temperature. Increasing temperature (or deposition rate) can lead to a transition from a Zone 1 film with voids (tension) to a smooth Zone T film (compression).

CONCLUSION

The stress, morphology, and resistivity of copper sputtered onto Kapton polyimide substrates at various pressures and deposition rates have been studied for 2500 A thick films. Our results are similar to results for other metal-substrate combinations. For a fixed deposition rate of 2 Å/sec, as sputtering pressure increases, the film initially in compression makes an abrupt transition to tension at 2.5 mTorr. A maximum stress of 0.30 GPa is obtained at an argon pressure of 3.3 mTorr. The stress decreases to zero at higher pressures. Correspondingly, the resistivity remains roughly constant at three times bulk resistivity until 3.3 mTorr, reaching a shallow minimum at 2.5 mTorr. At 25 mTorr, the resistivity climbs to 30 times its bulk value.

Our results may be understood in terms of the Thornton-Movchan-Demchishin diagram for Zone T film-growth at low argon pressure and low T/T_m; zone 1 columnar growth occurs at higher pressures (see Figures 3a, 3b and 6a.). The 2.5 mTorr point represents a bulk-like copper film with minimum stress and resistivity. Atomic peening at low argon pressures densifies the film, leading to compressive stress and a resistivity that is larger than the bulk value because of impurities and degradation. At higher pressures, the film begins to become columnar; stress becomes tensile and resistivity increases. At 3.3 mTorr, the tensile stress approaches the tensile strength of copper and cracks may propagate to reduce the stress, thereby causing the resistivity to increase.

We have also observed that deposition rate as well as working gas pressure can be used to drive a tensile film into compression.

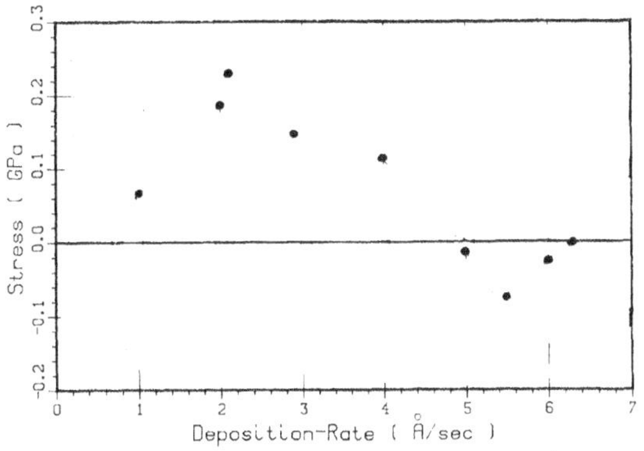

Fig 5. Stress as a function of deposition rate for a fixed argon gas pressure of 3.0 mT.

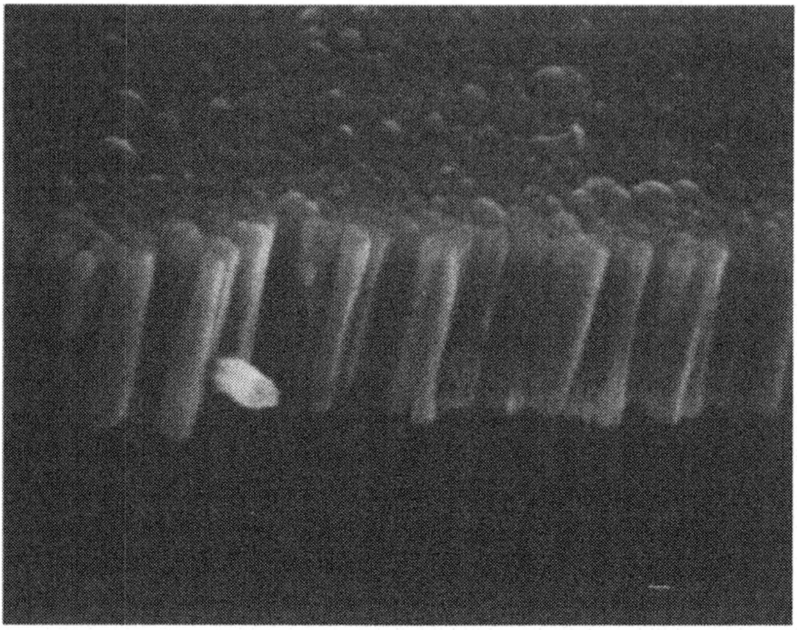

Fig 6a. Columnar structures in a 7500 Å copper film deposited on glass at 10 mT and a deposition rate of 2 Å/s. Photo of copper on glass illustrates better the columnar morphology.

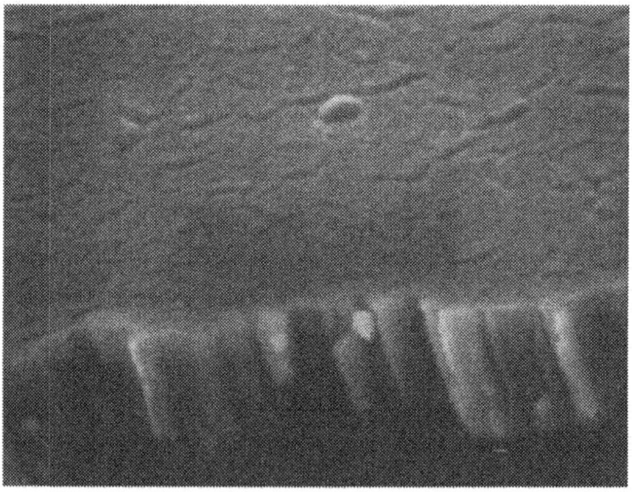

Fig 6b. Faded columnar structures in a 7500 Å copper film deposited on glass at 10 mT and a deposition rate of 4 Å/s.

For a fixed argon pressure of 3.0 mT, a transition from tension into compression occurs at a deposition rate of about 4.8 Å/sec. (This has also been observed with Molybdenum.) This result is again consistent with the TMD diagram under the assumption that an increase in deposition rate can produce an effect which is analogous to an increase in substrate temperature. For a fixed pressure, a low deposition rate leads to a Zone 1 tensile film with columnar structure. At a higher deposition rate, a Zone T compressive film results.

ACKNOWLEDGMENTS

We would like to thank Dr. Arthur Z. Kovacs, Head of the Physics Department, and Dr. John D. Paliouras, Dean of the RIT College of Science, for their continued interest and support. We are also indebted to Dr. Tom Frederick and Mrs. Sandy McCormack who helped us with the SEM; Mr. Bill Van Derveer provided valuable technical assistance; and Mrs. Karen Carr typed many drafts of this paper. Finally, Dr. David W. Hoffman, Ford Scientific Research Labs, helped us with useful suggestions and encouragement during many stages of this research.

REFERENCES

1. A. Entenberg, V. Lindberg, K. Fletcher, A. Gatesman, R. S. Horwath, J. Vac. Sci. Technol. A 5, 3373 (1987).
2. J. A. Thornton, Ann. Rev. Materials Sci., 7, 239 (1977); F. M. d'Heurle, Metall. Trans., I 725 (1970).
3. J. A. Thornton, in "Deposition Technologies for Films and Coatings", R. F. Bunshah, editor, pp. 211-219, Noyes, Park Ridge, NJ, 1982.
4. D. W. Hoffman and J. A. Thornton, Thin Solid Films, 45, 387 (1977).
5. D. W. Hoffman and J. A. Thornton, J. Vac. Sci. Technol, 20, 355 (1982).
6. D. W. Hoffman and J. A. Thornton, J. Vac. Sci. Technol., 17, 380 (1980).
7. J. A. Thornton and D. W. Hoffman, J. Vac. Sci. Technol., 14, 164 (1977).
8. D. W. Hoffman and J. A. Thornton, Thin Solid Films, 40, 355 (1977).
9. J. A. Thornton, J. Tabock, and D. W. Hoffman, Thin Solid Films, 64, 111 (1979).
10. J. A. Thornton, J. Vac. Sci. Technol., 11, 666 (1974).
11. A. Brenner and S. Senderoff, Journal of Research NBS, Research Paper 1954, 42, 105 (1949).
12. "Kapton Polyimide Film: Summary of Properties", product information bulletin E-72087 (January 1985) supplied by Du Pont Company, Polymer Products Department, Wilmington, Delaware.
13. A. G. Blachman, Metallurg. Trans, 2, 699 (1971).
14. "American Institute of Physics Handbook", Third Edition, D. E. Gray, editor, McGraw Hill, New York (1972).
15. R. W. Hoffman, in "Physics of Thin Films", Volume 3, G. Hass and R. Thun, editors, p. 211, Academic Press, (1966).
16. S. Craig and G. L. Harding, J. Vac. Sci. Technol., 19, 205 (1981).
17. D. W. Hoffman and Peters, "Proc. Ninth Int. Vac. Congress and Fifth Int. Conf. on Solid Surfaces", Madrid, Spain, 415 (1983).

METAL INTERLAYERS IN POLYMER FILMS: A SURVEY OF DEPOSITION PROCESSES, MORPHOLOGY, PATTERNING METHODS AND PHYSICAL PROPERTIES

S. Mazur*, L.E. Manring, M. Levy[1], G. T. Dee, S. Reich[1], and C.E. Jackson

Central Research and Development Department
E.I. du Pont de Nemours & Co., Inc.
Wilmington, DE 19880-0356

Metal interlayer deposition comprises a new family of wet-chemical processes by which a dense, electrically continuous metal film can be grown within the bulk of a pre-existing polymer film. The present article reviews three generic forms of the process, namely: electroless, polymer-mediated electrochemical, and carrier-mediated electrochemical deposition. Experimental examples are given with particular emphasis on polyimides as the matrix material. Mechanistic principles which govern the kinetics and morphology of the deposition are described.

These processes make possible wholly new strategies for fabricating metallized patterns as well as more complex contoured structures. Some detailed examples are given, and the virtues and limitations discussed. Finally, we review fundamental electrical and mechanical properties of polyimide films containing metal interlayers, and, in particular, the influence of post-deposition annealing on these properties.

I. INTRODUCTION

Organic polymer films laminated with metal foil or coated with metal film are employed in a wide range of practical applications. Typical fabrication processes include pressure lamination, adhesive bonding, screen printing, sputtering, vapor deposition, and electroless deposition (with or without subsequent electroplating). Over the past few years, we have been exploring a new family of methods: interlayer deposition processes. These differ substantially from conventional methods, especially in that they allow a metal layer to be introduced within the bulk of a pre-existing polymer film.

Metal interlayer deposition takes advantage of the fact that many polymer films can be swelled by solvent without loss of mechanical integrity and that the swollen film becomes quite permeable towards small molecules and ions. Thus, by simultaneously exposing opposite surfaces of a solvent-swollen film to independent sources of metal ions and reducing agents (or electrons) respectively, a reaction zone may be established within the bulk of the film wherein metal ions are converted to zero-valent metal.

These principles have been exploited in an electroless scheme and two different electrochemical variations. The electroless scheme requires a free-standing film interposed between two reagent solutions[2]. The electrochemical schemes involve a film supported on a cathode which serves to generate, in situ, reducing agents[3] or, in the case of an electroactive polymer, mobile electrons[4]. While there are important operational differences among these schemes, they share a common formal mechanistic description. Most importantly, in all three it is possible to establish a kinetic steady-state such that metal accumulates in the reaction zone, ultimately forming a dense, continuous mass[5].

The objective of this paper is to summarize aspects of metal interlayer deposition which may be relevant to potential applications, particularly electrical applications. These include the applicable range of materials, process conditions, and dimensions; methods for creating two- and three-dimensional metallized patterns; and lastly the influence of experimental variables on interlayer morphology and physical properties.

II. VARIATIONS OF THE PROCESS

Figure 1 illustrates three alternative deposition schemes and the parameters which govern them. They differ with respect to reagents and detailed mechanism. However, they share the common feature that a second-order homogeneous reaction is coupled to diffusional fluxes of two different reagents. Consequently, we can draw some useful generalizations regarding the spatial distribution of metal throughout the film and time-evolution of this distribution[7]. In each scheme are indicated those parameters most important in determining the distributions.

The Electroless Scheme, 1-A, requires a free-standing film and the process is stoichiometric in both metal salt and reducing agent. D_m and D_r represent the diffusion coefficients for the reagents in the film. (In general, these are at least two orders of magnitude lower than the corresponding coefficients in solution.) K_m and K_r represent equilibrium sorption coefficients for partition of reagents between film and solution. Finally, the homogeneous reaction is governed by a second order rate constant k.

Carrier-Mediated Electrodeposition (scheme 1-C) employs a reversible redox reagent as an "electron carrier". This small molecule is included in the electrolyte in its oxidized form, O. The carrier diffuses to the polymer/cathode interface where it is converted to the reduced form, R, which diffuses away from the cathode towards the reaction zone. Thus the flux of R in this process is governed by two independent experimental parameters: the concentration of O and the potential applied to the cathode. The carrier functions catalytically, being reduced and reoxidized at equal rates at the cathode surface and reaction zone respectively. In those cases where the metal ions can be provided by a sacrificial anode, then the only reagent consumed will be the anode.

Polymer-Mediated Electrodeposition (Scheme 1-B) can be viewed as a special case of Carrier-Mediated electrodeposition, where the polymer itself plays the role of electron carrier. Mechanistically there is a big difference since small molecule carriers function primarily via mass transport (molecular diffusion) whereas polymer-mediated transport occurs primarily via redox-exchange reactions among proximal reactive groups. Nevertheless, transport in both electrochemical schemes can be

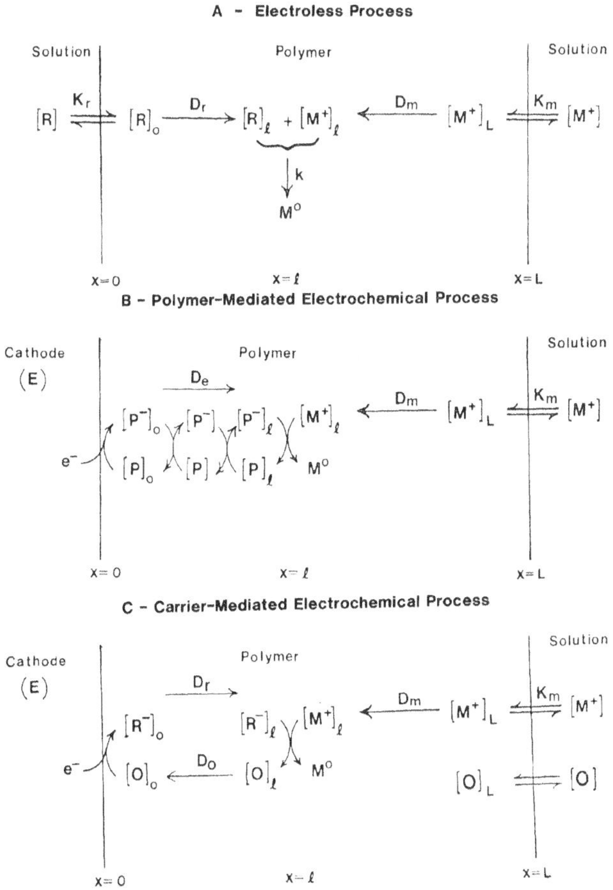

Figure 1. Schematic diagrams for three different versions of metal interlayer deposition processes. Reagents are designated R for reducing agent, M^+ for metal ions, M° for zero-valent metal, and O for electron carriers in their oxidized form.

formally described in terms of a single parameter, D_e, representing the effective diffusion coefficient for electrons. (For the Carrier-Mediated scheme, D_e depends explicitly upon the concentration of O). By virtue of this simplification, the electroless and electrochemical processes become formally equivalent. That is, the electrons function as any other reagent except that their concentration (at the polymer/cathode interface) is determined by the potential applied to the cathode.

The formal equivalence of the three processes is supported by many quantitative correlations. For example, values of permeability coefficients for the various reagents determined from the rate and location of interlayer deposition were found to be in good agreement with values measured in independent experiments.[2,3,4,7] Also, a more sophisticated model including the effects of electric field migration on electrodeposition[8] does not significantly improve the correlations with experiments.

III. MATERIALS CRITERIA

Certain criteria have emerged from the examples studied to date regarding viable combinations of polymer, metal and process conditions (see Tables I and II). Perhaps the single most influential factor is swelling of the polymer. In general, the greater the extent of swelling, the greater will be the permeability of reagents and deposition rate. However, such correlations tend to be highly non-linear. Thus for neutral polymers, there appears to be a lower limit somewhere in the range of 15-20% swelling (by volume) below which the permeability of most metal salts are simply too low to be useful. (A minimum "practical" deposition rate might be represented by a flux of about 1 nanomole/cm^2 sec.) However, this single parameter is not, in itself, sufficient. Thus, some metal salts displayed remarkably low permeability even for polymer/solvent combinations where swelling is in excess of 80%.

The swellability of polymer films can be strongly influenced by their degree of crystallinity and by more subtle aspects of thermal history. Table I lists examples of common neutral polymers which have been employed for interlayer deposition. Attempts to employ highly crystalline fluoropolymers composed primarily of tetrafluoroethylene were unsuccessful. However, in the case of poly(vinylidene fluoride) a small amount of crystallinity was found to be advantageous[3] because it suppressed dissolution of the sample in good swelling solvents such as N,N-Dimethylformamide (DMF) mixtures with acetonitrile.

Polyimides derived from pyromellitic dianhydride (PMDA) and various aromatic diamines have been of particular interest[16] because of their importance as high-temperature dielectrics in various electrical applications. These polymers are not readily processible; consequently, films must be prepared via dehydration of a precursor film (polyamic acid). Typical thermal dehydration conditions result in materials whose swellability is too low for purposes of interlayer deposition. However, it was discovered[4,9] that chemical dehydration carried out at temperatures below 100°C provides film with substantial degrees of swelling in acetonitrile (AN) (ca. 45%) and DMF (ca. 80%). In the swollen state, high permeabilities are observed for a wide variety of reagents (see Table II) and moreover, the films are electrochemically active by virtue of the elecron accepting ability of the pyromellitimide group.[9,17] Thermally and chemically imidized samples differ remarkably little in their infra-red spectra or degree of crystallinity.[10] It remains unclear what is the key factor which mediates the dramatic difference in solvent sensitivity.

Table I. Examples of Polymers Which Can Be Employed as Matrices for Metal Interlayer Deposition.

Polyimides derived from pyromellitic dianhydride (PMDA) and various aromatic diamines, including 4,4′-oxydianiline, 1,3- and 1,4-phenylenediamine
Cellulose
Poly(vinyl chloride)
Poly(vinyl fluoride)
Poly(vinylidene fluoride)
Poly(acrylonitrile)

Table II. Reagents for Metal Interlayer Deposition and Representative Examples of Their Permeabilities in PMDA/ODA Polyimide (PI) or Poly(vinyl chloride) (PVC).

Permeant	Permeability or Diffusion Coefficient (cm^2/sec)*	Redox Potential (V versus SCE)
Metal ions:		
Ag^+	3×10^{-8}	+0.32
"	1.5×10^{-7} (PVC)	"
Cu^+	10^{-9} (PI)	+0.1
Au^+	10^{-7} (PI)	+1.1
Hg^{+2}	--	0.3
Reducing Agents and Electron Carriers:		
BH_4^-, $BH(OCH_3)_3^-$	1.3×10^{-8} (PI)	<0
Naphthoquinone	10^{-7} (PVC)	-0.60
"	" "	-1.10
Benzil	--	-1.04
H_3C-N+ $-CO_2CH_3$	4×10^{-8} (PI)	-0.85
PMDA/ODA#	7×10^{-10}	-0.73
" "	--	-1.13

* All values in acetonitrile solutions at ambient temperature.
\# Values corresponding to the polymer-mediated mechanism.

Table II summarizes some examples of transport parameters for reagents (metal salts, reducing agents, electron carriers and electrons) in various solvent-swollen polymers. In the case of the polyimide derived from 4,4'-oxydianiline (PMDA/ODA) the data include examples of all three process types. It should be noted that the parameters in this table vary over several orders of magnitude. For reasons discussed in the following section, these parameters are relevant not only to the rate of interlayer deposition but also to the spatial distribution of deposited metal.

IV. MECHANISM AND MORPHOLOGY

Mechanistic understanding of interlayer deposition has developed from theoretical modelling supported by experiment. Both the rate of these processes and the morphology of the precipitate can be predicted from knowledge of the concentrations, and transport and reaction rate coefficients. We will review here only certain qualitative generalizations. Quantitative details and supporting data can be found in the primary references.[2,4,7]

In all experimental systems investigated to date, metal precipitates initially in the form of microparticles (from tens to hundreds of Angstroms in diameter). As the process continues, the volume fraction

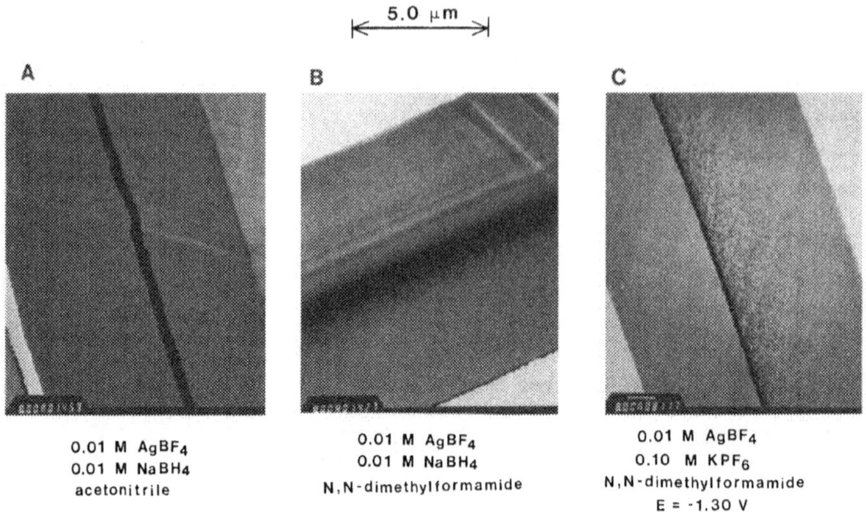

Figure 2. Transmission electron micrographs of cross-sections showing different morphologies for Ag interlayers in polyimide film. A and B were produced by the electroless process and C by polymer-mediated electrodeposition. (Reprinted from Reference 7, courtesy of the Journal of Physical Chemistry).

of these microparticles increases until a dense packing is created. The local volume fraction of metal can become as high as 0.72 (as much as 95% by weight). Once an interlayer of maximum density is established, it continues to grow in thickness until some other event intervenes. (Typically, the metal will ultimately penetrate the surface of the polymer, forming dendritic nodules at isolated locations).

The details of this morphology, particularly the spatial distribution of microparticles is subject to considerable variation depending upon the particular combination of reagents and conditions employed. Some representative examples are illustrated in Figure 2. In fact, interlayer morphology contains considerable mechanistic information. In addition, morphology governs many of the materials properties and is, therefore, relevant to practical considerations as well.

Figure 3 summarizes, in rough chronological order, the mechanistic steps from the initial encounter of reagents to the formation of a dense metallic interlayer. It is apparent that the overall kinetics is composed of both fast and slow components. For example, diffusion, homogeneous reaction, nucleation and growth represent relatively rapid events which occur in parallel. On a much slower time scale, as the volume fraction of microparticles increases, the overall mechanism evolves from the homogeneous to the heterogeneous mode. These two pathways may or may not overlap for any significant time interval.

It is useful to distinguish three different time domains:

1 - Pre-Steady-State Regime: The period prior to establishment of stable (time-independent) concentration gradients for reagents. (Heterogeneous reaction is presumed unimportant at this early stage).

2 - Steady-State/Homogeneous Regime: Characterized by constant diffusion fluxes coupled to purely homogeneous reaction. (Volume fraction of microparticles is still relatively low.)

3 - Steady-State/Heterogeneous Regime: Characterized by constant diffusion fluxes coupled to purely heterogeneous reaction. (Maximum volume fraction of microparticles.)

The first two regimes can be described by differential equations representing the effects of diffusion and bimolecular reaction on the position-dependent concentrations of each reagent (reducing species, metal ions, and zero-valent metal).[7] Numerical solutions to these equations yield predictions for the spatial distribution of metal at all times prior to the onset of heterogeneous reaction. This problem is greatly simplified by the fact that nucleation and growth are sufficiently fast relative to reaction that, to first approximation, these steps can be combined, (i.e., one assumes that metal precipitates as fast as it is produced). In the limit of this approximation it has been possible to obtain satisfactory agreement between experimental and theoretical metal distributions with a minimum spatial resolution on the order of 0.1 μm. For length scales smaller than this, the morphology is dominated by the details of particle size, shape, and clustering, features which are clearly governed by the missing steps. Most of our efforts have, therefore, focused on distributions within the range from 0.1 μm to the thickness of the film, L (typically on the order of 10 μm).

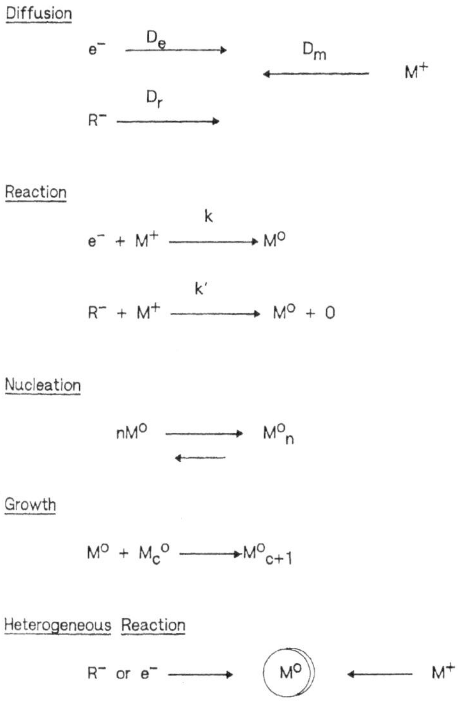

Figure 3. Summary of mechanistic steps in the formation of dense metal interlayer.

At any given time, all reaction occurs within a volume element (lamina of the film) which we shall describe as the "instantaneous reaction zone" or IRZ. The final distribution of metal across the thickness of the film reflects the following characteristics of the IRZ.

Location of the Instantaneous Reaction Zone

The center of the IRZ corresponds to a plane at which the fluxes of metal ions and reducing agent (or electrons) are equal. In the steady-state regime, each of these fluxes must also be equal to the product of the diffusion coefficient and the gradient of concentration between the IRZ and the film surface. Therefore, the location of the steady-state IRZ can be calculated from knowledge of the diffusion coefficients and concentrations. It follows that one can easily regulate the steady-state location simply by changing the concentration of either reagent[2] (or equivalently, by changing the electrochemical potential[3,4]). Taking advantage of this principle, metal interlayers can be grown at locations ranging from one surface of a film to the other, and multiple interlayers can be grown within a single film[3].

In the pre-steady-state regime, concentration gradients are non-linear and it is not possible to derive a general analytical solution. However, the earliest point in the reaction corresponds to the first encounter between the fluxes of reducing agent and metal ion. It can be shown[7] that this location of first encounter is not sensitive to the reagent concentrations. Consequently, the initial location of the IRZ need not be the same as its steady-state location. That is to say, during the pre-steady-state regime, a displacement may occur (vide infra).

Width of the Instantaneous Reaction Zone

In the limit of an extremely large bimolecular rate constant, each metal ion would react on its first encounter with a reducing species so that the IRZ effectively corresponds to a lamina of thickness equal to a single molecular diameter. However, for most reagents, reaction will not occur on first encounter. Many metal ions and reducing agents diffuse a finite distance past the location of equal fluxes before they finally react. In this case the width of the IRZ is determined by the imbalance between the local reagent fluxes and the local reaction rate.

The IRZ width cannot be determined analytically for any general condition. However, for the particular circumstance $D_e = D_m \equiv D$ and $[M+] = [R-] \equiv c$, numerical results can be generalized in terms of a dimensionless reaction rate parameter: kcL^2/D, where L is the polymer film thickness. For example, numerical results[7] show that kcL^2/D must exceed 20,000 in order to limit the width of the steady-state IRZ to less than 10% of the overall film thickness. (Rigorously, the IRZ has no discrete boundaries. The "width" discussed here represents limits within which some arbitrarily large fraction (>0.99) of the reaction occurs.)

The dramatic difference in microparticle distributions illustrated in Figures 2-A and 2-B are consistent with this analysis, since the rate constant for BH_4^- reduction of Ag^+ is much slower in DMF than in AN.

Displacement of the Instantaneous Reaction Zone

A broad distribution of microparticles can result either from a wide IRZ whose location remains constant throughout the process or else from a narrow IRZ whose location changes during the course of the process. Displacement of the IRZ can be the result of some extrinsic effect such as systematic variation of the reagent concentrations at the external surfaces of the film. For example, a broad interlayer can be produced in the electrochemical processes by slowly changing the applied potential[4]. However, displacement can also be the result of circumstances intrinsic to the deposition mechanism.

As previously indicated, the location of the IRZ at the first instant of encounter between reagent fluxes, ℓ_o, need not coincide with the location at steady-state, ℓ_{ss}, even when the surface concentrations are rigorously constant. ℓ_o and ℓ_{ss} are expressed as distance from the cathode/polymer interface or from the interface with reducing agent solution). In particular, ℓ_o/L is uniquely determined only by the ratio of transport coefficients[7]. For the case of polymer mediated electrodeposition:

$$\ell_o/L = \left(1 + \sqrt{\frac{D_m}{D_e}}\right)^{-1}$$

Thus for any given ratio D_m/D_e, ℓ_{ss} can equal ℓ_o only for a unique choice of concentration ratio [P-]/[M+]. From Table II we note that D_e for PMDA/ODA polyimide is much lower than $D_m K_m$ for any of the metal ions. As a consequence, ℓ_o always occurs very near the cathode surface for polymer-mediated electrodeposition in this material. The microparticle distribution shown in Figure 2-C is caused by displacement of the IRZ from this initial location to a steady-state location near the mid-plane of the film. The numerical model provides a good qualitative account of this effect.[7]

In situations where a diffuse distribution resulting from IRZ displacement is undesirable, experimental counter-measures are generally available. For example, in the case of polymer-mediated deposition of Ag in PMDA/ODA, it was found that a nearly symmetrical, narrow distribution can be achieved by delaying the introduction of metal salt into the electrolyte until a few seconds after initiation of the electron flux. By imparting this "head start" to the electrons one can ensure that the location of initial encounter will coincide with the steady-state location.

Some observations regarding the heterogeneous reaction regime are worth noting. Assuming reaction kinetics are not unusually slow, this mechanism becomes important once the probability of encounter between reagents and microparticles becomes significantly higher than the probability of encounter between the reagents. The heterogeneous reaction may be viewed as a kind of electroplating process where the cathode (dense interlayer) functions as a bipolar electrode, charged by reducing agents at one side and discharged by metal ions on the opposite side. In fact, the same morphology has been obtained by making direct electrical contact to a pre-formed interlayer, followed by electroplating to further thicken it.

Nevertheless, the dense interlayer morphology differs considerably from electroplating onto an exposed metal surface in the same electrolyte. The latter process produces coarse dendritic structures composed of crystallites several μm in diameter. By contrast, the

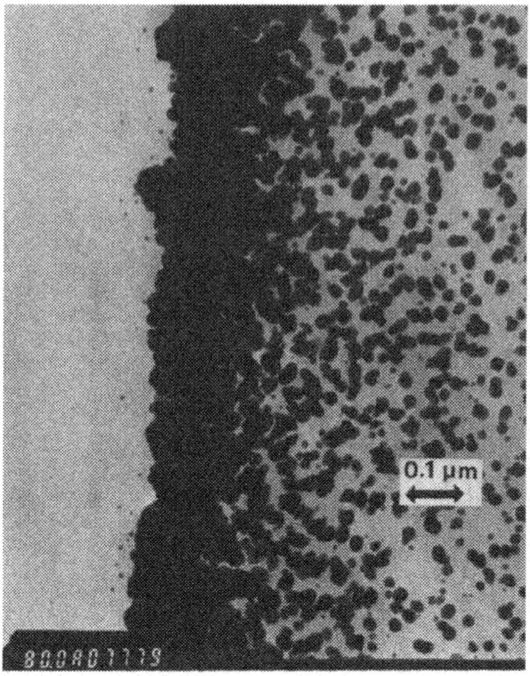

Figure 4. Transmission electron micrograph of Ag interlayer in polyimide film showing spherical shape of the microparticles and indications of columnar structure in the dense regions formed under heterogeneous reaction conditions.

heterogeneously grown interlayer is made up of densely packed crystallites with dimensions on the order of 10 nm. These differences may be a reflection of the influence of slow mass transport on heterogeneous nucleation and growth kinetics within the film. In several experiments, electron micrographs have indicated a columnar character to the heterogeneous growth (see Figure 4).

V. PATTERNS

Electrical circuitry represents one of the most important areas of application for metallized films.[11] These applications require metallized patterns in one or more layers parallel to the film surface. Often the outermost layer must be protected by an additional layer of polymer dielectric. Common methods for fabricating the metallized patterns are either selective etching of metal from a continuous metal film (a "subtractive" process), and screen-printing of conductive inks onto a polymer substrate (an "additive" process).

A number of different methods have been explored for fabricating patterned interlayers. They are "additive" in the sense that the metallized pattern is created directly in a single step. These methods fall into two generic categories:

Patternwise Regulation of Surface Concentrations:

The supply of reagents to one or both sides of the polymer film is regulated. The use of patterned cathodes[4], photosensitive cathodes[12] and ion barrier masks exemplify this approach.

Patternwise Regulation of Film Permeability

As discussed earlier, polymer film permeability is a function of swellability. Therefore, any treatment of the film which strongly alters its permeability can, in principle, form the basis for a patterning method. Two examples which have been demonstrated effective for PMDA/ODA polyimide are selective thermal treatment and incorporation of a photopolymerized network.[13]

A - Patterned Cathodes

Figure 5 shows an electron micrograph of an interlayer grown in a film on the surface of a patterned cathode. The original location of the edge of the cathode is discernible as a shallow step on the lower surface of the film. The interlayer was grown approximately 4 μm from the cathode/polymer interface. Two features are noteworthy. The edge of the interlayer extends beyond the edge of the conductive region of the cathode and, the interlayer edge curves back towards the cathode surface. These features are consequences of the fact that the concentration profiles for metal ions and electrons are not symmetric (due to the asymmetry in surface concentration distributions). Consequently, reagents diffuse laterally (in the plane of the film) as well as transversely. The numerical model for coupled reaction and diffusion quite accurately reproduces these features[7] as well as the systematic decrease in volume fraction of microparticles. No new principles are required to understand these effects.

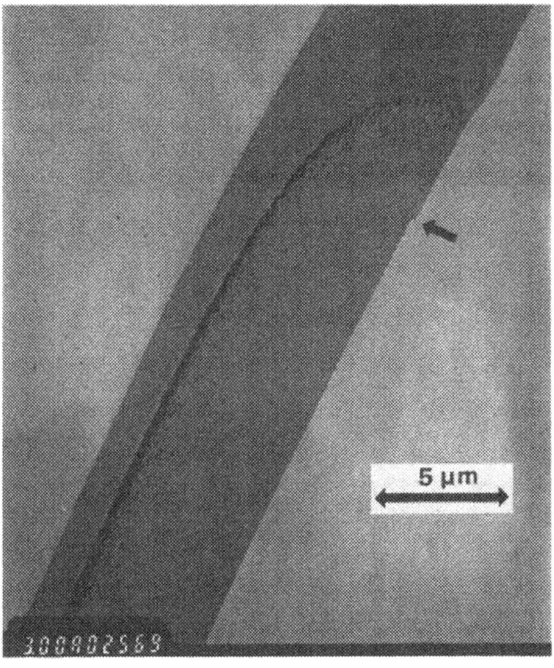

Figure 5. Transmission electron micrograph of a Ag interlayer grown in polyimide film using a patterned cathode. The cathode consisted of a glas plate wih a thin (0.5 μm) chromium pattern on its surface. An edge of the chromium pattern was originally located at the step indicated by the arrow in the figure.

This edge curvature represents an unusual kind of three-dimensional structure which cannot readily be obtained by conventional metallization techniques. However, it constitutes a problem for applications requiring complete insulation of the interlayer. The broadening effect does not represent an absolute limitation on pattern resolution, because the extent of the broadening is proportional to the distance of the interlayer from the cathode surface. Resolution of 2 μm features has been achieved for interlayers grown near the cathode surface.[4]

B - Photosensitive Cathodes

A somewhat more interesting case is represented by a continuous cathode (i.e., unpatterned) whose electrical properties can be modulated with light.[12] Since the interlayer itself would prevent any light from reaching the cathode from the direction of the electrolyte solution, it is necessary to illuminate the cathode from behind. The principle is illustrated schematically in Figure 6. Both photoconductive and photovoltaic effects can be exploited. In either case, in order to retain resolution, the surface in contact with the polymer must have very low dark conductivity. This requirement is simpler to satisfy in the case of a photoconductive cathode, so the remaining discussion is limited to this case.

The process illustrated in Figure 6 functions as follows. The cathode consists of a glass substrate covered with a thin layer of transparent conductive material (tin oxide) which constitutes the primary electrode, followed by a second thin layer of the photoconductor. The polymer film is cast on top of the cathode as usual and immersed in electrolyte. Electrical contact is made to the primary electrode at the edge of the plate. A potential E_a is applied, sufficiently negative to drive the deposition process for a normal cathode. However, in the present situation the electrical resistance of the thin photoconductor layer in the dark is very high so that the potential which appears at the polymer/photoconductor interface, E, is shifted by Ohm's law positive with respect to E_a. When this shift is

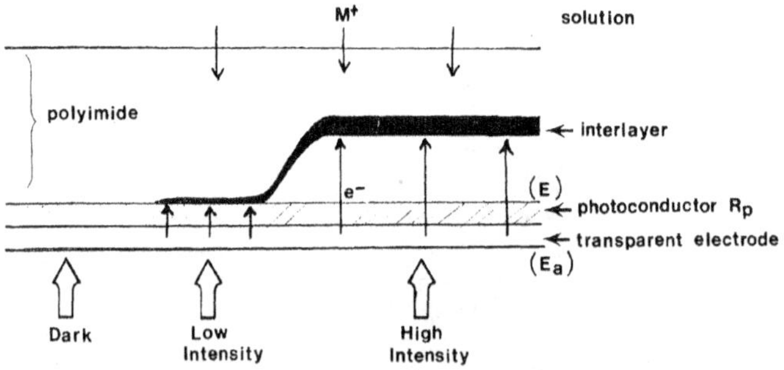

Figure 6. Schematic illustration of the use of photoconductive cathode for deposition of a contoured interlayer. A photoconductive thin film (shaded region) is interposed between a transparent primary electrode (tin oxide coated glass) and the polymer film. The potential E which determines the rate and location of interlayer deposition depends both upon the applied potential E_a, and on the light intensity.

sufficiently large, a negligible amount of current will flow, i.e., the process is "shut off". In order to achieve the necessary voltage drop across the photoconductor (a few hundred millivolts across 0.5 μm) its dark resistivity must be on the order of 10^9 ohm cm. When the photoconductor is illuminated through the back of the cathode, this resistivity drops so that E more closely approaches E_a and the process is turned on. In fact this negative shift in E increases with increasing light intensity and the deposition current increases correspondingly. Since the location of the steady-state IRZ is determined by the electron flux, the light intensity also determines the transverse location of the interlayer.

The model for steady-state polymer-mediated electrodeposition[4] can be modified to include the additional constraint of Ohm's law:

$$I \times A \times R_p = E - E_a$$

where I is the current density, A the cathode area, and R_p the resistance of the photoconductive layer. Results of some calculations for typical deposition conditions (10 μm film, 0.01 M Ag+) are shown in Figure 7. Note that for a given value of E_a, the current is characterized by two plateau regions. The first corresponds to direct electroplating of metal at the cathode/polymer interface at potentials more positive than the redox potential of the polymer. For lower values of R_p, the polymer-mediated process becomes effective. The second plateau is reached when the polymer is fully reduced at the cathode surface. For reference, typical values of R_p are indicated for a 0.5 μm amorphous hydrogenated silicon (a-Si:H) thin film over a range of light intensities. (Conductivity generally varies as the logarithm of light intensity for such materials.)

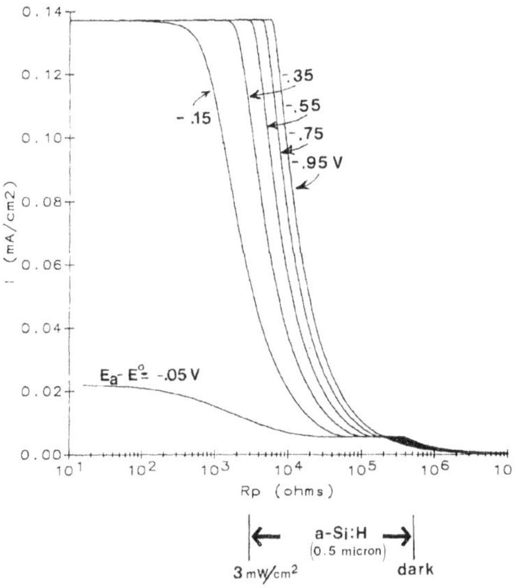

Figure 7. Calculated response of a photoconductive cathode in conjunction with a 10 μm polyimide film with 0.01 M Ag+. Current density, I, depends both upon the applied overpotential $E_a - E^\circ$), where E° is the redox potential of the polymer and on the resistance of the photoconductor R_p. Non-ohmic effects are neglected.

This method provides some unique capabilities. In typical experiments, the master pattern consisted of a black and white negative whose image was projected onto the back of the cathode. Variations in optical density of the negative provide variations in the rate and transverse location of interlayer growth. Thus, the interlayer can be a contoured structure with variable depth and thickness. By depositing in sequence two or more interlayers from different negatives, it is possible to generate complex multilayer structures. Examples of these capabilities are illustrated in Figure 8.

There are, however, two problems which restrict the utility of this method. First, as in the case of patterned cathodes, the edges of the interlayer always curve back towards the cathode/polymer interface. This means, for example, that two interlayers with different patterns will often contact one another along the edge of the uppermost layer. Secondly, pinholes in the photoconductor result in unregulated metallization at local points in the film. The pinhole density of typical thin-film semiconductors is rather high (on the order of ten per cm^2 for glow-discharge deposited a-Si:H) so that only very small defect-free samples were obtained.

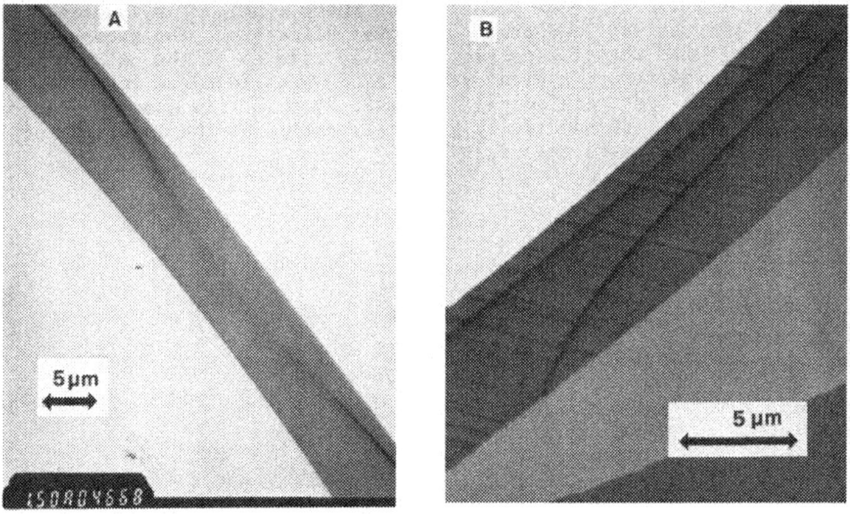

Figure 8. Transmission electron micrographs of complex structures (Ag in polyimide) grown by means of a photoconductive cathode (a-Si:H). Sample A was produced by illuminating through an optical ruling (62.5 µm pitch). Sample B consists of two different layers deposited by projecting two images in sequence at different light intensities. The upper layer (higher light intensity) is continuous within the microtomed section, while the lower level (lower light intensity) has reached an edge.

C - Photo-Polymerized Networks

Methods based on patternwise control of permeability have the inherent advantage of avoiding the edge curvature caused by unsymmetrical concentration profiles. One such method involves the use of a photopolymerized network to control the swellability and permeability of PMDA/ODA polyimide films.[13]

The scheme is summarized in Figure 9. The polyimide films are derived from dehydration of a polyamic acid precursor. In the present instance polyamic acid was combined with a multifunctional photopolymerizable monomer (tetraethyleneglycol diacrylate, for example) and an appropriate combination of photosensitizer and initiator. On exposure to light, the photo-monomer polymerizes to create a network (presumably interpenetrating with the polyamic acid). Next, the film is immersed in the conversion bath containing acetic anhydride and pyridine which converts the precursor to polyimide.

The photopolymer network is influential in two respects. First, it restricts the swelling of polyamic acid in the conversion bath so that conversion in the photolyzed regions is greatly retarded relative to that in unexposed regions. Secondly, it restricts the swelling of polyimide under deposition conditions (acetonitrile solutions). Since the polyamic acid itself swells relatively little in acetonitrile, the resulting permeability of the photolyzed regions is greatly diminished. Unreacted photomonomer is extracted from the unexposed regions during conversion, so that normal permeability can be obtained in these

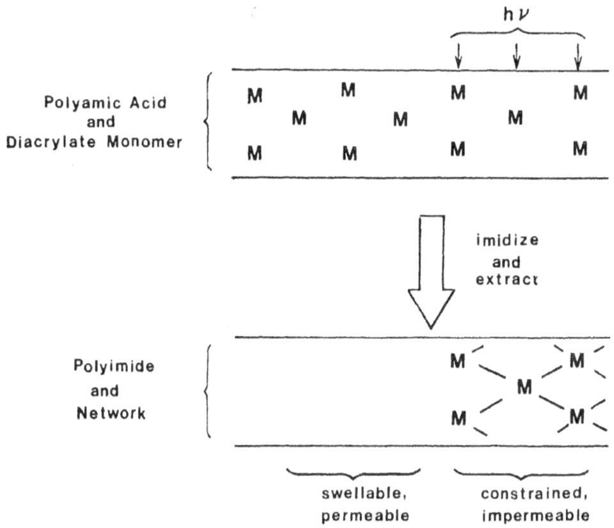

Figure 9. Schematic diagram for the use of a photo-polymerized network to regulate the swellability and permeability of a polyimide film for purposes of creating patterns. M represents the diacrylate monomer.

regions. Conditions, employing as little as 10% diacrylate by weight, provide excellent contrast. For example, unexposed regions exhibited swelling in acetonitrile on the order of 40% by volume while the photolyzed regions of the same sample swelled by less than 1%. With films of this type it has been possible to deposit interlayers as thick as 5 μm without incorporation of any metal in photolyzed regions. Figure 10 shows a portion of a flexible circuit pattern fabricated in this manner.

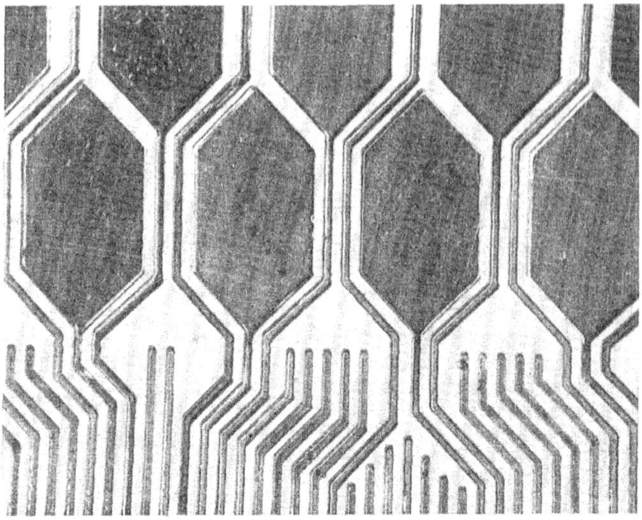

Figure 10. Optical micrograph of a fine-line flexible circuit pattern (Ag in polyimide) produced by means of a photo-polymerized network. The metallized lines are 50 μm wide and approximately 5 μm thick. The film is 40 μm thick overall.

VI. PHYSICAL PROPERTIES

A very broad spectrum of physical properties must be evaluated in considering any practical applications. For the materials and structures fabricated by metal interlayer deposition, this characterization remains incomplete at the present time and will need to be examined much more extensively.

A - D.C. Electrical Properties[14]

For interlayers situated more than about 0.1 μm beneath the external surfaces of the film, the bulk electrical conductivity of the samples is generally indistinguishable from that of the neat polymer. The electrical properties of the interlayer itself are only manifest when contact is made directly to it by etching away a portion of the polymer overlayer. This was accomplished using either an oxygen plasma or, for more robust structures, wet chemical etching. (For example, polyimide samples with interlayers thicker than about 1 μm could be etched using alcoholic KOH solutions.) D.C. conductivity measurements were made in the plane of the interlayer using a standard four-probe configuration. Some results are summarized in Table III.

In all cases the as-deposited interlayer exhibited conductivity lower than that of the corresponding bulk metal by at least a factor of 500. However, on heating the samples, even to rather modest temperatures, the conductivity was found to increase irreversibly. The effect is illustrated by Figure 11 for the case of Ag in PMDA/ODA polyimide. (Similar behavior was observed for Au.) Following exposure to

330°C, the conductivity and temperature coefficient of conductivity approached values characteristic of the bulk metal.

Studies[14] of crystallographic and morphological changes which occur at these temperatures indicate that the microparticles are sintering. Despite the high-melting points of these metals, sintering occurs at relatively low temperatures apparently because of the small size of the microparticles. (The microparticle dimensions approximately double in the process.) The low initial conductivity may reflect the fact that the as-formed interlayer is composed of densely packed spheres which make only point-contacts with one another. Sintering is believed to occur by surface diffusion which does not need to proceed very far in order to greatly improve inter-particle contacts. Consistent with these ideas is the observation that dense interlayers which have not been annealed can be completely dispersed as colloidal metal upon dissolution of the polymer matrix. By contrast, dissolution of the matrix from annealed samples leaves the interlayer intact as a self-supporting thin film.

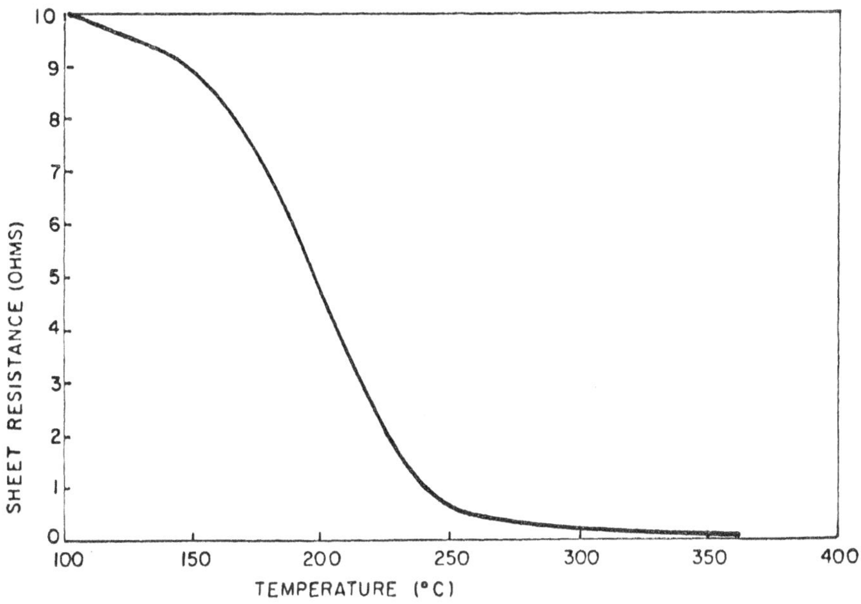

Figure 11. Effect of annealing (sintering) on the sheet resistance of 1 μm Ag interlayer in polyimide film (Reprinted from Reference 14. courtesy of the Journal of Applied Physics).

The temperatures required for sintering exceed the limits for the majority of common thermoplastic resins; however, they are compatible with certain high-temperature polymers such as aromatic polyimides. It is a convenient coincidence that virtually the same annealing conditions (e.g., 350°C, 15 minutes) are required to achieve the full dielectric properties of polyimides which have been prepared by chemical dehydration. Mechanical properties are also influenced by annealing.

Table III. D.C. Electrical Properties of Metal Interlayers.

Metal* (polymer)	Interlayer Thickness, μm	Conductivity $(ohm\ cm)^{-1}$	TCR* $oK^{-1} \times 10^{-3}$
Ag (PI)	0.19	650	0.42
" "	0.90	1700	0.62
" ", annealed	0.44	6.7×10^5	3.6
" "	diffuse	5	-
" ", annealed	"	700	-
Ag (control)	-	6.8×10^5	4.1
Au (PI)	0.5	24	-
" " . annealed	"	1.8×10^5	-
Au (control)	-	4.2×10^5	-
Cu (cellulose)	-	1.2×10^4	-
" " , annealed	-	2.4×10^4	-

Unless otherwise noted, all values represent as-deposited samples.
* Temperature coefficient of resistance measured between -100 and 25°C

Mechanical Properties

Tensile properties for polyimide films, with and without interlayers, with and without annealing are summarized in Table IV. Several points are worth noting. First, the mechanical properties of experimental polyimide films prepared for this process fall short of those typical of commercial high-performance film. This is not surprising in view of the fact that the experimental samples are virtually unoriented. Secondly, we note that the incorporation of a dense metal interlayer greatly enchances the modulus of the film. This is to be expected in view of simple mechanical models for composite materials[15] (see Table IV), and, for the very high volume fractions characteristic of dense interlayers. This effect is insensitive to the continuity of the metal.

Finally, the most important consequence of annealing on mechanical properties is the decrease in ultimate elongation. This is consistent with the fact that annealing sinters the particles into a continuous (thought not fully dense) metallic mass. Thus, the elongation of typical cold-rolled Ag is similar, on the order of 6%. Microscopic examination of the failure surfaces shows considerable differences between unannealed and annealed samples. The former show evidence of plastic deformation of the matrix and uniform separation of the microparticles. By contrast, annealed samples appear to fail in the metal layer itself. Clean cracks which develop in the interlayer propagate across the entire film thickness.

The relatively low elongation of the annealed films is reflected in flex properties as well. For films thicker than about 15 μm, it is necessary to maintain the interlayer near the midplane of the film in order to maintain stability in flexure (creases or bends of radius less than 0.1 mm). A simple geometrical analysis shows that the surface of such a film sustains elongations exceeding 6%.

Table IV. Tensile Properties of the Interlayer Structures in PMDA/ODA Polyimide Film.

Sample*	Modulus (GPa)	Max. Elongation (%)	Strength (MPa)
Kapton® 100-H (reference)	3.17	45	180
Chemically converted, unannealed	3.03	29	90
Chemically converted, annealed	2.75	72	130
With Ag interlayer (8 μm), unannealed	6.00	13	69
Composite theory# for discrete particles	7.38	--	--
With Ag interlayer (8.5 μm), annealed	6.21	5	76
Composite theory@ for continuous layer	7.65	--	--

* All samples are 0.001" films, results are averages of five 0.88" x 0.19" specimens strained at 0.2"/min.

\# Calculated from the volume fraction of particles and thickness of the interlayer and the modulus and Poisson ratio for each of the components (Reference 15).

@ Calculated from thickness of the interlayer assuming bulk Ag modulus.

VIII. SUMMARY

As a practical tool, metal interlayer deposition is still in its infancy. Further advances are needed to increase the scope and versatility of the process itself, and many of the critical physical properties remain to be defined. While certain capabilities, such as the formation of contoured surfaces, appear to be unique, it remains to be seen how these processes may best be exploited.

REFERENCES

1. Permanent address: Department of Materials Science, Weizmann Institute of Science, Rehovot, Israel.

2. L.E. Manring, Polymer Communications, 28, 68-71 (1987); L. E. Manring and S. Mazur, U.S. Pat. #4,692,360.

3. M. Levy, L.E. Manring, and S. Mazur, J. Electrochem. Soc., 135, 2479 (1988); M. Levy, U.S. Pat. #4,668,354.

4. S. Mazur and S. Reich, J. Phys. Chem., 90, 1365 (1986); idem, in "Integration of Fundamental Polymer Science and Technology", L.A. Kleintjens and P.J. Lemstra, Editors, Elsevier, 265-72, (1985); and S. Mazur, U.S. Pat. #4,512,855.

5. Similar processes involving precipitation of other inorganic phases ($BaSO_4$, AgCl, and Prussian Blue) have been studied.[6] However, these processes are intrinsically self-limiting because the precipitated phase is impermeable to both reagents. By contrast, metal interlayer deposition is never self-limiting because the precipitate remains "permeable" to electrons.

6. Some representative examles are: P. Hirsch-Ayalon, Rec. Trav. Chim. Pays-Bas, 75, 1065 (1956); K.F. Mueller, Science, 225, 1021 (1984); L.E. Manring and S. Mazur, J. Phys. Chem., 90, 3269 (1986); K. Honda, H. Hayashi, K. Chiba, Chem. Lett. (Chem. Soc. Jap.), 191-194 (1988).

7. G.T. Dee, L.E. Manring, and S. Mazur, J. Phys. Chem., 91, 6699 (1987).

8. G.T. Dee and S. Mazur, unpublished results (1984).

9. S. Mazur, P.S. Lugg, and C. Yarnitzky, J. Electrochem. Soc., 134, 346 (1987).

10. S. Mazur, L. E. Manring and K. H. Gardner, unpublished results (1985).

11. S. Gurley, "Flexible Circuits, Design and Applications", Electrical Engineering and Electronics, Vol. 20, Marcel Dekker, New York (1984).

12. H. Hadley and S. Mazur, U.S. patent #4,657,833.

13. L.E. Manring, U.S. patent #4,774,157.

14. S. Reich, S. Mazur, P. Avakian and F.C. Wilson, J. Appl. Phys., 62. 287 (1987).

15. We are indebted to Dr. Ron Kander (Du Pont) for these calculations based on the Halpin-Tsai equations.

16. K.L. Mittal, editor, "Polyimides: Synthesis, Characterization and Applications, "Vols. 1 and 2, Plenum Press, New York, 1984.

17. R.C. Haushalter and L.J. Krause, Thin Solid Films, 102, 161 (1983).

PART II. SPECTROSCOPIC INVESTIGATION OF INTERFACIAL INTERACTIONS

SURFACE SPECTROSCOPIC TECHNIQUES APPLIED TO METALLIZED PLASTICS

N.J. DiNardo

Department of Physics and Atmospheric Science
Drexel University
Philadelphia, PA 19104

To ultimately understand macroscopic adhesion in metallized plastics, fundamental metal-organic bonding chemistry at metal-polymer interfaces must be probed. Experimental surface spectroscopies have generally produced information for well-defined systems in surface chemistry. New applications of surface techniques to complex systems such as metal-polymer interfaces have recently produced exciting new developments in studies related to adhesion technology. Recent studies of electronic and vibrational structure and the morphology of metal-polymer interfaces are reviewed. Due to the high degree of interest in the particular system, we concentrate upon the metal-polyimide interface as a means to describe the scientific approach. The general applicability of these techniques and future directions are discussed.

INTRODUCTION

The field of surface science has advanced significantly over the last several years and great progress has been made in understanding interfacial properties of materials. Advances in surface spectroscopies, surface structural measurement techniques, and theoretical capabilities have all contributed to this progress. In addition, coupling experiments with theoretical calculations has provided new insights into the chemical and physical phenomena relevant to interfacial growth. These advances have provided guidance to technologies that require interfacing materials for development of new processing and fabrication strategies.

The great potential to apply scientific discoveries in surface and interface science to microelectronics device and processing technology has been realized in many ways. Much of the scientific work has been focused on clean and adsorbate-covered metal and semiconductor surfaces. Surface chemists direct their attention towards catalytic processes and surface modification through chemistry. Surface physicists and materials scientists involved in microelectronics technology have concentrated much effort in understanding the growth and control of semiconductor surfaces and the formation of metal-semiconductor interfaces. Issues of fundamental science have driven much of the research effort, yet the effort has also led to many new ideas and methods for applications to technological areas.

We identify another area of surface and interface research which has applied established surface science methods to advanced technological requirements -- coupling surface chemical and structural investigations to macroscopic adhesion at metal-polymer interfaces. Experiments on metal-polymer systems are designed in an inverse fashion to those which most surface scientists study. Usually, an adsorbate or thin film on a metal or semiconductor surface is studied. Here we discuss samples which may be insulating polymers with evaporated metals as "adsorbates." The goal is to relate macroscopic adhesion to microscopic chemical interactions between metallic species and functional groups comprising the relevant monomer.

This paper provides a review of the recent applications of some common and not-so-common surface science methods to study the formation of metal-polymer interfaces. Our goal is to consider how surface science methods have been used to characterize microscopic phenomena at metal-polymer interfaces. The paper focuses on a limited number of systems in order to emphasize various aspects of this scientific approach. We will concentrate primarily on thin film substrates of the polyimide PMDA-ODA because of intense technological interest in this material. The general applicability of these surface techniques to other metal-polymer systems can be extended straightforwardly.

We will compare metals which possess contrasting adhesion characteristics -- Cr and Cu (or Pd) -- on polyimide. We briefly sketch results obtained for other metal-polyimide systems for completeness. In the following sections we discuss:

(1) scientific interest in metal-polymer interfaces;
(2) experimental details;
(3) studies of electronic structure;
(4) studies of vibrational structure;
(5) microstructural measurements;
(6) summary and future directions.

SCIENTIFIC INTEREST IN METAL-POLYMER INTERFACES

Metal-plastic/metal-polymer adhesion has particular importance in advanced materials technologies.[1-3] Interfacial integrity and stability is a major concern in areas ranging from microelectronics device and packaging technology to protective layers for space environments and the manufacture of decorative coatings. In relation to advanced microelectronics device technology, an intense effort, particularly at IBM, has been directed to apply the methods of surface science to such problems.

The relation of macroscopic properties of adhesion between two media to microscopic interfacial properties is a fundamental issue in adhesion science.[4] Macroscopic testing procedures provide trends in adhesion characteristics between two media including systematic measurement of adhesion degradation or enhancement under various conditions. Chemical models and studies in solution provide a first order basis to explain such behavior on a molecular level. However, it is crucial to view such phenomena on a microscopic scale in the solid state, as well. Adhesion, being an interfacial phenomenon, has been addressed recently from a fundamental point of view utilizing surface science techniques directed at probing interfacial phenomena on an atomistic level.

The initial work at applying surface spectroscopies to metal-polymer interfaces began in the mid-1970's[5,6] and developed[7,8] through measurements of metal and polymer atomic core levels. Measurements of core-level electron binding energies have provided a fingerprint of the local chemical environment of individual atoms at the interface to explain metal-polymer reactions. X-ray photoelectron spectroscopy (XPS) is the technique of choice for measurements of core electron binding energies. Clark[5] and others performed extensive XPS studies on clean polymer surfaces, relating local bonding to electron binding energies and evaluating the mean free paths of electrons at the polymer surface.[6] Burkstrand[7,8] extended such work to study the metal-polymer interface. His XPS studies of Cu, Ni, and Cr films on various polymer surfaces established trends relating macroscopic metal bonding strength to microscopic chemical reactivity. The observed trends correlated bonding strength to the presence of Oxygen in the polymer. The formation of metal-oxygen complexes was proposed as the mechanism for enhanced adhesion. His data also suggested that the trend of increasing adhesion strength as a function of metallic species progresses as follows: Cu $\rightarrow$ Ni $\rightarrow$ Cr. This was related to the increasing electronegativity of these metals as well as to their ability to form oxygen complexes. In addition, it was proposed that arene-containing polymers might increase adhesion by the formation of π-bonding complexes. Related to this suggestion, a model proposed by Chou and Tang[9] has been successful in determining the bonding strength of various metals to polyimide in terms of the thermodynamic ability to form metal oxides. While it is not based on microscopic theory, this theory provides a criterion with which to predict reactivity. Successful empirical models have provided a basis from which to begin experimental investigations elucidating microscopic bonding mechanisms.

The properties of polyimide include good thermal stability (to $T \sim 400°$ C) and mechanical stability, low dielectric constant, and good chemical resistance. Polyimide has important applications in device packaging and as thin insulating films in microelectronics devices.[1-3] If a polyimide film is to be used as a thin-film insulator between thin-film metallic conductors, a strong and contiguous metallic film is desired. No diffusion through or delamination from the insulating film over the range of temperatures used in device processing or operation should occur. In order to better define and elucidate the microscopic basis of the characteristics of these metal-polymer interfaces, a focused scientific effort has been assembled at several research institutions, particularly IBM. This research has begun to provide a fundamental understanding of metal-plastic adhesion properties which will undoubtedly be applied to future materials applications.

EXPERIMENTAL DETAILS

Most experiments in surface science are performed in an ultrahigh vacuum (UHV) environment to maintain cleanliness at the monolayer level over extended periods and to avoid interactions of particle beams with ambient gases. One requirement for polymer samples then is that they be UHV-compatible with a low vapor pressure at the temperatures of interest (generally at and above room temperature). If polymer films are formed from solvent suspensions, methods to remove the solvents and other adsorbed species are required. This includes both *ex-situ* and *in-situ* curing. Metallization of polymer samples irreversibly alters them. In order to process several samples in a surface spectroscopy system, a sample transfer mechanism must be implemented. We have designed and used such a device[10] which allows rapid sample introduction. Intermediate processing can be performed in an auxiliary chamber; a wide range of sample cooling and heating in the main chamber is also possible with this design.

We first discuss PMDA-ODA (pyromellitic dianhydride-oxydianiline) samples (Figure 1) which are typically prepared by spinning dilute solutions of polyamic acid onto solid substrates.[11] Si wafers have generally been used in order to maximize thin-film uniformity. In many cases, metals are first evaporated onto the wafers in order to provide a conductive backing with which to heat the polymer films resistively. Film thickness is typically measured by ellipsometry; most films studied are on the order of 30-300Å thick. Two curing processes follow spinning -- a-curing at ~85° C to drive off the solvent and c-curing at T>250° C for cycloimidization. The repeat units for polyamic acid and PMDA-ODA are shown in Figure 1a.

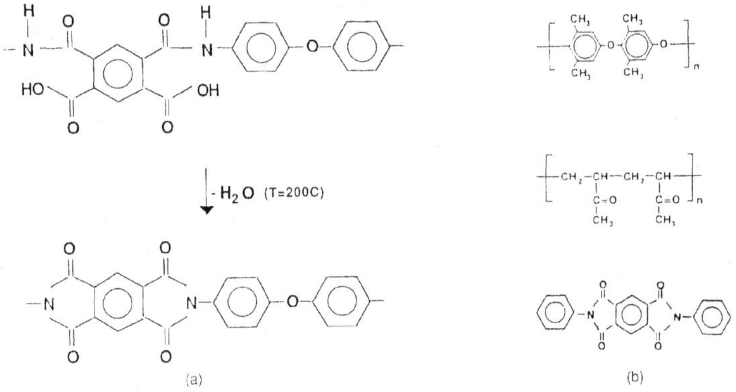

Figure 1. (a) The repeat unit of polyamic acid (upper) which is cycloimidized upon annealing to form the PMDA-ODA polyimide (lower); (b) Structures of various model compounds: top - PMPO {poly(methylphenylene oxide)}, center - PVMK {poly(vinylmethyl ketone)}, bottom - PMDA-model {N,N'-diphenylpyromellitimide}

The a-curing step is performed in a N_2 ambient, while c-curing can be performed *ex-situ* or *in-vacuo*. Annealing *in-vacuo* would follow c-curing anyway to remove adsorbed water vapor and other contaminants.[12] Formation of polyimide films by vapor deposition polymerization has also been reported recently.[13] This approach avoids problems that might occur in processing due, for example, to the presence of residual solvents. Upon curing, these films possess excellent properties and are comparable to the solvent-prepared films. No detailed surface science studies of the metallization of vapor-deposited films have yet been reported.

The structure of PMDA-ODA has been characterized by x-ray diffraction[14] and infrared/Raman spectroscopies.[15] Molecularly-ordered polyamic acid converts to the polyimide below the glass transition temperature indicating that large spatial rearrangement is not required for chain closure.[14] X-ray studies show that there is no long-range crystalline order in the annealed films. The plane of the ODA part is rotated with respect to the plane of the PMDA part by ~60° and the ODA arene rings are rotated by ~45° with respect to each other. The reduced orbital overlap then separates the PMDA unit from the ODA unit electronically. Vibrational studies also find that conformational changes in the polymer under stress modify the degree of conjugation in the ODA group.[15] We note that angle-resolved photoemission applied to studies of polystyrene has also concluded that the chains are randomly-oriented.[16] Azimuthal-angle-dependent measurements are therefore of little use on such polymer films. Other polymer samples which have been investigated have been prepared by numerous means such as sectioning bulk pieces.[7]

Resistively-heated, well-outgassed evaporation sources are generally used for the deposition of metal films. Metal coverages are generally measured with a quartz crystal microbalance calibrated by Rutherford backscattering spectroscopy. Other means of coverage measurement involve using electron emission intensities of the deposited metallic species.

Figure 2a presents a schematic representation of a surface spectroscopy experiment. Operating in UHV and using a sample transfer device, experiments are performed in the main chamber which houses the experimental apparatus. In order to enhance surface sensitivity, the experimental philosophy has been to use probes which interact strongly with matter, such as charged particles, so that the characteristically short mean free paths will only probe the surface region. Electron mean free path varies with electron kinetic energy, E_K. The "universal curve" for solids,[17] which also applies to polymers,[6] shows electron mean free paths of ~30Å at $E_K \sim 10$eV, 5Å at $E_K \sim 100$eV, and 10Å for $E_K \sim 500$eV. We note that grazing incidence/exit further enhances the surface sensitivity since the electron traversal distance perpendicular to the surface is minimized. Insulating substrates present difficulties for charged particle spectroscopies since the surface electric potential is not easily controlled. Maintaining a large occupation of conduction states in a thin sample is a means to reduce charging effects and this scheme has been demonstrated even for low electron energy spectroscopies.[18]

The information of interest comes from emission experiments (e.g. photoemission spectroscopy) or scattering experiments (e.g. electron or ion scattering). Schematics of these processes are shown in Figure 2b. The kinetic energy of charged particles can be analyzed with electrostatic electron energy analyzers, such as with a Cylindrical Mirror Analyzer (CMA) or a Hemispherical Analyzer.

In photoemission, light incident (photon energy = $h\nu$) on a sample excites electrons into the vacuum with kinetic energy, $E_K = h\nu - E_B - W$, where W is the work function and E_B is the electron binding energy. Thus, depending on the excitation energy and the spectral region of interest, one can measure the binding energies of core electrons or valence electrons. Core electron photoemission is usually referred to as XPS (X-ray Photoelectron Spectroscopy), or ESCA (Electron Spectroscopy for Chemical Analysis; UPS (Ultraviolet Photoelectron Spectroscopy) refers to valence photoemission. Often synchrotron radiation is used allowing continuous variability of incident photon energies. Since photon energy is directly related to the photoelectron kinetic energy from a given state, surface vs. bulk sensitivity can be adjusted exploiting the relation of mean free path to electron kinetic energy.

In Electron Energy Loss Spectroscopy (EELS), monochromatic electrons of energy E_i are incident upon the sample. The electrons are channeled through an electrostatic monochromator stage to provide uniform energy to the incident beam. The incident electrons can excite vibrational modes or electronic transitions giving up an energy, E_{loss}, in the process. Thus the electrons analyzed have an energy $E_K = E_i - E_{loss}$. For vibrational spectroscopy, E_{loss} is less than ~0.4 eV and the resolution requirements are rather severe (<10 meV).

In Medium Energy Ion Scattering Spectroscopy (MEIS), ions with energies on the order of 100 keV are incident upon the sample and lose energy by the Rutherford Scattering process. Ion scattering spectra can be analyzed to obtain detailed information on elemental composition, absolute concentration, and depth distribution at the interface region.

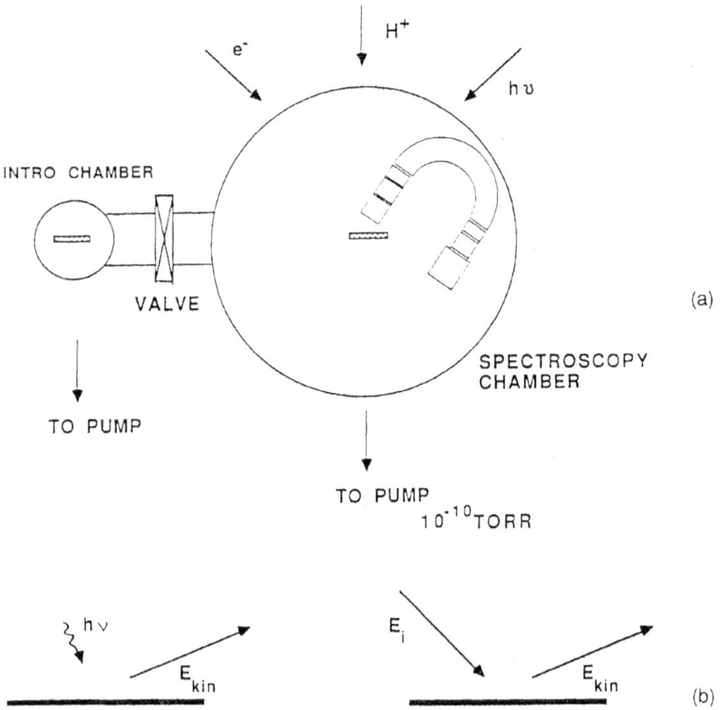

Figure 2. (a) General scheme for surface spectroscopy experiments; (b) Kinematic schemes for emission spectroscopies and scattering spectroscopies.

We will discuss each technique in more detail later as we address specific experimental studies. For more details, the reader is referred to excellent reviews and monographs on each topic: XPS,[19] UPS,[20] EELS,[21] and MEIS.[22]

STUDIES OF ELECTRONIC STRUCTURE

A major part of the surface science effort in metal-polymer adhesion has centered around studies of the electronic structure of polymer surfaces and of polymer-metal interfaces using photoemission spectroscopy. Two reasons exist for the widespread use of this approach. First, a direct signature of the occurrence of a chemical interaction should be observed by modifications in the electronic energy levels involved in the interaction with respect to those of the unreacted substrate and metal. Second, photoemission spectroscopies are widely used as analytic tools in surface science and much background information for analogue molecular species and gas-phase compounds already exists. In addition to obtaining electron binding energies, angle- and energy-dependent measurements can also provide structural information concerning the interface exploiting the mean free path of electrons in the medium.

Synchrotrons or x-ray/ultraviolet laboratory sources supply the photons and electrostatic energy analyzers are used to obtain energy distribution curves. The observation of core level binding energies by photoemission is the most widely-used elemental fingerprint in surface science. For polymers, the monomer contains atomic species which can exist in several different chemical environments. Core level binding energies associated with atoms in different environments may, therefore, contain several components that are shifted with respect to each other. The identification of the local environment of each atom by core level spectroscopy is important to study site-specific reaction chemistry upon surface metallization. The resolution of most reported XPS studies is between about 0.4 eV and 1.2 eV depending upon source energy linewidth and electron energy analyzer resolution. The problem of peak deconvolution in separating peaks related to specific atomic components in a polymer often arises since core levels may shift by less than 1 eV and there may be several different environments for a given atomic species. Reference data from isolated functional groups and analogue compounds as well as theoretical calculations of model compounds help the deconvolution process and the subsequent analysis considerably.

Binding energies of the valence electrons which take part in chemical bonding can also be analyzed. Photoemission spectra for these lower binding energies do not carry so much element-specific information as do the core levels. However, these spectra measure energies of the bonding electrons in the surface region. For polymers, σ and π bonding orbitals may be manifest as bands due to the delocalized bonding interactions which occur. Various functional groups can be identified particularly if theoretical calculations are available. One of the major difficulties in these studies is that the lack of crystalline order severely broadens the spectra since the bonding levels are so sensitive to the local bonding configuration. This covers the band effects that might otherwise be present. Of additional importance, metal s and d electron emission appears as metals are deposited allowing the analysis of the development of metallic emission at the Fermi level.

Core Level Photoemission Spectroscopy of Polyimide

The study of metal-polyimide reactions is a two-step process - (1) understand the clean polyimide surface, assigning peak positions and relative intensities in accord with the energetic shifts associated with particular atomic sites and relative populations of atomic environments, respectively; (2) study the changes that occur upon growth of a metallic interface on the polyimide surface. To explain the data for metal-polyimide reactions, two models have been developed which we will discuss: formation of π-complexes or electron transfer with possible fragmentation.

Typical clean polyimide core level spectra are illustrated from the data of Haight, et al.[23,24] in Figure 3. Theoretical calculations have established assignments for the spectral features observed and these are denoted with the deconvoluted peaks.[25,26] These spectra were deconvoluted by fitting the data with a number of Gaussians each with FWHM ~ 1.1-1.3 eV, the experimental energy resolution. The monomer contains 22 carbons, 5 oxygens, and 2 nitrogens. In the C(1s) spectra, we observe four components originating from carbon atoms in five different chemical environments -- carbonyl carbons (Ⓒ=O), nearest-neighbor carbonyl carbons on the central PMDA ring (-Ⓒ-C=O), carbons attached to nitrogens, (Ⓒ-N), ODA ring carbons ($Ⓒ_{ODA}$), and ether carbons (Ⓒ-O-Ⓒ). The peaks in the experimental spectra are assigned by comparison to calculated spectra[23,25] (see also Figure 6). The highest binding energy peak is attributed to the carbonyl carbons and is well-separated from the other features. The large relative shift in binding energy for these carbons occurs due to charge transfer through the delocalized π-system to the electron-withdrawing carbonyl oxygens (C=Ⓞ) and to the (Ⓒ-C=O) carbon atoms. Thus the carbonyl carbon (Ⓒ-O) is in an electron-deficient environment increasing the electron binding energy. The O(1s) spectrum has two contributions, the lower binding energy component from the carbonyl oxygens, where there is a high electron density. The two nitrogens sit in identical chemical environments so only a single peak is observed.

Broad structure is observed at higher binding energies in the clean spectra. This originates from "shake-up" processes where valence electrons are excited into unoccupied states during core-excitation. Energy conservation dictates that the emitted core electron will have lower kinetic energy; the auxiliary excitation can be viewed as inelastic scattering of the exiting core electron. Thus the corresponding emission appears at higher binding energy. The unoccupied valence levels are attributed to π^* levels of the PMDA unit ($\pi \rightarrow \pi^*$ transition energy ~4 eV) since π levels of the ODA unit have been calculated[12,23] to be ~5 eV lower in energy than those for the PMDA unit. The expected trend that the intensity of each core level component scales with the relative number of atoms in the corresponding environment holds fairly well after accounting for shake-up contributions. Reported discrepancies in relative intensities have been explained by invoking polymer degradation[26] and/or the presence of isomeric species at the surface.[27]

The metal-polyimide interaction is considered by comparing coverage-dependent data for Cr and Cu, strongly- and weakly-bonded metals on polyimide, respectively. The data will be discussed by considering studies of model compounds and theoretical studies that analyze various spectral changes which occur due to metal deposition.

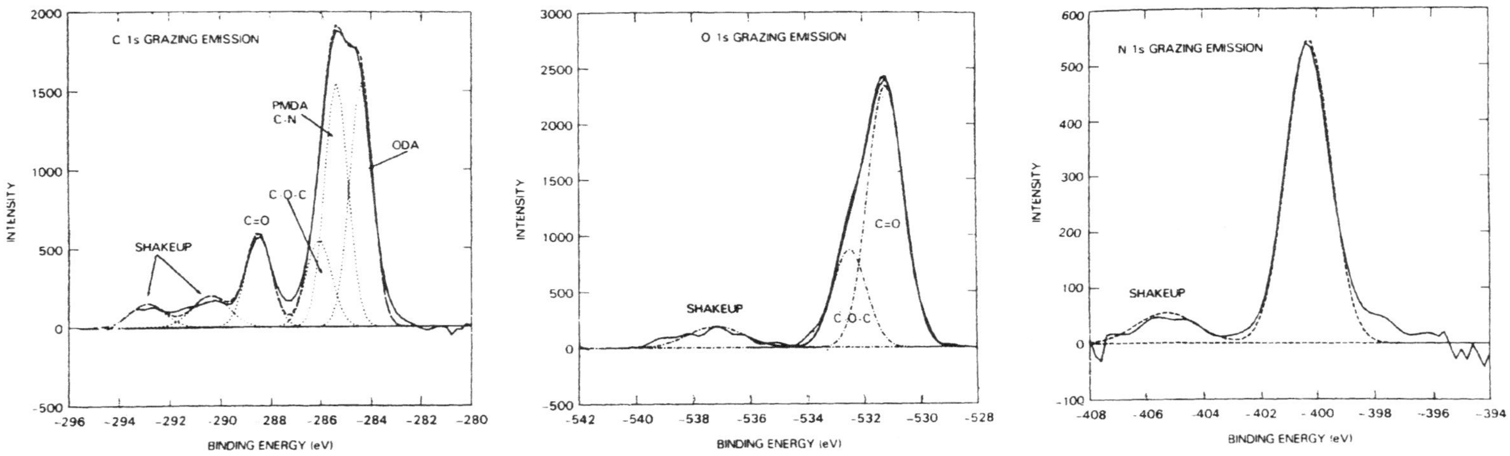

Figure 3. Core level photoemission spectra of PMDA-ODA in the C(1s), O(1s) and N(1s) spectral regions (Haight, et al.[23]).

Figure 4 presents the data of Haight et al.[23] for coverage-dependent core level binding energies for Cr deposited on polyimide at room temperature. Less than 1 Å Cr (1 monolayer Cr ~ 2.6 Å) modifies emission of the (Ⓒ-C=O) and the (Ⓒ=O) C(1s) components. The C=Ⓞ O(1s) level is preferentially attenuated and the Cr(2p) core level exhibits an increase in binding energy of ~2 eV with respect to metallic Cr. Since the initial action involves carbons associated with the PMDA part and not the ODA part, the initial guess is that it is the PMDA part of the monomer which is reactive to the metallic species. The shake-up intensity is found to be reduced upon Cr deposition. This is consistent with the assertion that the PMDA unit is involved in the bonding and the π* orbitals become occupied due to metal → PMDA charge transfer.

Higher Cr coverages continue to attenuate the C(1s) emission both due to further chemical interactions and inelastic electron scattering in the developing Cr layer. A new feature is observed at a lower binding energy. The O(1s) behavior is consistent with that of the C(1s) -- the carbonyl contribution is modified initially and another peak appears at lower binding energy for higher Cr coverages. The nitrogen atoms terminate the PMDA unit and the N(1s) signal is only found to be attenuated at low coverages until another peak arises at higher coverages. The observed behavior suggests a two-step Cr growth process -- at very low coverages and at higher coverages.

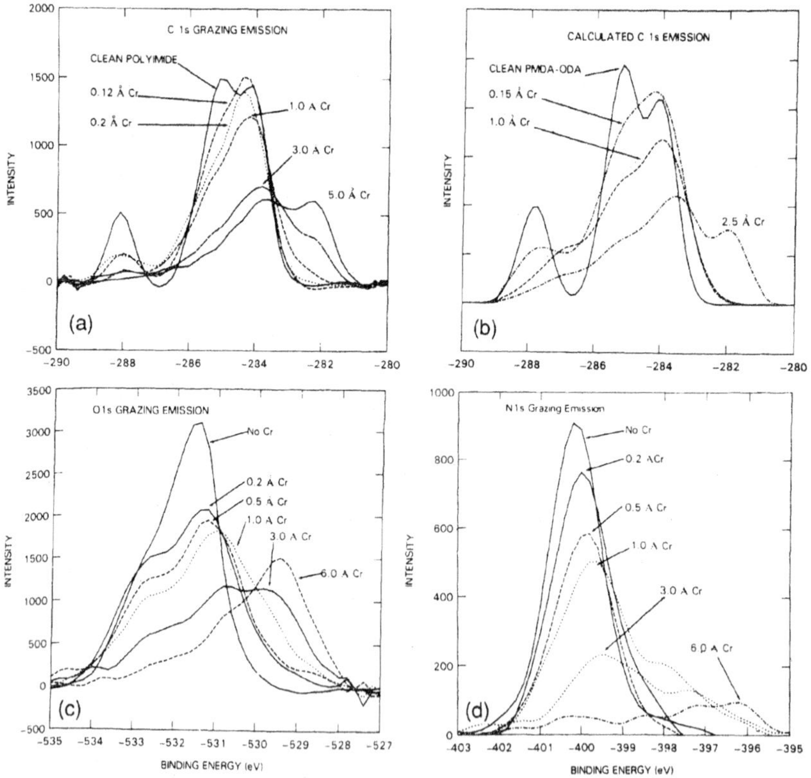

Figure 4. Chromium coverage-dependent core level photoemission spectra for the Cr/PMDA-ODA system in the C(1s), O(1s) and N(1s) regions; upper right panel presents calculated C(1s) emission for similar Cr coverages (Haight et al.[23]).

C(1s) spectra for Cu deposition on polyimide are shown in Figure 5. The only changes which are observed are a uniform reduction in signal with some preferential attenuation of the (-Ⓒ-C=O) and in the (Ⓒ=O) levels spectra. This "chemical" attenuation occurs to a far lesser degree than for Cr. On this basis, the conclusion has been made that the interaction which occurs with polyimide is weak but is also centered on the PMDA part. Higher Cu coverages did not produce any additional spectral features indicating a single-step process.

The authors have suggested a two-step reaction sequence for Cr/polyimide.[23] The first step involves immediate changes in the PMDA part. At high coverages, new low binding energy features emerge for C, N, and O. For Cu/polyimide, weak chemical interactions were observed. We consider experiments on model systems and theoretical calculations next.

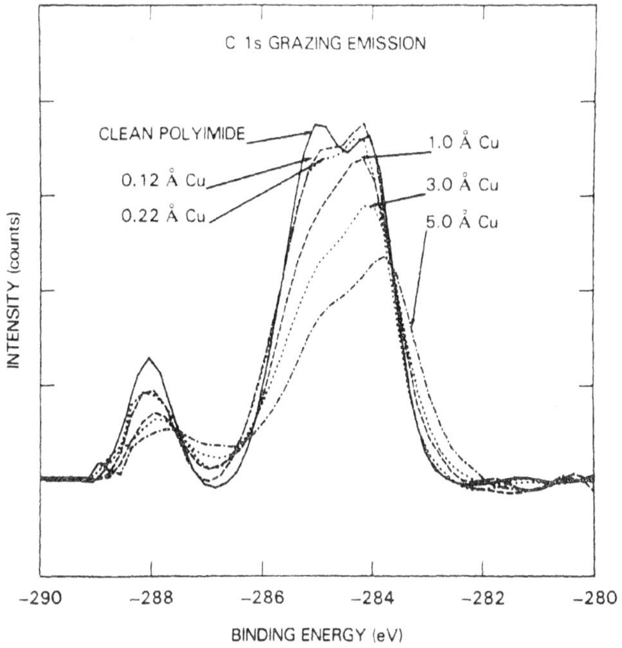

Figure 5. Copper coverage-dependent core level photoemission spectra for Cu/PMDA-ODA in the C(1s) region (Haight, et al.[23]).

Photoemission Studies of Model Systems

PMDA and ODA can be considered to be chemically separable in the PMDA-ODA polyimide.[23] Steric effects limit the overlap between the delocalized π-systems of the individual parts of the monomer. This assumption of separability has prompted an empirical approach[28,29] studying model compounds which contain a subset of the PMDA-ODA functional groups in order to identify possible mechanisms for metal-polyimide interactions. The model compounds are polymers in the solid state which provide a better correspondence to the actual situation than species in solution where kinetics are markedly different, although the energetics of individual interactions may correspond well. These experiments have met with good success at explaining at least initial metal-polyimide interactions. In addition, model studies of charge-transfer reactions on polyimide have been performed by Cesium adsorption.[30]

XPS studies have been performed for the three "model" compounds shown in Figure 1b:

"PMDA"	N,N'-diphenylpyromellitimide	
"C-(C=O)-C"	poly (vinyl methyl ketone)	PVMK
"ODA"	poly (2,6-dimethylphenylene oxide)	PMPO

The PMDA-model polymer contains the central arene of PMDA-ODA with terminal arenes. PVMK contains carbonyl species bonded in a saturated chain. PMPO essentially consists of the ODA-unit with the bridging ether-oxygen species.

C(1s) core level spectra of the clean PMDA-model compound showed components due to the carbons of (1) the carbonyl, (2) the central phenyl ring, and (3) the terminal phenyl ring consistent with PMDA-ODA binding energies.[25] Cr was found to preferentially modify the carbonyl C(1s) emission. Cu, on the other hand, was found to interact only with the terminal arenes, those exclusive to the model compound.

PVMK contains the carbonyl unit but lacks the delocalized π-system of PMDA. The PVMK C(1s) spectrum contained two components - one from carbonyl carbons and the other from aliphatic carbons. There is only one O(1s) component (from the carbonyl). Cr deposition was found to attack the carbonyl component as evidenced in both C(1s) and O(1s) spectra. Higher coverages resulted in the growth of a low binding energy peak. This behavior was taken as an indication that the Cr reaction on polyimide takes place at the carbonyls exclusive of the arene.

Clean PMPO produces two C(1s)-derived components in XPS spectra: (1) the ether carbons and (2) the arene and methyl carbons. The interaction of Cr with PMPO was found to have a weak effect at low coverages. The C(1s) peaks are attenuated almost uniformly indicating the lack of a strong reaction. Whatever reaction occurs also affects the ether carbons. Again, a new peak at lower binding energies appears at higher metal coverages. Cu also exhibited weak reactivity with the ether functional group.

Studies of the reduction of polyimide by deposition of the alkali metal Cs supported the assertion that the initial reaction of Cr with polyimide involves charge transfer to the PMDA unit. The XPS experiments of Clabes, et al.[30] showed only Cs-induced changes to PMDA C(1s) and O(1s) core levels. Angle-dependent measurements indicated the intercalation of Cs into the substrate with polyimide beneath the surface having been reduced. The reaction is consistent with reduction of polyimide to $(PMDA\text{-}ODA)^{-1}$.

These model experiments have provided complementary information to those on polyimide. In a simplified picture, Cr appears to react strongly with PMDA-model and PVMK in a two-step sequence. Cu reacts weakly with the PMDA-model polymer at sites unrelated to the polyimide.[29] Both react weakly with the ODA-model polymer. These observations are consistent with observations on PMDA-ODA. At high coverages, either the same reaction mechanism occurs for both PMDA-model and PVMK (bonding at the C=O site), or two separate reaction mechanisms occur (π-complex for PMDA-model and local bonding for PVMK).

Theoretical Models of Polyimide-Metal Interactions

We outline theoretical work which has treated the interaction of Cr with the polyimide PMDA-ODA in a π-complex model. We will discuss and compare the work of Rossi, et al.[25] and Haight, et al.[23] for experimental core electron binding energies. The theoretical work has been based upon consideration of charge transfer into the delocalized π-system of the PMDA unit of the monomer positioning the Cr above the phenyl ring producing spectral results which compare well with experiment. The results of the *ab-initio* calculations of Rossi et al.[25] are reproduced in Figure 6 and those of Haight, et al.[23] were given in Figure 4. Both compared calculated spectra for clean and Cr-reacted surfaces. Rossi et al. limited their calculation to the initial reaction (low coverage); Haight, et al. considered both initial and high coverage Cr using a sequential bonding scheme.

Calculations of core level shifts correlate well with experimental results for the initial Cr/polyimide interaction manifest by significant changes for the C(1s) binding energies of the (Ⓒ=O) and the (Ⓒ-C=O). Rossi et al.[25] first considered Cr bonding to benzene and then the PMDA-model compound was used for the model studies. In analogy to bis-arene complexes, they placed the Cr above the arene ring to form a $Cr(C_6H_6)$ complex and above the central benzene ring of the PMDA model. For this geometry, they found that the benzene π* orbitals interact most strongly with d_{xy} and $d_{x^2-y^2}$ Cr orbitals. For the PMDA model compound, the strongest interaction was with the d_{xy}. This interaction results in charge transfer throughout the delocalized π-system. Charge accumulates at the -Ⓒ-C=O and at the C=Ⓞ sites. The electron density at the carbonyl carbon remains nearly constant; this is consistent with the character of the PMDA-model a_2 level. The calculated electron binding energies of the (-Ⓒ-C=O) and the (Ⓒ-O) C(1s) were upshifted due to increased electron density in the vicinity of these atoms. Calculated spectra of Rossi, et. al. for a 50-50 mix of reacted and unreacted Cr/polyimide provided the best fit of the XPS data.[25]

Haight, et al.[23] obtained comparable results for the initial formation of the Cr-polyimide chemical bond using the bonding geometry of Rossi, et al. The analysis was extended to consider the high coverage Cr regime where experimentally a new lower binding energy peak was observed in C(1s), O(1s), and N(1s) spectra. They observed that the occurrence of metal-metal interactions at higher metal coverages may initiate the formation of a laminar Cr film now placing Cr atoms in proximity to the ODA complex. While strong reactions occur in the PMDA unit, they found that stable Cr/ODA complexes can also form at these higher coverages. In the calculation, complexing of the Cr to the ODA shifted the levels of both the aromatic and aliphatic C(1s) levels down ~2 eV in binding energy. The ether O(1s) levels were found to shift as well. In the model PMPO studies discussed above, we recall that Cr reacted weakly in the vicinity of the ether oxygen further providing support that reactivity with the ODA part can occur. The N(1s) shift is not fully explained by this model but may occur due to a proximity effect. With proper consideration of the attenuation of photoelectron emission from the polymer due to inelastic scattering in the developing Cr layer, the new C(1s) levels obtained from theory agree well with the data in energy position and relative intensity (Figure 4).

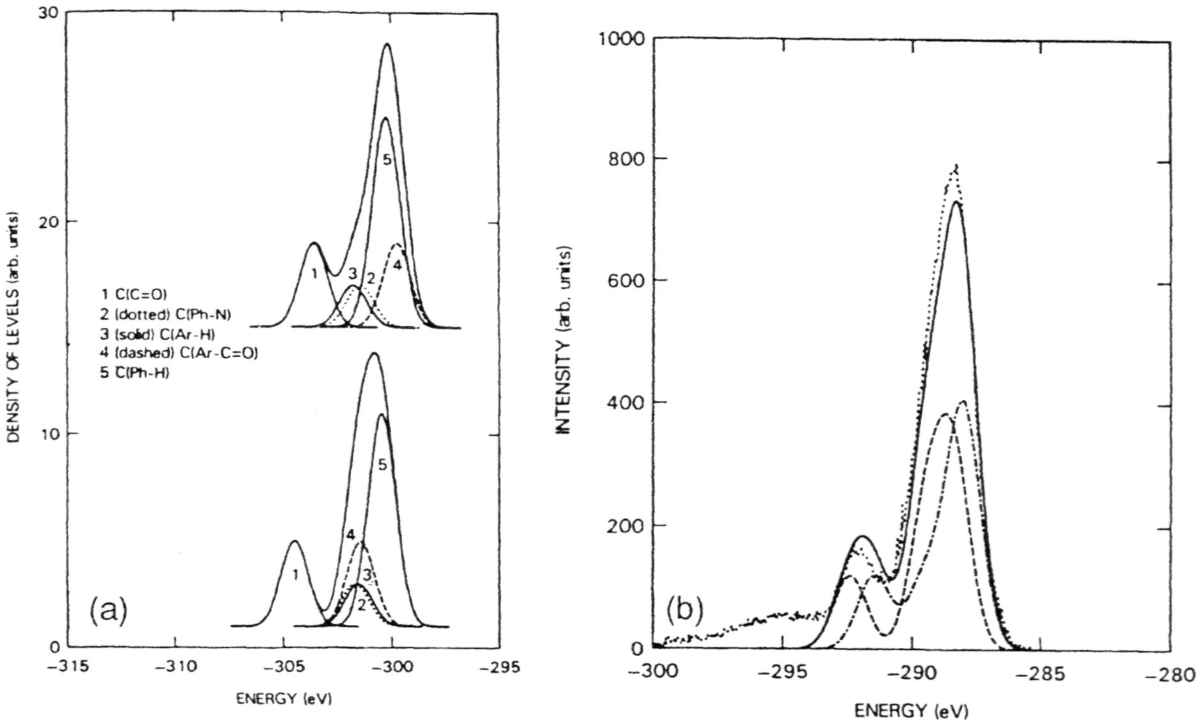

Figure 6. (a) Calculations of Rossi, et al.[25] for clean and Cr-reacted PMDA-model compound; (b) Comparison of calculation of a 50/50 mixture of Cr-reacted and unreacted PMDA-model compound of Rossi, et al.[25] with XPS C(1s) spectrum of Sanda, et al.[29] for PMDA-model compound.

These theoretical studies[23,25] proposed the formation of a π-complex to explain the data for initial coverages. In this model, no fragmentation occurs and delocalized bonding throughout the PMDA essentially allows considerable redistribution of charge throughout the PMDA. The high coverage data was also explained by complex formation, although the origin of the N(1s) shift at high coverages was not fully explained.

The data suggest that the chemical interaction is particularly strong for Cr. The picture of the initial chemical interaction of Cr with polyimide derived from these model studies involves charge transfer from the metal to the PMDA group. The points of agreement center on the strong reaction with the PMDA part and delocalized bonding through the π-system. Theoretical modeling has only considered the symmetric metal site above the phenyl ring which is successful at reproducing experimental spectra. Other chemical evidence[31] would suggest against the formation of such a π-complex and would suggest that electron transfer occurs at an alternative site in greater proximity to the carbonyl.

It has thus been proposed that the high-coverage low binding energy feature originates from either (1) Cr bonding in the ODA environment[23] or (2) fragmentation of the polyimide and the formation of Chromium oxides, nitrides, and carbides.[28] From their studies of model compounds and comparison with binding energies for core emission from metal oxides, nitrides, and carbides, Jordan, et al.[28] have proposed that these spectral features develop from the decomposition of the polyimide during reaction. In support of such a mechanism, it is noted that the feature is common to both the PVMK and the PMDA-ODA model. PVMK lacks the π-system of PMDA. In order to have similar characteristics, it was asserted that charge transfer does not occur by bonding above the phenyl ring but through bonding to the C=O. Bonding of Cr at or near the carbonyl oxygen has not been considered theoretically, yet this is a reasonable site for attack and subsequent fragmentation on the basis of many chemical reactivity studies. That similarities exist in the spectra for these two species is not conclusive evidence for either mechanism. It must be noted that these species must all form simultaneously to explain the data.

Haight et al.[23] asserted that their theoretical calculations indicate that the lower binding energy feature arises from interactions with the ODA. In fact, model studies do show that weak interactions with the PMPO occur causing changes in the vicinity of the ether, neither proving nor disproving the site that they chose for the calculations. As experimental support for a weak interaction, they found that annealing the film reduced emission from this component. A weak reaction and reversibility would not be expected if bond scission and the formation of new compound(s) occurred. It should be noted, however, that kinetic constraints may be lifted upon annealing, resulting in related spectral modifications. The origin of the low binding energy feature in the N(1s) emission also remains in question in this model.

The high Cr coverage interface is an important issue since macroscopic adhesion involves a thick interfacial region. The reaction mechanism must still be established in order to provide a valid model for the bonding interaction and to relate the microscopic chemical properties to macroscopic mechanical properties of the interface.

Valence Photoemission Spectra of Polyimide

Valence electron binding energies of clean polyimide and metal-polyimide surfaces have been studied experimentally. For clean polyimide, broad spectral features originating from the close-lying σ and π molecular orbitals of PMDA-ODA have been found. Figure 7a presents the clean valence spectrum of Haight, et al.[23] before metal deposition with the origin of the broad features observed. Calculations of valence electron binding energies[32] predict much more structure. ~2 eV broadening must be added to compare theory[32] to experiment,[12,23] as shown in Figure 7b. If only instrumental resolution was considered, considerably more structure should have been observed in the clean spectrum. Two possibilities were suggested to account for this spectral broadening -- energy dispersion in (structurally-ordered) films or structural disorder.[32] Brédas and Clarke[32] reported calculations of the PMDA-ODA band structure where the maximum valence bandwidths are found to be about 0.2 eV or less. Dispersion of the valence levels cannot be invoked to account for the observed magnitude of broadening. Structural disorder has been invoked to explain the observed broadening. Being of a greater magnitude than the observed bandwidths, this disorder can be taken as localized to individual PMDA-ODA units.[33] UPS spectra exhibit less structure for films prior to annealing *in-vacuo*.[12] This has been attributed to a combination of new features and increased disorder from thermally-activated adsorption of H_2O.

Although broad valence features are generally observed, it has been shown that UPS can still be applied to metal-polymer interfaces providing valuable information in at least two regards: (1) bonding to the polymer units can be related to preferential changes in spectral features and (2) the development of metallic electron emission provides information on the growth of the interface. Comparisons of UPS spectra for Cu and Cr on polyimide are consistent with the main conclusions of XPS studies. In particular, large changes observed in the -10 eV feature of clean polyimide for small Cr coverages indicates that chemical interactions with the π-system of the PMDA unit occur. Further deposition of Cr gives rise to the development of emission at the Fermi level and uniform attenuation of the polyimide features (with nearly complete attenuation of the PMDA π molecular orbital peak). The formation of a uniform metallic film of Cr reacted with the polymer surface is concluded. In contrast, only small changes in the -10 eV feature occur for Cu deposition, suggesting that a weak chemical interaction occurs. Cu s-states at the Fermi level were observed to develop slowly with an equally slow attenuation of the bulk polyimide features. Such studies[12] concluded that diffusion of weakly-interacting Cu into the bulk can occur, in agreement with microstructural measurements.[34]

Spectra of Electronic Excitations in Polymers

Interest in absorption spectroscopy has developed recently as a means to analyze the unoccupied states of clean and reacted polymers. Such experiments have been performed on model systems by uv-visible absorption[35] and has been extended to surface studies by investigating the core $\rightarrow$ bound transition region with Near-Edge X-ray Absorption Fine Structure (NEXAFS)[36] and valence $\rightarrow$ bound excitations with Electron Energy Loss Spectroscopy (EELS).[37] These states are the origin of the shake-up satellite structure in XPS. The *ab-initio* calculations which have been described could not produce the spectrum of unoccupied states accurately.[25] However, the spectroscopy of unoccupied states of model compounds can offer another means to understand the bonding interaction for metal-polymer systems.

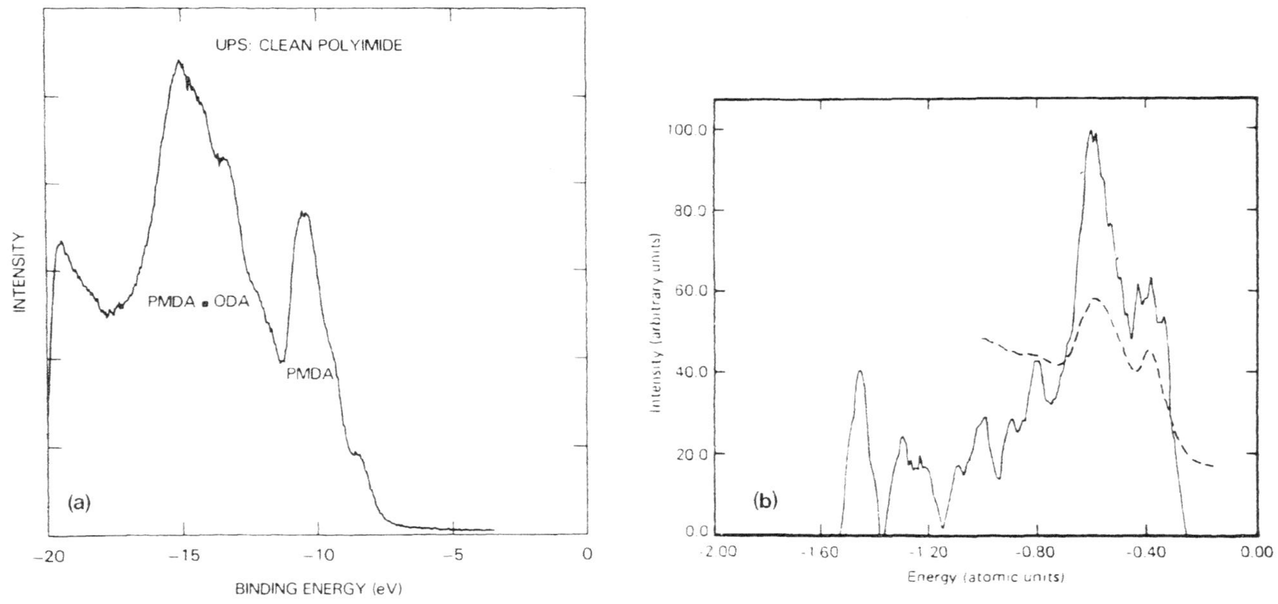

Figure 7. (a) Valence level photoemission spectrum for clean PMDA-ODA (Haight, et al.[23]); (b) Calculated valence binding energies for PMDA-ODA using 2 eV broadening; dotted curve depicts the experimental data of Hahn, et al.[12] (Brédas and Clarke[32]).

NEXAFS measurements on PMPO, PVMK, a PMDA-model compound, and PMDA-ODA have recently been reported.[36] Absorption spectra were obtained by measuring the secondary electron yield as a function of photon energy near the C(1s) binding energy (the K-edge) using synchrotron radiation. Assignments for clean PMDA-ODA were made on the basis of the model compounds studied. The delocalized π^*-orbitals are most straightforwardly assigned while σ^*-resonances are difficult to separate for complex polymer structures.[36]

Cr-polymer interface formation was also investigated in this set of studies. Consistent with XPS studies, Cr interacts with the PMDA unit of polyimide. The results indicated that no reaction occurs for PMPO and that Cr interacting with PVMK involves the carbonyl group. For Cr/polyimide, the carbonyl π^*-resonance is the most severely quenched C(1s) $\rightarrow \pi^*$ excitation supporting the view that the carbonyl is strongly involved in bonding. Furthermore, this observation was taken as support of the previous view of Jordan, et al.[28] that Cr bonds at the carbonyl oxygen rather than forming a Cr-arene complex. Calculations must be performed to determine whether such a π-complex could produce such a result.

We note that Pireaux, et al.[37] have reported the electronic excitation spectrum of clean polyimide using EELS. They have reserved interpretation of the observed features due to the lack of analogous information for model compounds.

Photoemission Studies of Complementary Systems

The goal of this paper is to indicate how a surface science approach can be employed to elucidate local chemical interactions as they contribute to metal-polymer adhesion. We are thus considering model systems for which several points of view have been developed. There has been a parallel effort to study several other metals to elucidate bonding characteristics in light of intense study of Cr. It is found that behavior of other metals not only produces different degrees of interaction, but sometimes different types of interactions. Generally it is found that Cr, Ti, Al, and Ni react strongly with polyimide while Cu, Ag, Au, and Pd are weakly interacting. There is general consistency with the model of Chou and Tang[9] which is based upon trends in electronegativity. For completeness, we briefly outline results for other interfaces providing the reader references from which to obtain details of the studies. Since systematic studies have not been performed as for Cr, only similarities and differences with respect to Cr and Cu films can be noted.

Aluminum-polyimide interfaces have been the subject of numerous studies and new results using vibrational spectroscopy have been reported.[38] Atanasoska, et al.[39] found that the carbonyl C(1s) peak is again preferentially attenuated upon initial Al exposures but never vanishes. All the polyimide features exhibit little attenuation. The Al(2p) is initially shifted and then soon forms into characteristic metallic emission. The transparency of the polyimide and the rapid formation of metallic Al features is indicative of cluster growth at the surface. The initial chemical reaction is proposed to occur in the vicinity of the carbonyl sites (at the carbonyl oxygens) with the strong Al shift related to charge withdrawal by the carbonyl oxygen. Both terminal and bridging Al species are proposed to occur without polyimide fragmentation. Reactions at elevated temperatures have also been studied.[40]

Titanium-polyimide interfaces follow a Stranski-Krastinov growth mode -- completion of the first monolayer is followed by island formation. Focusing on <1 monolayer, XPS data[41] has shown changes in the C(1s) intensity in a stepwise fashion. Initially, the carbonyl C(1s) is attacked indicating a reaction with the carbonyl. This is followed by the appearance of a low binding energy feature interpreted as due to the formation of a surface carbide. Related changes in the O(1s) and Ti(2p) have been interpreted as being due to the formation of a C-O-Ti complex. Auger spectroscopy data correlated with XPS to support the occurrence of a carbidic C species due to increased Ti density and increased charge transfer as the monolayer develops.

A study of Cobalt deposition on polyimide[42] concluded that uniform layer growth and a reaction sequence notably different than for Cr occurs. It was found that Co immediately forms low binding energy emission on the main C(1s), O(1s), and N(1s) peaks. The degree of broadening of the features indicates several inequivalent interaction sites. Although there was evidence of immediate interaction with the carbonyl carbons, a layer of polyimide fragments also begins to grow at the onset of deposition saturating upon completion of the interface. Co is carried by the PMDA fragments. Coverage-dependent measurements further suggest that the fragments apparently remain near the surface as the Co film grows beneath. Thus two components need to be separated in this model -- the bonding at the Co film-polyimide interface, presumably bonded to decomposed polyimide chains, and the Co-polyimide fragments above, which are unrelated to adhesion.

In light of applications to semiconductor technology and the study of bonding mechanisms with polymers in general, it is of interest to ask what differences might occur at semiconductor-polyimide interfaces with respect to metal-polyimide interfaces. Atanasoska, et al.[43] have investigated the formation of the Ge-polyimide interface where the possibility of local bonding chemistry might be expected to a greater degree than for a metal film. Again, the carbonyl group of the PMDA is initially affected in the spectra. Ge(3d) and Ge(2p) core level spectra exhibit a shift of ~3 eV for low coverages, indicating a strong interaction with the substrate. This is viewed as due to the formation of a Ge-O-C complex. Multiple bonding of Ge to PMDA has been proposed at higher coverages[43] resulting in modifications of bond orders and shifts of electron binding energies throughout the PMDA unit.

STUDIES OF VIBRATIONAL STRUCTURE

The analysis of the vibrational modes of molecules and solids is an important means to determine structure and interatomic bonding potentials in materials. One distinct advantage of vibrational spectroscopy for adsorbed species is the ability to identify vibrational modes of the species with reference to gas phase species. Shifts in vibrational energies can often be attributed to the molecule-surface interaction, and the absence of a mode generally corresponds to bond scission occurring upon adsorption.

Infrared and Raman spectroscopy can be used to measure vibrational modes of gas, liquid, or solid phase materials. Difficulties arise when optical techniques are applied to surfaces due to the lack of surface selectivity. Both the large sampling depth and problems arising with the incorporation of these techniques with ultrahigh vacuum has limited the general applicability of optical techniques. Electron spectroscopy, having been well-established in surface science, has also been applied to surface vibrational spectroscopy in high-resolution EELS. The advantage of wide spectral range allows measurements to extend from the vibrational regime to electronic excitations with the same instrument.[37,44]

High-resolution Electron Energy Loss Spectroscopy (EELS) provides a means to directly observe surface chemical bonding by analysis of the vibrational modes of surface species.[21,44] Electrons can couple to excited intermolecular vibrational states in a variety of ways. The mechanism most generally invoked, especially for work on single crystals, is the dipole scattering mechanism. This is the analogue to infrared absorption. The process is dominated by small momentum transfer, implying the forward scattering of electrons. The backscattered electrons which are detected are then either weakly backscattered or forward scattered and reflected by the substrate. For single crystals, the latter process dominates and dipole scattering intensities peak for a specular scattering geometry. Since this scattering/reflection is a two-step process, dipole scattering scales with the reflected elastic intensity. Image effects of a metallic surface give rise to dipole selection rules such that modes parallel to the surface are screened and are weak or not observed with dipole scattering. This rule allows the orientation of a molecule on a single crystal surface to be deduced immediately.

Impact scattering and resonance scattering are two other vibrational excitation processes. Impact scattering is treated in multiple scattering formalisms which include vibrational excitations of the medium. Resonance scattering represents a process by which a short-lived negative ion state is formed. De-excitation to the ground electronic state can be to an excited vibrational state showing up as an energy loss in the spectrum. This process has an energy dependence representative of the negative ion state formed. Both mechanisms have angular dependences governed by the local structure. For polymers, where crystalline order is not generally observed, specular scattering is not a relevant consideration to deduce dipole scattering. The scattering mechanisms are rather distinguished by the energy dependence of inelastic scattering.

To date, energy resolution on the order of ~2 meV has been reported.[45] This clearly approaches that of optical techniques. However, the diffuse nature of scattering from polymer surfaces has created limitations in resolution for polymer surfaces, partly due to inhomogeneity and partly due to limitations of the electron optics for diffusely scattered electrons.

The novel application of EELS to detect interfacial vibrational structure of organic substrates has been demonstrated recently by a few groups. One of the earliest reports involved studies on conducting BF_4-doped polypyrrole.[46] Broad vibrational bands enveloping the bulk vibrational spectrum of the polymer were observed. A limitation of the technique applied in this way became evident -- closely-spaced vibrational excitations observed routinely with infrared or Raman spectroscopy are not easily resolved as individual spectral features due to the inherently poorer resolution of EELS. Following this work, EELS was applied to fatty-acid Langmuir-Blodgett films deposited on Ag substrates.[47] Three spectral regions were distinguished: 90-200 meV including various bending and scissoring modes; 200-220 meV containing the C=O stretch; and the 350-400 meV region including the CH stretching region. With an energy resolution of ≥10 meV (80 cm^{-1}), the 100-200 meV region was observed as a broad band from which individual vibrational losses were resolved as peaks and shoulders. The behavior of the carbonyl stretching band(s) is particularly interesting. Generally strong in infrared spectra and found at ~210 meV, the carbonyl loss was observed as a weak shoulder at ~206 meV. The weak signal was interpreted as being consistent with the orientation of the carboxylic units towards the substrate and screened from the electric field of the impinging electrons. The energy shift was related to charge transfer due to the delocalized interaction of the molecules with the substrate. The strength of the loss provided a qualitative feel for the sampling depth of EELS while the shift indicated the ability to observe chemical reactivity.

We have applied EELS to thin polyimide films[49,50] as well as to polymethyl methacrylate (PMMA) and various Langmuir-Blodgett films. Two specific goals were defined: (1) to demonstrate the feasibility of analyzing the clean PMDA-ODA surface with EELS and (2) to study the formation of the polyimide-metal interface by analysis of the surface vibrational structure of polyimide upon metal deposition. We note that a calculation of the vibrational modes of the PMDA-ODA polyimide has been reported recently.[51]

An EELS vibrational spectrum of the clean polyimide surface before and after c-curing is presented in Figure 8. The sample was prepared[51] by spinning polyamic acid on a gold-coated Si substrate[52] with partial curing *ex-situ*. The thickness of the films was estimated to be ~50 Å by ellipsometry. Samples were inserted into the vacuum system via the sample transfer stage,[10] and curing was completed *in-vacuo*. Ultraviolet photoemission spectroscopy was used for comparison to other measurements[12] ensuring comparable surface conditions and good film uniformity. The vibrational assignments were based upon a comparison to the infrared studies of Ishida, et al.[15] on polyimide and other model compounds.

We separate the spectra into three regions. The region between 100-200 meV contains contributions from the imide functional group and from various bending modes. We identified the following features in this region: the C_6H_2 wag (106 meV), imide-OC stretch (141 meV), (C-O-C) asymmetric stretch (155 meV), and imide CN stretch (170 meV). The vibrational band in the 210-220 meV region contains the carbonyl stretching modes. The mode observed here at 212 meV is the asymmetric carbonyl stretch, the stronger infrared-active mode. The third region contains CH stretching modes in the 360-380 meV range. Other vibrational modes could not be identified due to limited resolution and/or low cross-sections; in most cases, the strongest bulk infrared-active modes were correlated with the EELS data. Our major focus was placed on the carbonyl stretch, however. We also noted other spectral differences that were observed for as-introduced films, before cycloimidization (c-curing). Generally the spectra are broader due to greater inhomogeneity and the presence of additional bands. In addition, contributions of O-H and N-H bands were observed.

A subsequent EELS study of the polyimide surface by Pireaux, *et al.*[37] corroborated our vibrational assignments. Improved instrumental resolution and a refined annealing recipe allowed improvements in the identification of vibrational features as peaks and shoulders within the vibrational bands, particularly in the 100-200 meV range, as shown in Figure 9. In these studies and in complementary studies of the linear saturated polymers, polyethylene and hexatriacontane,[53] impact energies

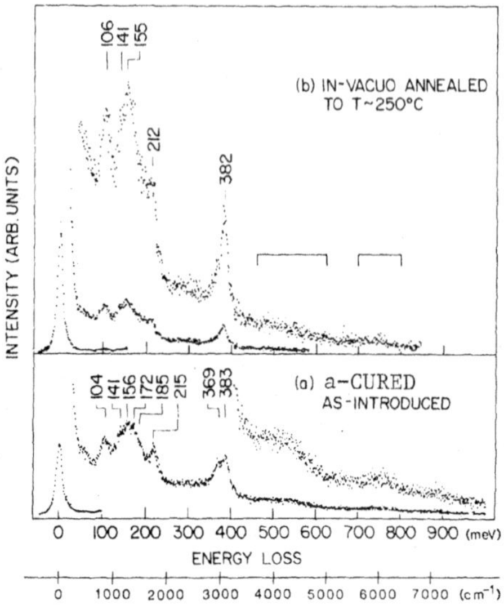

Figure 8. EELS spectra of surface vibrational losses of PMDA-ODA (a) before (a-cured) and (b) after (c-cured) cycloimidization (DiNardo, et al.[49]).

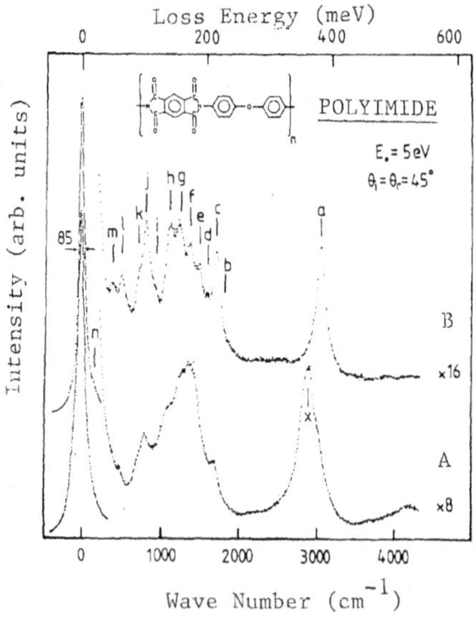

Figure 9. EELS spectra of PMDA-ODA of Pireaux, et al.[37] (A) as-introduced and (B) after heating *in-vacuo* at 530K for one hour. Peak assignments are provided in Reference 37.

and scattering angles were varied to distinguish inelastic electron scattering mechanisms. The dipole scattering contributions were evaluated by measuring the ratio of the loss intensity to the elastic intensity for various modes as a function of impact energy. It was concluded that, while dipole scattering contributions are indeed present, the possibility of resonance and/or impact scattering could not be excluded. Based upon ideas from dipole interaction for infrared-active modes, the orientation of polymer chains was addressed. Much of the directional information was deduced from non-dipolar components of the inelastic scattering. Their angle-dependent measurements produced the assertion that the orientation of C=O groups is different from that of CH groups.

Success in resolving and assigning major vibrational surface bands of clean polyimide with EELS allowed us to study the reaction of Cr and Pd with polyimide by EELS.[54] The main goal was to obtain information on the location of the reaction by analyzing modifications in the vibrational structure of clean polyimide upon reaction. Cr is known to be reactive from XPS measurements; Pd was considered to be less reactive by consideration of XPS studies of Ni and Cu on polyimide.[55] From our surface vibrational measurements, we concluded that Cr-polyimide reactions involve the carbonyl group in agreement with core level studies and that Pd interactions are significantly weaker. The fact that the carbonyl vibrational bands are well-separated from other polyimide vibrational bands provided the crucial advantage in this study.

Figure 10 presents the comparison between a typical clean polyimide spectrum ("a") and metal-polyimide spectra. Both EELS and UPS spectra are shown. For Pd, curve (b) represents ≤5 monolayers; for Cr, curve (b) represents ~1 monolayer Cr and curve (c) represents ~5 monolayer Cr. Cr coverages were estimated by comparing the development of Cr d-electron emission intensity with previous correlations of valence and core level intensities by Hahn, *et al.*;[55] Pd coverages were estimated in a similar manner.

Throughout the range of coverages (b-e), Pd was observed to uniformly reduce the intensity of the polyimide vibrational losses without itself introducing any new spectral features. This behavior continued until Pd began to form a continuous conducting film. The elastic peak width increased and a large loss background related to a continuum of quasielastic losses[56] was observed. This signaled the formation of a two-dimensional conducting layer; the broadening is related to the excitation of collective surface modes within a 2-d conducting film. Analysis of the broadening can provide direct information on the connectivity of the film.[54,56]

The situation was quite different for the Cr/polyimide system. At the start of Cr deposition, a chemical reaction was evident as deduced from the downshift of the carbonyl peak from 212 meV to ~185 meV. The remaining vibrational bands underwent uniform attenuation as the surface became covered (b - e). Continuum broadening again occurred due to formation of a conducting film. The shift is consistent with a reduction of the CO bond order due to charge transfer from Cr into carbonyl antibonding orbitals. We have attributed the shift to metal → polymer electron transfer in the delocalized bonds of the PMDA part in accord with theoretical calculations.[23,25] Due to the delocalized nature of bonding in the PMDA component, we can only conclude that the interaction is in the PMDA part but cannot deduce the site without a more detailed vibrational analysis, requiring better energy resolution. Bonding of Cr above the arene could result in a

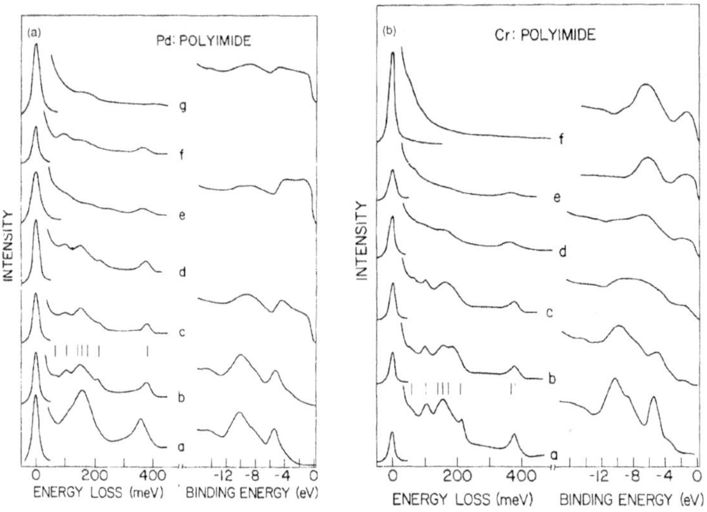

Figure 10. (a) Palladium coverage-dependent EELS/UPS spectra for sequential Pd exposures onto polyimide at room temperature (DiNardo, et al.[54]); (b) Chromium coverage-dependent EELS/UPS spectra for sequential Cr exposures onto polyimide at room temperature (DiNardo, et al.[54]).

reduction of bond order as well as bonding at the carbonyl. Clabes, et al.[30] have reported FTIR vibrational measurements for neutral and monoanionic solutions of a PMDA model compound. They observed a downshift in the CO stretch upon one-electron reduction of the monomer, which provides some consistency with our results. In our studies, further deposition of Cr was found to further attenuate the 185 meV feature (curve c) suggesting possible bond scission or a further reduction in bond order due to the metal-carbonyl interaction at higher metallic densities.

These studies allowed us to identify the local environment of the Cr - PMDA-ODA bonding interaction. They are in agreement with photoemission studies which concluded that the Cr bonds to the PMDA part. The downshift in the C-O vibrational frequency can be explained by the π-complex model[23,25] as delocalized electron transfer to an antibonding orbital of the carbonyl. These measurements also agree with infrared studies of Clabes, et al. for electron transfer from Cs to model systems at the carbonyl site.[30] This leads to the assertion that similarities may exist in spectroscopic observations for either π-complex formation or local bonding at the carbonyl oxygen since both invoke charge transfer. At higher coverages, we conjectured that fragmentation of the carbonyl might occur. We note that fragmentation of the polyimide has been concluded in an EELS study of the Al/polyimide system.[38] The one piece of information which we have not extracted is the actual bonding site. Higher resolution is required to determine the structure of all vibrations and their modification upon metal deposition for this to be accomplished. The possibility remains that local disorder and other effects might present other difficulties; however, ongoing EELS studies are proceeding in this area.[38] Improvements in infrared spectroscopy of surfaces will be a great advantage for observing metal-polymer interfaces in their solid state environment rather than in solution with high resolution. In addition, EELS studies in progress[57] promise to allow an understanding of scattering mechanisms and geometries such that the parameters available can be exploited to extract a maximum amount of information about the surface region.

MICROSTRUCTURAL MEASUREMENTS

There are two aspects of the microstructure of the metal-polymer interface which should be considered in order to understand the metallization process. The first involves the structure of polymer chains at the surface and the role of the substrate to define this topology. Are the polymer chains oriented in an ordered fashion; how do they sit? The second concerns the nature of the interface formed upon metal deposition. Is the interface sharp; does metal adhere to the surface or does the metal agglomerate; do metal atoms diffuse? Most experimental work has centered about the latter question which is the subject of this section.

Transmission Electron Microscopy (TEM), Medium Energy Ion Scattering (MEIS), and XPS studies have provided data concerning the structure of the interface between deposited metals (both reactive and non-reactive) and polyimide surfaces. The trend which is observed is that weakly-reacting metals agglomerate rather than form as a contiguous metallic film. Deposition at elevated temperatures with low fluence can cause metal clusters to form deep below the polyimide surface.

For TEM investigations, samples prepared *in-vacuo*, removed from the chamber, sectioned, and ion milled to ~50 μm thickness have been studied.[58] We first consider the Cr/polyimide interface, which exhibits a strong chemical interaction with the polyimide surface. TEM indicated that Cr deposited at room temperature or at ~300°C results in sharp, uniform metal films atop the polymer surface (see Figure 11). While atomic-scale intermixing with the substrate cannot be observed with TEM, no Cr precipitates were observed within the polymer film. This supported a picture where relatively strong chemical bonding stabilizes the reaction with the surface and uniform films can form.

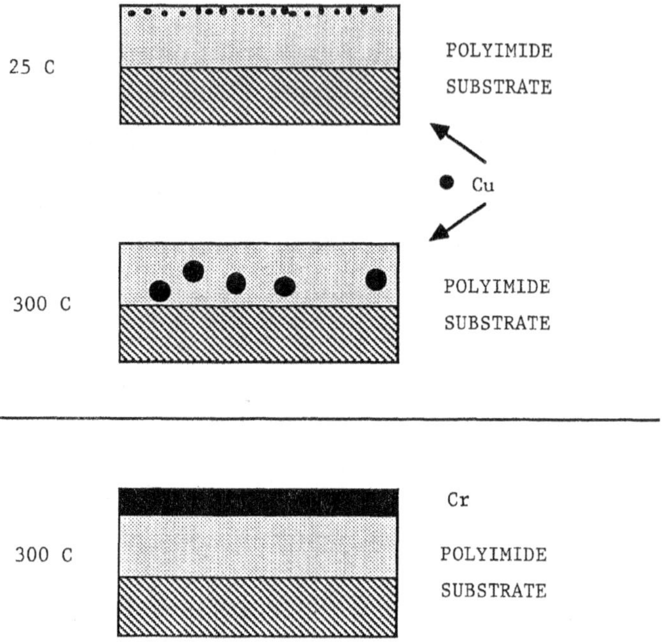

Figure 11. Sketches of TEM images obtained for Cr and Cu evaporated at room temperature and at elevated temperatures for low fluence (see Reference 58).

In marked contrast to the Cr behavior, Cu deposition was found to result in metallic agglomerates forming beneath the polyimide surface. TEM images showed that Cu, deposited at low fluence, migrates into the bulk polyimide forming nearly spherical single-crystal grains[58] as sketched in Figure 11. These appeared at various distances beneath the surface depending on the deposition temperature. This observation supported the view that Cu weakly interacts chemically with polyimide, whereby atomic diffusion and Cu-Cu interactions dominate over surface chemical bonding. For room temperature deposition, the Cu precipitates were small and were found just below the polyimide surface; >250°C depositions resulted in larger spheres which agglomerated below the surface to the polymer-substrate interface.

Using higher metal fluences resulted in the formation of Cu films at the surface. However, subsequent annealing or higher temperature deposition caused the Cu to form islands on the surface with no diffusion into the material. High deposition rate observations indicated again that Cu-Cu interactions dominate over Cu-polyimide surface interactions and weakly-interacting Cu atoms diffuse within the polymer until they bond to other Cu atoms to form a precipitate. The precipitates thus do not diffuse.

The results for Cu were supported by novel MEIS studies of the Cu/polyimide interaction. MEIS is a surface technique in which ions of energy ~50-200 keV incident upon the surface can backscatter through the Rutherford scattering process. The energy of a backscattered ion that has been elastically scattered by a target atom is determined by the kinematics of the scattering process. The backscattered ions were detected with an (angle-resolving) electrostatic analyzer. Since the cross-section and energy losses for ion interactions with various elements is well-known, this information could then be used for analysis of interfacial structure. In addition, ions suffer inelastic collisions as they traverse the substrate composed of atoms lighter than Cu. If the Cu is buried, this additional energy loss will be resolved and the depth beneath the surface be deduced. A depth resolution of the order of 5Å can be achieved. Positions, widths, and integrated intensities of the loss peaks provide a structural picture of the interface through a spectroscopic rather than a microscopic method.

For single crystal surfaces, precise beam-sample alignment allows ions to channel down atomic rows while backscattered ions in well-defined trajectories are blocked by the outermost crystal layers on their way to the detector. Quantitative analysis of the energy and angular characteristics of the scattered ions then provides structural information concerning the position of the surface layers with respect to the bulk. For the metal-polymer case, structural order is not the issue. Rather, MEIS data are analyzed to distinguish thin uniform films from metallic clusters through the width of the energy loss peak. If the metallic species exists beneath the polymer surface, the entire loss peak is shifted down in energy.

Figure 12 presents the data of Tromp, et al.[34] They have shown that Cu deposited at room temperature produces a broad Cu peak of a well-defined shape for all coverages. The width and integrated area of the energy loss peak immediately suggested that the Cu grows in clusters rather than as a uniform thin film. Subsequent annealing shifted the peak to lower energy but did not change its shape indicating diffusion of the Cu cluster just beneath the surface (or an overgrowth of polyimide). The energy shift is caused by inelastic scattering by the polyimide while retention of the peak shape is evidence that the Cu depth distribution remained the same. Data for Cu depositions at higher temperatures showed (1) a downshift of the ion energy maximum, (2) extensive broadening of the energy loss peaks, and (3) reduced intensity of the loss peaks. Such observations are consistent with diffusion deep inside the film and the formation of Cu clusters within the bulk polyimide. The energy downshift again resulted from inelastic scattering from polymer layers atop the copper. The broadened peaks indicate that Cu was distributed over a range of depths as temperature was increased. The reduced intensity results from the development of larger grains of Cu deep inside at higher temperature.

The MEIS and TEM observations were explained by a model where weakly-interacting Cu atoms arrive at the surface and diffuse beneath at distances that depend upon temperature. Beneath the surface, individual atoms meet and clusters nucleate to form metallic grains. This is consistent with the TEM results of LeGoues, et al.[58]. Higher deposition temperatures increased the size of the metallic clusters due to the increased mobility of Cu atoms. Diffusion of clusters at elevated temperatures was found to be negligible. Higher deposition rates resulted in formation of metallic clusters at the surface because nucleation and agglomeration occur before diffusion of Cu atoms into the polymer. Thus a metallic "film" of such a weakly interacting metal can be formed. The instability of such a film was assessed by elevating the temperature. At such elevated temperatures, surface diffusion was preferred over reaction with the substrate and metallic islands were formed. Apparently, once metal atoms begin to nucleate, diffusion into the bulk is greatly reduced and only surface migration dominates.

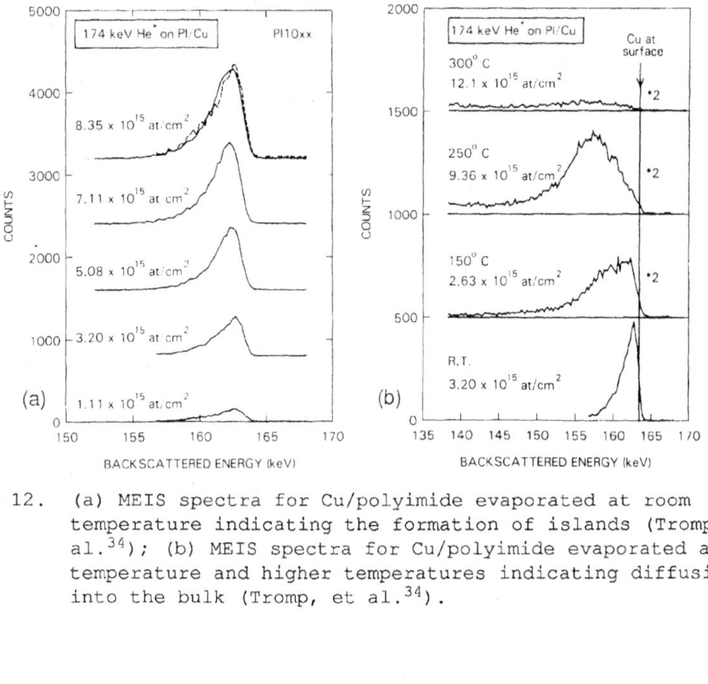

Figure 12. (a) MEIS spectra for Cu/polyimide evaporated at room temperature indicating the formation of islands (Tromp, et al.[34]); (b) MEIS spectra for Cu/polyimide evaporated at room temperature and higher temperatures indicating diffusion into the bulk (Tromp, et al.[34]).

TEM images for Ni and Al interacting with polyimide have indicated that chemical interactions diminish the degree of intermixing compared with Cu but not to the extent of strongly-interacting Cr.[58] Monte-Carlo simulations for these systems, where different polymer-metal interaction energies served as the input, supported the modes of formation which were found experimentally. Such studies of interfacial microstructure are a valuable complement to spectroscopic information for they provide the more macroscopic pictures of the microscopic surface chemistry which occurs in the metallization process. We draw comparisons with electron spectroscopy results for Cu/polyimide to Cr/polyimide. The observed microstructure of the Cu/polyimide system supports the spectroscopic observations that polymer/metal chemistry plays at most a minor role in this system. On the other hand, the formation of sharp Cr/polyimide interfaces correlates with the stronger chemical interaction that Cr has with the polymer.

Interfacial microstructure can also be obtained with EELS and with photoemission. EELS is generally limited to the observation of a continuous two-dimensional conductor from the continuum broadening of the spectra for which the temperature-dependence can be analyzed.[56] In principle, if the resolution was high enough, analysis of all vibrational modes would provide direct information on bonding sites for metals on polymer surfaces. XPS (and UPS), provide a direct means to obtain the growth mode of a metal film. Variation of photon energies causes an equivalent change in the kinetic energy of the photoemitted electron. With synchrotron radiation, changing the kinetic energy of the photoemitted electron permits control over the mean free path to enhance or diminish near-surface contributions. In addition, varying the angle of exit can provide another means to vary the surface sensitivity. Besides these, the most common means to assess the growth mode is to monitor the film and the substrate core level intensities as a function of metal coverage or exposure. Laminar growth of a film will be characterized by a linear increase in the metal core level coupled with a linear decrease in the substrate intensities. Laminar growth followed by island formation will result in two linear segments in the intensity vs. coverage curves with a discontinuity in slope at one monolayer. Island growth will show slow nonlinear behavior in the metal or substrate uptake curves with bare, unreacted substrate clearly visible to very high coverages. In the discussion of the growth of Al films at the polyimide surface by Atanasoska, et al.,[39] we recall that observations of the latter type led to the conclusion of island growth at the surface.

The introduction of new techniques in surface science, notably Scanning Tunneling Microscopy,[59] has provided researchers with the ability to assess real-space surface structure with atomic resolution. In addition, spatially-localized spectroscopic data can be obtained from current vs. tunneling bias (I vs. V) measurements at various locations on the surface. Structural information at the highest resolution is generally obtained with ordered conducting substrates. However, it is significant that STM has been shown to provide structural information on ordered polymer films.[60] Furthermore, spatial variations in occupied and unoccupied states of disordered polymer films have recently been studied.[61] In addition, the recent coupling of FTIR (Fourier Transform Infrared Spectroscopy) and STM in the study of polymer surfaces has recently been reported;[62] this is an ideal configuration with which to relate geometric information with related vibrational excitations of the substrate. It is expected that STM will be applied to metallization of plastics in the future providing a host of new insights by spatial imaging the growth of thin metallic films and/or clusters and possibly relating this to spatially-localized properties of the electronic structure of the resulting interface.

SUMMARY AND FUTURE DIRECTIONS

We have reviewed several experimental and theoretical studies to provide a background in the use of surface science techniques for metal-polymer interfaces. The use of XPS and UPS, EELS, and TEM and MEIS are a relatively new means to look at such systems. The development of these techniques for polymer substrates brings a good deal of promise to future technological advances. The model systems which were considered -- polyimide substrates and metals which exhibit strong or weak macroscopic adhesion properties -- are prototypes for applications in the future and also prototypes with which to understand microscopic interactions now. The microscopic picture that has developed indicates that specific metal-polymer chemical reactions produce good macroscopic adhesion.

In many of these studies, the isolation of various reaction channels has been assumed, particularly for comparison of experiment to theory. That such an assumption can be made is supported by the reproducibility of results between groups and the consistency of interpretations between techniques. However, while primary reactions may be identified, it should be noted that many secondary reaction paths can also proceed in such complex systems. The result of such reactions may be contained in the experimental data especially at high coverages or under extreme processing conditions. Interpretation of experimental data should be made with this possibility in mind.

Two-step reaction processes are typically observed for a strongly reacting metal such as Cr. First, at the sub-monolayer level, there is a "local" interaction with one unit of the polymer. Then at and above the monolayer level there are further interactions. This produces the view that, for example, a good Cr interface first consists of a set of atoms bound locally at specific sites of the large PMDA-ODA unit cell -- at the PMDA. To these metal atoms are attached many other metal atoms to complete the film. Atoms in the monolayer metal film interact further with the polyimide surface probably causing fragmentation, creating new bonding interactions, and providing cross-links between fragmented polymer chains. While the occurrence of strong reactivity is related to macroscopic adhesion, the question remains as to which step is key to promoting macroscopic adhesion. By far the greatest attention experimentally has been placed on the first step at submonolayer coverages. While not yet fully understood, it appears to be the "cleanest" part of the reaction.

Electronic and vibrational spectroscopies of these metal films provide a truly microscopic view. The analysis, however, is complicated by the size of the polymer unit cell and the number of near-lying excitations which exist in each cell. It appears we are not yet at the point to conclude specific reaction sites to relate to adhesion. Photoemission, for example, has produced two models for Cr bonding -- the π-complex and electron transfer models. Both models are based upon known processes in organometallic chemistry. It is clear that there has been a growing sophistication in the analysis of the reactions on the model systems. Experiments on model compounds in the solid state, electrochemisty associated with various monomeric analogues, and good *ab-initio* theoretical calculations have come on line to analyze the processes associated with metal-polymer interaction.

Many aspects of the Cr interaction with polyimide have been settled. First, the initial reaction occurs with the PMDA unit. This has been established with both core level and valence level photoemission and with EELS. Unfortunately, the information which has been obtained is not site-specific enough. The delocalized valence orbitals distribute the bonding charge throughout the unit causing ambiguity in the analysis of the origin of carbonyl binding energy shifts. Similarly, the only changes observed in the EELS spectra deal with the carbonyl stretch and are explained on the basis of the delocalized electronic states. Higher resolution in EELS has the potential to contribute a great deal if other relevant structural modifications can be identified in the vibrational spectra. Model studies of charge transfer with Cesium deposition are very interesting. Although the system is quite different from Cr, the result is consistent with charge transfer in the vicinity of the carbonyl. The anionic species created is not expected to possess a localized Cs site such that infrared spectra only see a carbonyl vibrational shift reminiscent of the Cr/polyimide study. The spectroscopy of unoccupied states again is limited by the delocalized nature of the π^*-band.

Theoretical models have focused mostly on the symmetric geometries related to bis-arene species. These must be extended to proposed sites which are not symmetric, e.g., to bonding at the carbonyl oxygen to provide insights as to the compounds which might form, the resulting charge distribution throughout the PMDA, the conditions for fragmentation, and the relation to spectroscopic data. The calculations truly represent a microscopic picture which takes into consideration the chemistry associated with the particular metal atom and does not rely on less specific electronegativity arguments, etc. Theoretical modeling needs to be reconciled to a fuller degree with known electrochemical processes also. Extensions to cross-linking of polymer chains stabilizing bonding interactions must also be addressed.

Microstructural analysis provides information on the integrity of the films on a larger than atomic scale. The results are rather consistent with the spectroscopic studies -- if there is no strong bonding to the polymer surface, metals easily diffuse on the surface or into the bulk. The mobility of the metal atoms and the metal-metal interactions which prevail then cause agglomeration rather than laminar growth. On the other hand, strong metal-polymer interactions give morphologically flat, stable films. This data provides needed insight into the morphology of films during and after processing. This also guides scientists in technological problems such as the configuration of diffusion barriers. For example, Cr can be used as a stable intermediate onto which to deposit other metals which would otherwise weakly bond and diffuse into the polyimide.

CONCLUSION

The applications of surface science techniques on metal-polymer systems have seen a very productive beginning. These model studies have established a good experimental foundation, showing both advantages and limitations of each technique. Such investigations are required to proceed to systematic studies of new polymer-metal combinations, surface modification, and new processing environments which can only begin once the aspects of microscopic bonding have been understood. New insights will undoubtedly be obtained once the machinery is put into place to perform systematic studies of other metal-polymer systems -- theory and experiment -- where the organometallic chemistry can likewise be addressed on these solid state systems. We have seen that other metal systems exhibit many interesting phenomena for which the analysis is currently incomplete. Improved instrumental resolution, the possibility to incorporate atomic-resolution instrumentation into these studies, and increased interest of theoreticians in this problem will aid in creating a substantial impact in the future.

As this understanding grows, it appears that this application of surface science will provide an opportunity to develop new processes in the future for microelectronic device technology as well as other technologies based upon metal-polymer interfaces.

ACKNOWLEDGEMENTS

I wish to acknowledge the support of IBM Research and of Joe Demuth and Tom Clarke, my collaborators for the EELS work which was discussed above. In addition, I am indebted to J.G. Clabes, R.C. White, K.L. Mittal, and A.J. Smith for critical readings of the manuscript.

REFERENCES

1. W.D. Grobman, J. Vac. Sci. Technol. A3, 725 (1985).
2. A.J. Blodgett and D.R. Barbour, IBM J. Res. Develop. 26, 30 (1982).
3. A.J. Blodgett, Sci. Am. 249, 86 (1983).
4. K.L. Mittal, J. Vac. Sci. Technol. 13, 19 (1976).
5. D.T. Clark, CRC Crit. Rev. Solid State Mater. Sci. 8, 1 (1978).
6. D.T. Clark, H.R. Thomas, and D. Shuttleworth, J. Polym. Sci. Polym. Lett. Ed. 16, 465 (1978).
7. James M. Burkstrand, J. Vac. Sci. Technol. 16, 363 (1979).
8. James M. Burkstrand, J. Appl. Phys. 52, 4795 (1981).
9. N.J. Chou and C.H. Tang, J. Vac. Sci. Technol. A2, 751 (1984); N.J. Chou and C.H. Tang, in "Surface and Colloid Science in Computer Technology", K.L. Mittal, ed., pp. 287-298 (Plenum, New York, 1987).
10. N.J. DiNardo, J.E. Demuth, W.A. Thompson, and P.G. Ledermann, Rev. Sci. Instrum. 55, 1492 (1984).
11. Polyimide samples are typically fabricated from DuPont 5878 spun on a flat substrate.
12. P.O. Hahn, G.W. Rubloff, and P.S. Ho, J. Vac. Sci. Technol. A2, 756 (1984).
13. J.R. Salem, F.O. Sequeda, J. Duran, W.Y. Lee, and R.M. Yang, J. Vac. Sci. Technol. A4 369 (1986).
14. Nobuyuki Takahashi, Do Y. Yoon, and William Parrish, Macromolecules, 17, 2583 (1984).
15. Hatsuo Ishida, Stephen T. Wellinghoff, Eric Baer, and Jack L. Koenig, Macromolecules, 13, 826 (1980); Stephen T. Wellinghoff, Hatsuo Ishida, Jack L. Koenig, and Eric Baer, Macromolecules, 13, 834 (1980).
16. N. Ueno, W. Gädeke, E.E. Koch, R. Engelhardt, and R. Dudde, J. Elect. Spect. and Rel. Phen. 36, 143 (1985).
17. D.R. Penn, Phys. Rev. B13, 5248 (1976).
18. H. Liehr, P.A. Thiry, J.J. Pireaux, and R. Caudano, Phys. Rev. B33, 5682 (1986).
19. L. Ley and M. Cardona, eds., "Photoemission in Solids I and II" (Springer-Verlag, New York, 1979).
20. E.W. Plummer and W. Eberhardt, in "Advances in Chemical Physics", I. Prigogine and Stuart A. Rice, eds., Vol. 49, pp. 533 (Wiley, New York, 1982).
21. H. Ibach and D.L. Mills, "Electron Energy Loss Spectroscopy and Surface Vibrations" (Academic, New York, 1982).
22. Wei-Kan Chu, James W. Mayer, and Marc-A. Nicolet, "Backscattering Spectrometry" (Academic, New York, 1978).

23. R. Haight, R.C. White, B.D. Silverman, and P.S. Ho, J. Vac. Sci. Technol. A6, 2188 (1988); R. Haight, B.D. Silverman, R.C. White, P.S. Ho, and A.R. Rossi, Mat. Res. Soc. Symp. Proc., 108, 233 (1988).
24. R.C. White, R. Haight, B.D. Silverman, and P.S. Ho, Appl. Phys. Lett. 51, 481 (1987).
25. A.R. Rossi, P.N. Sanda, B.D. Silverman, and P.S. Ho, Organometallics 6, 580 (1987); B.D. Silverman, J.W. Bartha, J.G. Clabes, P.S. Ho, J. Polym. Sci.: Pt. A: Polym. Chem., 24, 3325 (1986).
26. P.L. Buchwalter and A.I. Baise, in "Polyimides: Synthesis, Characterization, and Applications", K.L. Mittal, ed., Vol. 1, p. 537 (Plenum, New York, 1984).
27. A. Toth, I. Bertoty, T. Szekely, J.N. Sazanov, T.A. Antonova, A.V. Shchukarev, and A.V. Gribanov, Surf. Interface Anal. 8, 261 (1986).
28. J.L. Jordan, P.N. Sanda, J.F. Morar, C.A. Kovac, F.J. Himpsel, and R.A. Pollak, J. Vac. Sci. Technol. A4, 1046 (1986); J.L. Jordan, C.A. Kovac, J.F. Morar, and R.A. Pollak, Phys. Rev. B36, 1369 (1987).
29. P.N. Sanda, J.W. Bartha, J.G. Clabes, J.L. Jordan, C. Feger, B.D. Silverman, and P.S. Ho, J. Vac. Sci. Technol. A4, 1035 (1986).
30. J.G. Clabes, M.J. Goldberg, A. Viehbeck, and C.A. Kovac, J. Vac. Sci. Technol. A6, 985 (1988).
31. M.J. Goldberg, J.G. Clabes, and C.A. Kovac, J. Vac. Sci. Technol. A6, 991 (1988) and references therein.
32. J.L. Brédas and T.C. Clarke, J. Chem. Phys. 86, 253 (1987).
33. C.B. Duke, Mol. Cryst. Liq. Cryst. 50, 63 (1979).
34. R.M. Tromp, F. LeGoues, and P.S. Ho, J. Vac. Sci. Technol. A3, 782 (1985).
35. M.J. Goldberg, unpublished results (1988).
36. J.L. Jordan-Sweet, C.A. Kovac, M.J. Goldberg, and J.F. Morar, J. Chem. Phys. 89, 2482 (1988).
37. J.J. Pireaux, C. Grégoire, P.A. Thiry, R. Caudano, and T.C. Clarke, J. Vac. Sci. Technol. A5, 598 (1987); J.J. Pireaux, C. Grégoire, M. Vermeersch, P.A. Thiry, and R. Caudano, Surf. Sci. 189/190, 903 (1987).
38. J.J. Pireaux, M. Vermeersch, C. Grégoire, P.A. Thiry, R. Caudano, and T.C. Clarke, J. Chem. Phys. 88, 3353 (1988).
39. Lj. Atanasoska, Steven G. Anderson, H.M. Meyer, III, Zhangda Lin, and J.H. Weaver, J. Vac. Sci. Technol. A5, 3325 (1987).
40. J.W. Bartha, P.O. Hahn. F. LeGoues, and P.S. Ho, J. Vac. Sci. Technol. A3, 1390 (1985).
41. F.S. Ohuchi and S.C. Freilich, J. Vac. Sci. Technol. A4, 1039 (1986).
42. Steven G. Anderson, H.M. Meyer, III, and J.H. Weaver, J. Vac. Sci. Technol. A6, 2205 (1988).
43. Lj. Atanasoska, H.M. Meyer, III, Steven G. Anderson, and J.H. Weaver, J. Vac. Sci. Technol. A6, 2175 (1988).
44. N.J. DiNardo, Ph. Avouris, and J.E. Demuth, J. Chem. Phys. 81, 2169 (1984).
45. H. Ibach, to be published.
46. W.D. Jennings, G.S. Chottiner, R.W. Hoffman, and I. Lundström, and W.R. Salaneck, Bull. Am. Phys. Soc. 29, 359 (1984).
47. J.H. Wandass and J.A. Gardella, Surf. Sci. Lett. 150, L107 (1985); Joseph H. Wandass and Joseph A. Gardella, Langmuir, 2, 543 (1986).
48. 1 meV = 8.065 cm^{-1}
49. N.J. DiNardo, J.E. Demuth, and T.C. Clarke, J. Chem. Phys. 85, 6739 (1986).

50. N.J. DiNardo, J.E. Demuth, and T.C. Clarke, J. Vac Sci. Technol., A4, 1060 (1986).
51. B.D. Silverman, to be published.
52. The gold film circumvented problems with a large loss background due to electron-induced excitations in the Si substrate.
53. J.J. Pireaux, P.A. Thiry, R. Caudano, and P. Pfluger, J. Chem. Phys. 84, 6452 (1986).
54. N.J. DiNardo, J.E. Demuth, and T.C. Clarke, Chem. Phys. Lett. 121, 239 (1985).
55. P.O Hahn, G.W. Rubloff, J.W. Bartha, F. LeGoues, R. Tromp, and P.S. Ho, Mat. Res. Soc. Symp. Proc., 40, 247 (1985).
56. B.N.J. Persson and J.E. Demuth, Phys. Rev. B30, 5968 (1984).
57. R.J. Purtell, private communication (1988).
58. F.K. LeGoues, B.D. Silverman, and P.S. Ho, J. Vac. Sci. Technol. A6, 2200 (1988).
59. G. Binnig and H. Rohrer, Ch. Gerber, and E. Weibel, Phys. Rev. Lett. 49, 57 (1982).
60. Tycho Sleator and Robert Tycho, Phys. Rev. Lett. 60, 1418 (1988).
61. Dawn A. Bonell and Marie Angelopoulos, to be published.
62. V.M. Hallmark, A. Leone, S. Chiang, J.D. Swalen, and J.F. Rabolt, Polym. Prepr. 28, 22 (1987).

A STATIC SIMS STUDY OF INTERFACES BETWEEN EVAPORATED METAL FILMS AND POLYIMIDES

W.J. van Ooij*, R.H.G. Brinkhuis* and J.M. Park**

*Department of Chemistry
Colorado School of Mines
Golden, Colorado 80401

**IBM Corporation
Endicott, New York 13760

Static SIMS has been used to characterize the surfaces of a variety of commercial polyimides and of the interfaces between a polyimide of the PMDA-ODA type and evaporated films of copper and chromium. The commercial materials were all contaminated with organic and inorganic materials: the most common one is a free anhydride. The spectra obtained from the metal-PI interfaces are initially different for Cu and Cr but become identical after a post-deposition annealing treatment. A possible mechanism for the metal-polymer interaction based on the SIMS results is presented.

INTRODUCTION

Polyimides form a class of very important engineering polymers because of their excellent thermal and chemical stability. They can be processed in the form of thin films or coatings. They are used extensively in microelectronic devices because of their dielectric properties. In such devices a metal-polyimide interface may be formed by vapor deposition of a metal film onto fully cured PI or by spin coating of the PI precursor on a metal substrate followed by curing. The former type of interface has been studied extensively by surface and microanalytical tools, such as XPS, UPS, AES and TEM.[1-3] The latter has not been studied in as much detail.[4,5] The peel strength of Cu films on fully cured PI is very low and that of Cr films is considerably higher. The peel strength of metal films is now generally believed to be related to a covalent interaction between the metal and the carbonyl groups in the PI surface. There is little or no interaction in the case of copper and a strong interaction between PI and metal such as Cr, Ti or Al. These conclusions are based on XPS studies of the interface between PI, usually of the PMDA-ODA type, and monolayers or less of the evaporated metals. Such thin films allow a non-destructive characterization of the interface due to the rather long sampling depth of XPS. In other studies, solutions of low-molecular weight compounds have been used as model systems for PI.[6] There is little or no information in the literature on the structure of the interface between PI and much thicker films such as Cu or Cr, other than TEM studies of cross sections.[7]

In this study we report on the characterization of the interface between copper and chromium films and PMDA-ODA type polyimide substrates following peeling of the films and analysis by static Secondary Ion Mass Spectrometry (SIMS). This analytical tool was selected for two reasons. First it has a much higher surface sensitivity than XPS. It essentially analyzes monolayers of all materials and detects all elements, including hydrogen, with high sensitivity. It is, therefore, more suited to determine the exact locus and chemical composition of a fracture interface than other surface tools. Secondly, static SIMS is potentially a very powerful tool for the identification of polymeric structures or of structural features such as unsaturation, branching, crosslinking and even surface orientation of polymeric molecules or segments.[8] Changes in the structure of PI due to metal evaporation or the presence at the surface of contaminants or of low-molecular weight components can therefore more easily be detected with static SIMS than with XPS.

The static SIMS spectra of polymers are sometimes fairly complex and cannot easily be predicted. Therefore, several workers have published spectra of many important polymers or classes of polymers with small structural differences.[9,10] Such spectra are not yet available for polyimides. For this reason we have recorded the SIMS spectra of a number of commercially available PI films of different compositions. These spectra are compared with that of the non-commercial spin-coated PMDA-ODA film which was used for the interface studies with Cu and Cr.

EXPERIMENTAL

The Cu-PI and Cr-PI samples were provided by IBM Corporation. They consisted of Dupont PMDA-ODA polyamic acid spin-coated onto a clean silicon wafer and thermally cured in a nitrogen atmosphere. Copper films of 50 μm or copper films of the same thickness with a 50 nm Cr interlayer were vapor deposited onto the PI film. Annealing of some of the Cu-PI samples was done at 300°C under vacuum for 2 hours. Just prior to analysis by SIMS the metal films were slowly peeled off the polymer at approximately 90° angle in ambient conditions. Visual inspection indicated that fracture was at the metal-polymer interface for both type of samples. Both sides of the interface were introduced into the UHV system of the SIMS equipment immediately after peeling and analyzed within 30 min. The uncoated PI film was rinsed in hexane and NMP before analysis, briefly dried in air and finally in the UHV of the instrument.

The commercial PI films that were analyzed were Kapton H®, HN® and V® from Dupont, APICAL® from Allied (all of the PMDA-ODA type), Novax® (Mitsubishi Chemical), Upilex R® and S® (Ube Industries) and Thermid MC-600® (National Starch). The last four are different from PMDA-ODA. Prior to analysis, these films were rinsed briefly in hexane and in NMP and dried in air.

SIMS analyses were carried out on two instruments. The spin-coated PI and the metal-PI interfaces were analyzed on a VG SIMSLAB instrument equipped with a 0-8000 amu quadrupole mass analyzer and a 10 keV, 200 nm Ga^+ Liquid Metal Ion Source. The spectra were acquired over an area of 2x2 cm^2. The total data acquisition time per sample area was 200 s. These conditions are in the static SIMS regime.[8] Charge neutralization for the polymer samples was done with a defocussed flood of electrons of 600 eV. In the positive mode the spectra were optimized by maximizing the integral yields of all ions over 100 amu. Variables in this procedure were the quadrupole extraction voltage, the electron beam current and the position of the quadrupole energy window. In the negative mode the spectra were recorded with samples held at target bias of -100V and the extraction optiwere operated in the zero extraction voltage mode. The spectra were optimized here by varying the bias voltage and the electron flood. The PI films were all analyzed on a model 6300 PHI SIMS spectrometer with a 0-255 amu quadrupole mass analyzer and a primary beam of 4.5 keV, 60 pA Xe^+ ions, which was focussed over a $2x2cm^2$ area. Data acquisition time here was 2-3 min per spectrum of 100 amu. The ion dose was $1\text{-}3x10^{12}/cm^2$, representing truly static SIMS conditions. Charge neutralization was done here with a defocussed source of electrons of 500 eV and a current typically a few times that of the ion beam. The electron beam was rastered over a $3x3\ mm^2$ area. Optimization of the spectra was done as described above.

RESULTS AND DISCUSSION

1. SIMS Spectra of Commercial PI Films

The spectra of the commercial materials are given ing Figures 1-11. Only the positive spectra are shown. The negative spectra did not show characteristic fragments. These results will only be discussed very briefly here. With the exception of Figure 11 (Upilex S®, the spectra are all similar. The aromaticity of the PI is indicated by fragments 91, 103, 105, 115, 128 and 141, which are all fragments containing an aromatic ring.[8] Oxygen-containing fragments are 29 (COH^+) and 43 (CH_3CO^+). Nitrogen manifests itself usually in even-numbered peaks[11], such as 42, 44, 56, 58, but none of the PI's forms these in high amounts. A detailed comparison of the spectra shows that they are all different, although for a particular material they were reproducible. We believe, therefore, that in all commercial materials the surface layers do not consist of the pure polymer, but there is evidence for contaminants such as inorganic ions, oxidized species, residual solvent, low molecular weight components and other extraneous materials, despite the cleaning step prior to the analysis.

Figure 1 (Kapton®) is typical of an aromatic hydrocarbon with a high degree of surface oxidation (fragments 29 and 43). It shows no evidence of nitrogen-containing components. Figure 2, another Kapton® sample, is highly contaminated with inorganic ions (Na and K) and it has an unusually high yield of an ion of 58 amu, which is probably $CH_3CO\text{-}NH_3^+$. We believe that this ion is indicative of the presence of polyamic acid, i.e. incompletely cured or hydrolyzed PI. The Kapton® of Figure 3 has an unusually high mass 44. This is the ion $CH_3CH_2NH^+$ which may originate from the solvent, DMF or DMAc. The other commercial PMDA-ODA type polyimides (Figures 4-7), all show a high yield of peak 149. This ion can be identified as $C_6H_4\text{-}(CO)_2OH^+$, i.e. an anhydride type ion. It is always formed from materials such as PET, phthalate plasticizers, etc. Since it is completely absent in Figures 1-3, we feel that it originates from an anhydride impurity or incompletely reacted component in the PI. This identification is consistent with fragments 104/105, which are also formed by phthalates.[12] Peaks 73 and 147 in Figure 4 indicate the presence of a siloxane and some of them form ion 58 (Figures 4 and 6), others do not. The sample of Figure 5 has a high yield of mass 113, which is also formed by some other materials. It is completely absent in others. Such a fragment cannot be a pure hydrocarbon fragment. It must be a highly oxygenated species and is, therefore, either a solvent residue or an alternative structure in the PI, e.g. an isoimide. Upilex R® is a BPDA polyimide. Its spectrum (Figure 10) is not different from the other PI's. It also has high 58 and 149 peaks as in several Kaptons. The spectrum of Upilex S®, however, is completely different. The few fragments ions that are formed can uniquely be identified as originating from CF_3 groups. So this material must have undergone some sort of treatment, e.g. in a plasma, which has completely fluorinated the surface.

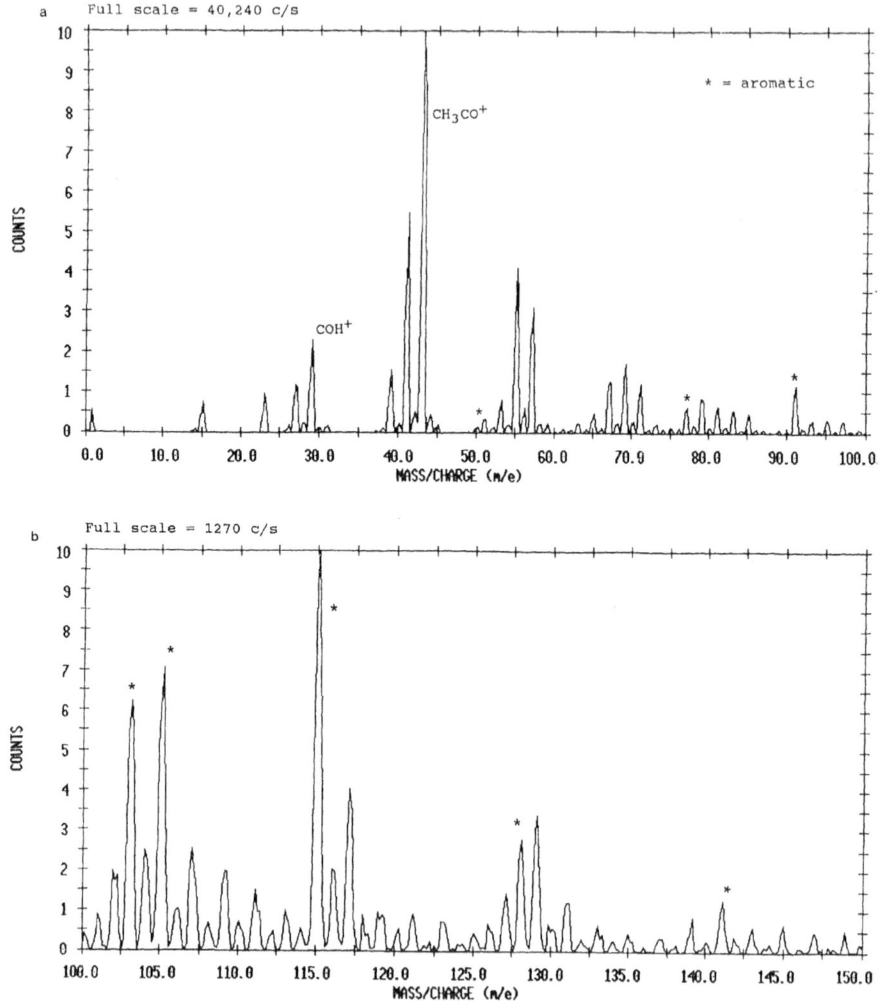

Figure 1. Positive SIMS spectrum of aged PMDA-ODA polyimide spin-coated and cured on silicon wafer; analysis by quadrupole SIMS using a 4.5 keV, 60 pA Xe^+ primary beam; acquisition time 3 min; in Figures 1-11 the analysis conditions were identical.

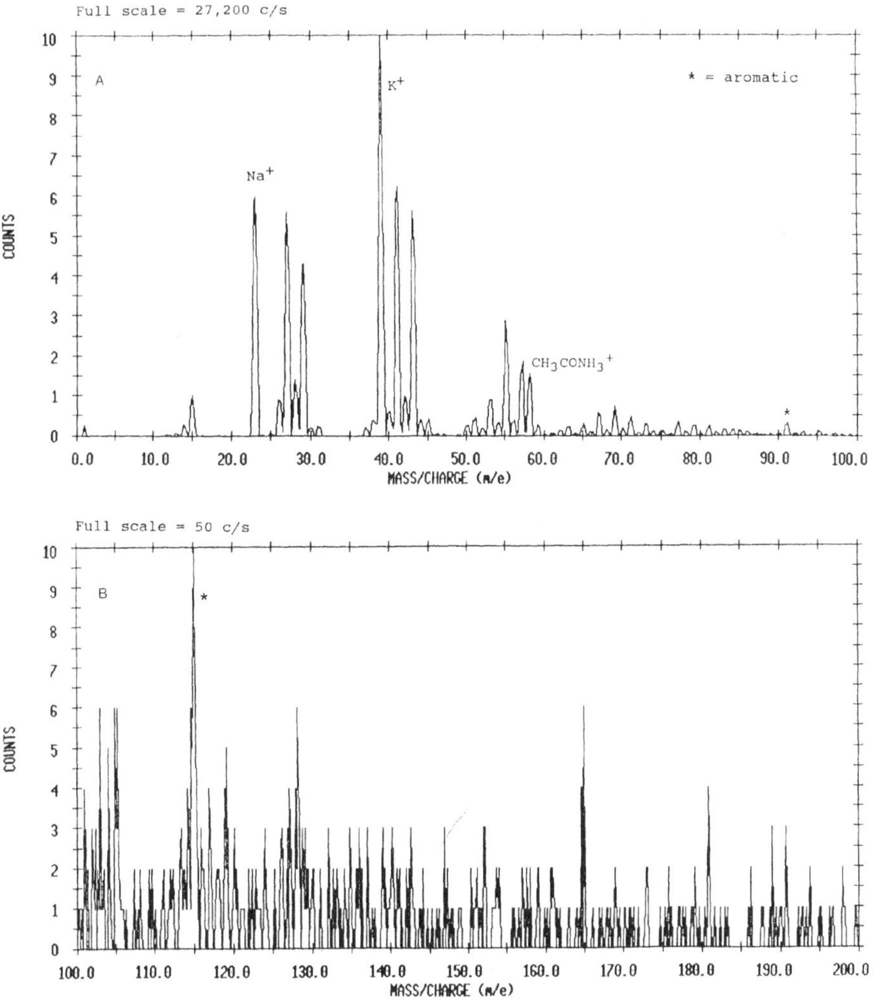

Figure 2. SIMS spectrum of Kapton© showing high concentrations of Na^+ and K^+.

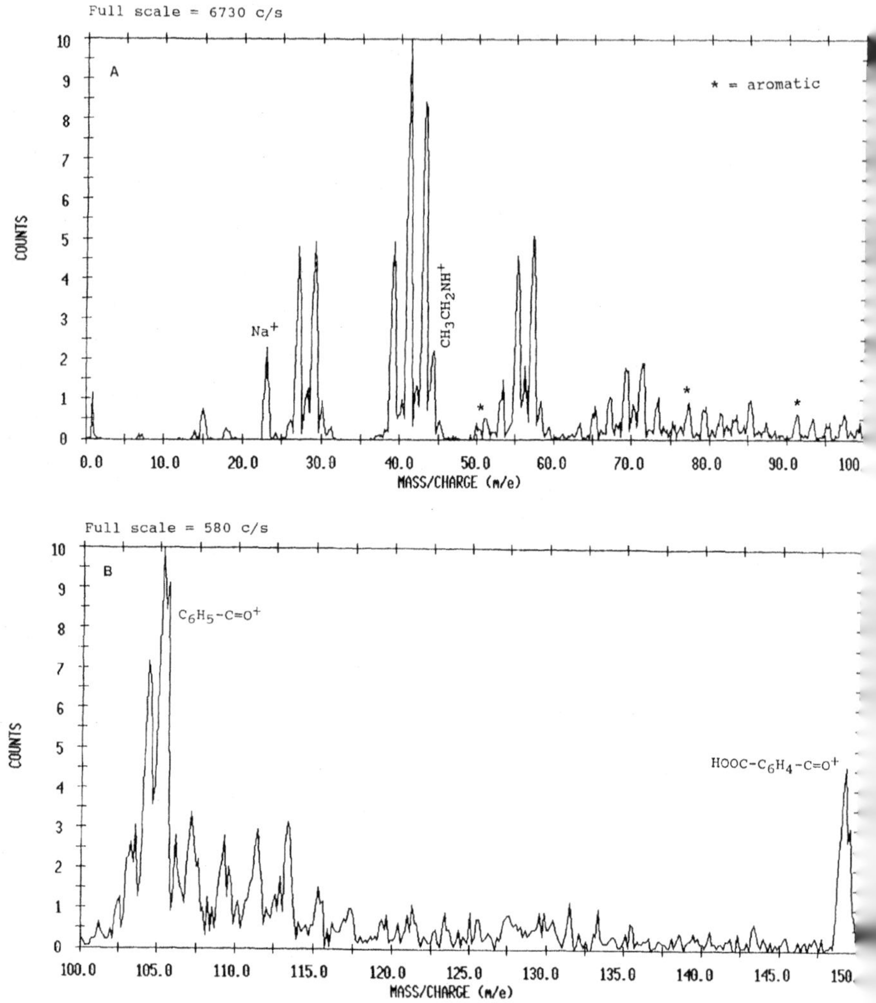

Figure 3. SIMS spectrum of a Kapton® sample showing high yield of ion a mass 44 at p-phthalate-type peaks at 105 and 149.

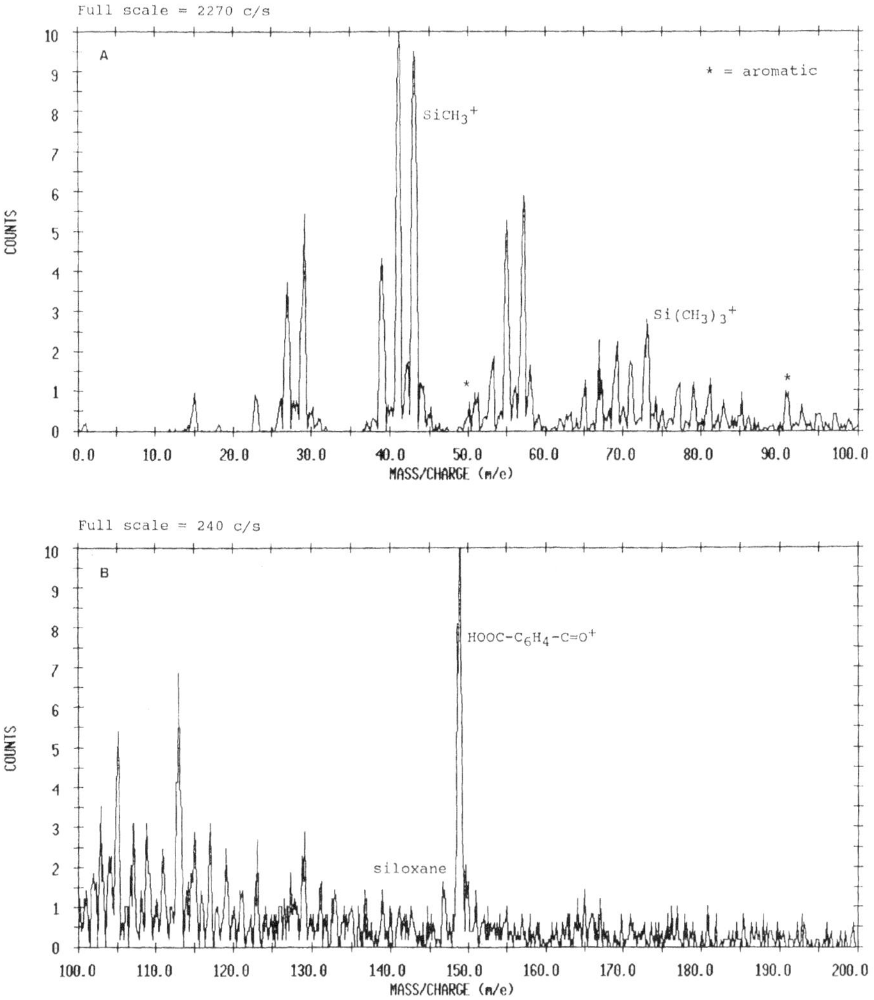

Figure 4. SIMS spectrum of Kapton-H® sample.

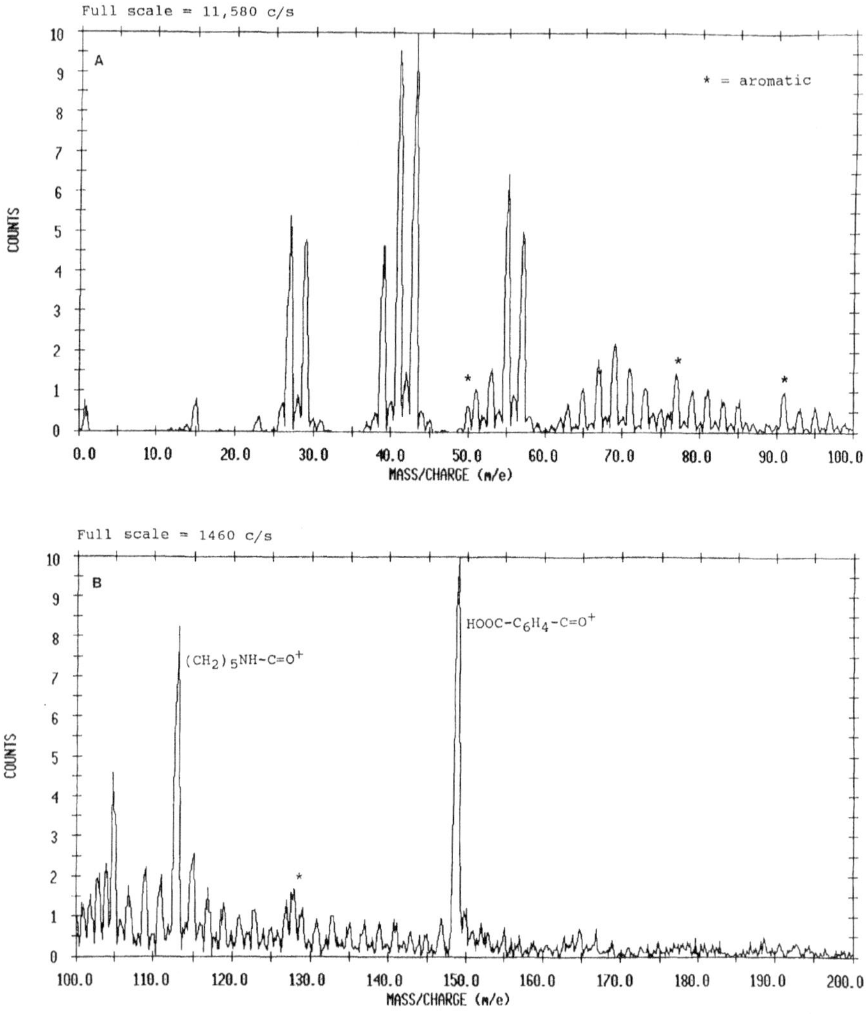

Figure 5. SIMS spectrum of Kapton-HN® sample.

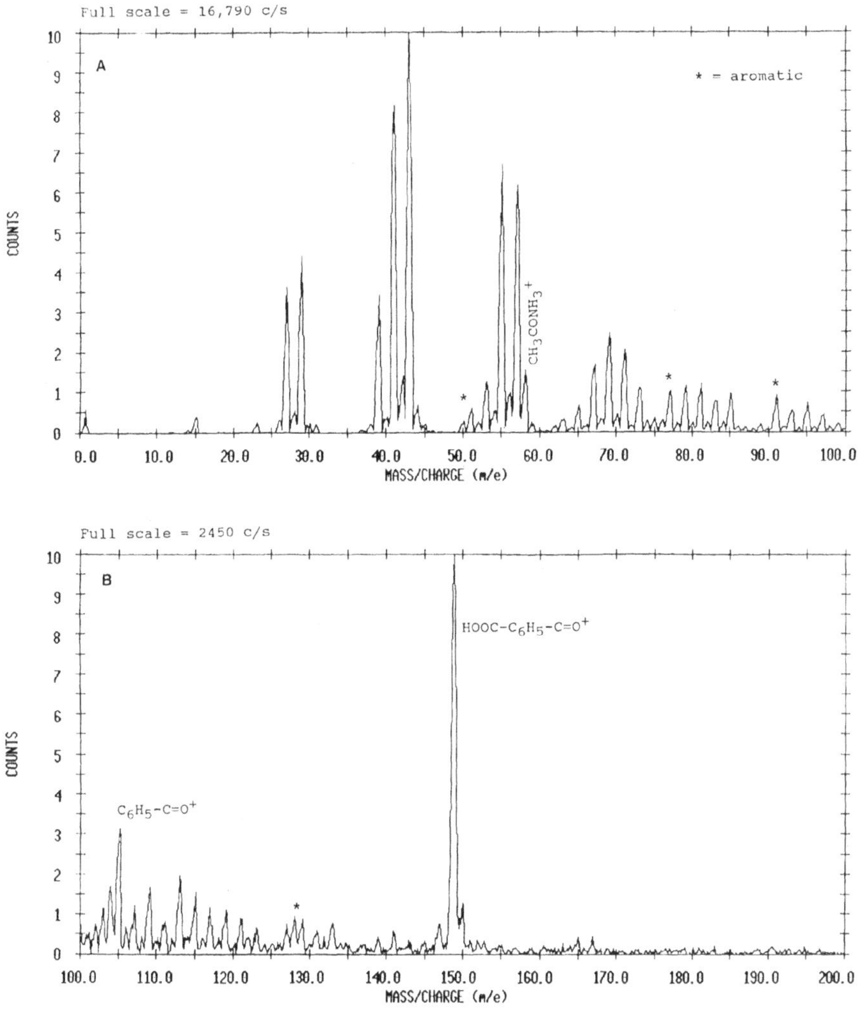

Figure 6. SIMS spectrum of Kapton-V® sample.

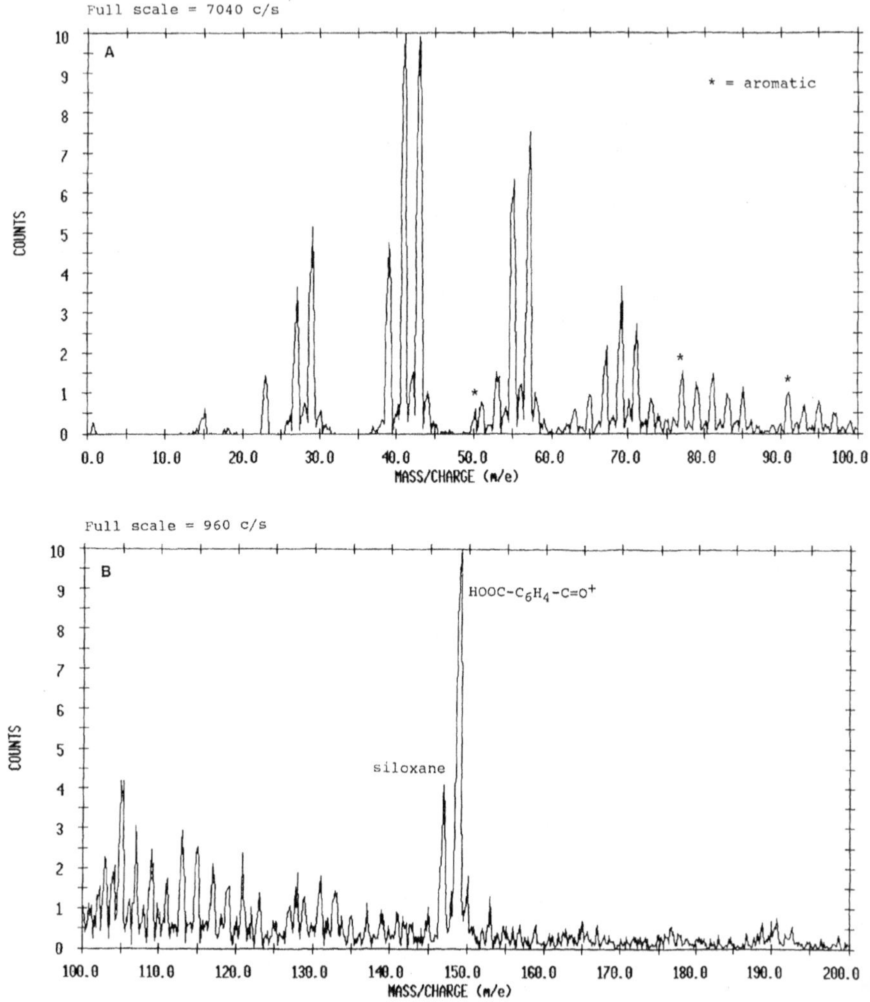

Figure 7. SIMS spectrum of APICAL® sample.

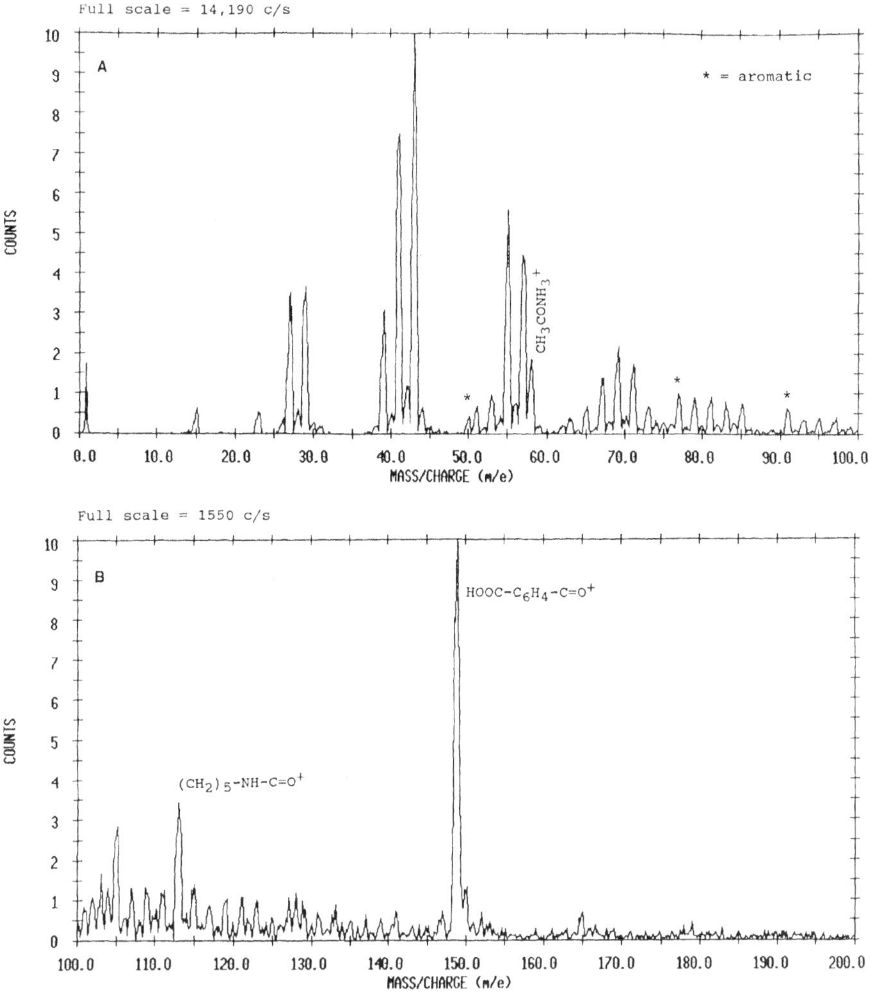

Figure 8. SIMS spectrum of NOVAX® sample.

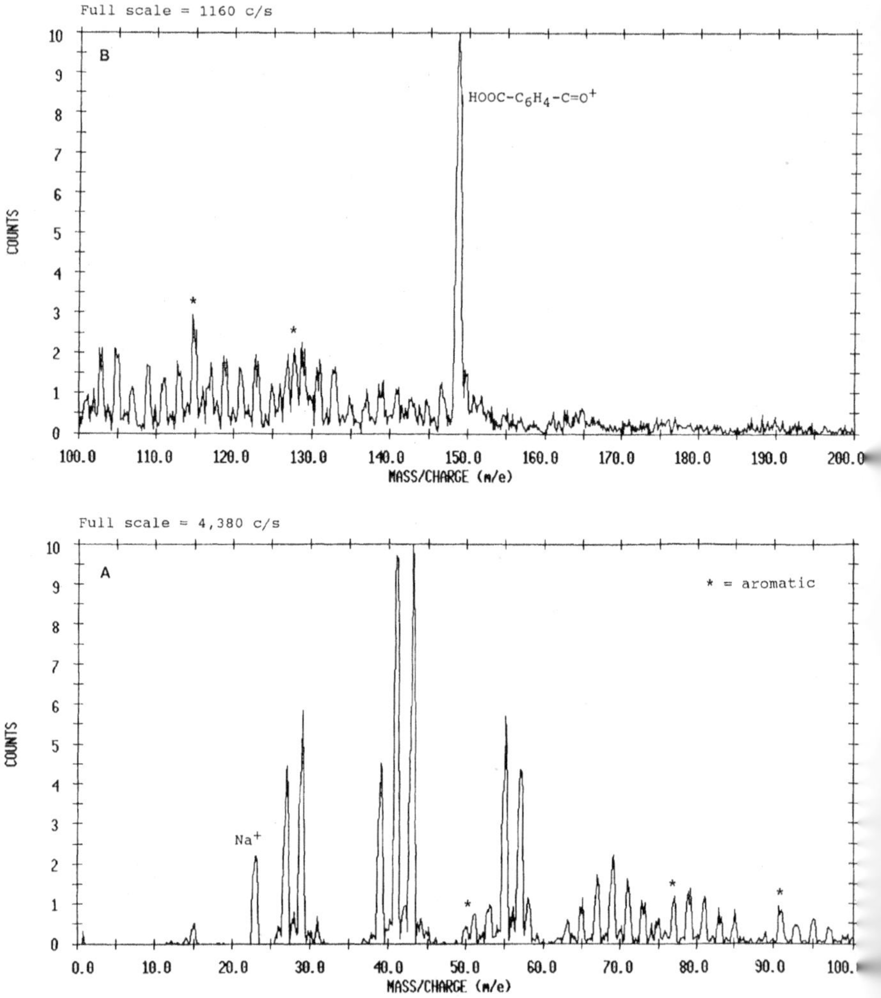

Figure 9. SIMS spectrum of Thermid-MC-600® sample.

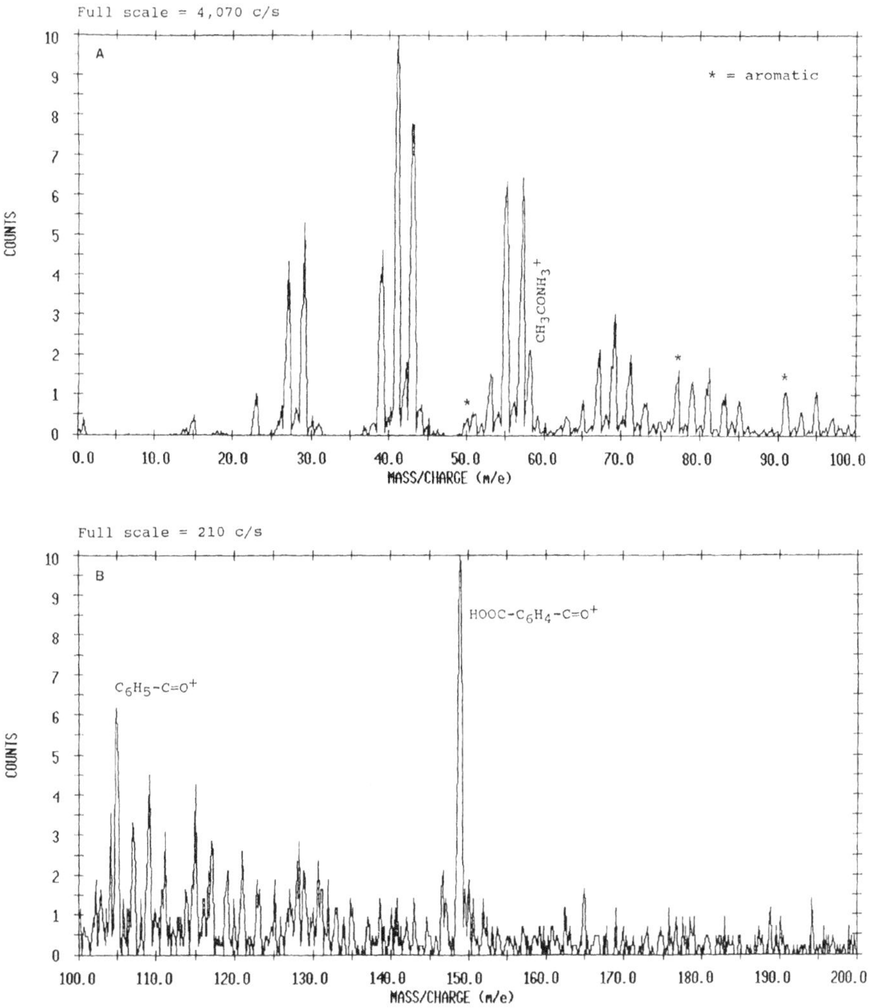

Figure 10. SIMS spectrum of Upilex-R® sample.

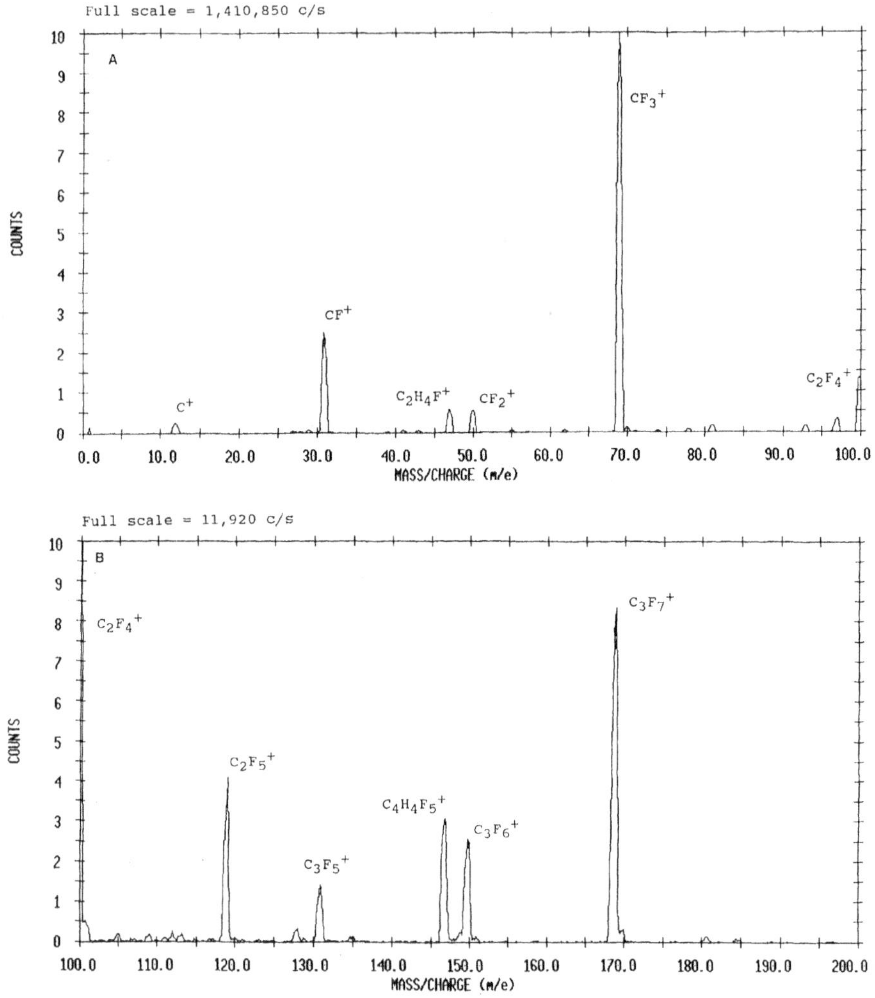

Figure 11. SIMS spectrum of Upilex-S® sample.

In summary, this brief comparison of SIMS spectra of a wide range of commercial films and spin-coated polyimides show that no two materials are alike. There is evidence for contaminants such as polyamic acid, solvents, inorganic ions, and other structures deviating from the expected one. In other materials, oxidation effects or deliberate surface treatments may be present. These results have important implications for the correct interpretation of adhesion phenomena of polyimides, but more work has to be done, e.g. using model systems, before the SIMS spectra can be completely interpreted.

2. Metal-Polyimide Interfaces

An obvious conclusion of the results presented above is that it is very difficult to interpret the SIMS spectra obtained from metal-polyimide interfaces if the surface composition of the polymer cannot be identified conclusively. However, this also applies to previously published studies by XPS. Although the XPS spectra may have been in agreement with the expected structure, a surface film of 1-2 monolayers with a composition or structure different from the bulk may have gone undetected in such work. The enormous surface sensitivity of SIMS allows the detection of such surface modifications; but, on the other hand, non-destructive interface analysis as in XPS is not possible with SIMS.

Table I. SIMS intensity ratios in PI, Cu-PI and Cr-PI.

ratio:	$\frac{CN^-}{CH^-}$	$\frac{CNO^-}{CH^-}$	$\frac{O^-}{CH^-}$	$\frac{CNO^-}{CN^-}$	$\frac{50+51}{55}$	$\frac{27}{29}$
PI	6.0	0.9	2.5	0.16	1.2	1.15
PI(Cu)	5.1	0.8	2.8	0.16	2.1	1.38
PI(Cr)	14.5	3.2	2.8	0.22	2.8	1.71
PI*(Cu)	13.6	3.1	2.8	0.23	2.7	1.69
Cu	6.0	2.4	2.0	0.40	0.3	0.74
Cr	10.8	4.9	2.4	0.45	0.5	0.94
Cu*	10.4	4.8	2.5	0.41	0.5	0.91

*annealed, 2 hr at 300°C

The results of the PI surface and PI-metal interface characterization are given in Figures 12-16 and are summarized in Table I. In this study the negative spectra are also presented. These spectra were recorded on a different instrument, using a Ga^+ source instead of a Xe^+ source used for the analyses of Figures 1-11. It should be emphasized that the type and energy of the primary beam in static SIMS does not have a qualitative effect on the spectra from polymers, although the ion yield is dependent on the primary beam conditions.[13] So, the spectra of the two series of experiments can be compared qualitatively in terms of fragmentation patterns. In the spectra obtained with the Ga^+ source, intense peaks due to reflected primary ions are observed, the Xe^+ ions (131 and 132 amu) do not show up in the spectra.

A comparison between the SIMS spectra of the PI surface and those of Figures 1-11 shows that they are similar but not identical. The spectra of Figure 12 were obtained repeatedly, even on different instruments, so they do represent a particular material. The most striking difference from the spectra of Figures 1-11 is the low yields of ions 29 and 43, which most likely originate from the carbonyl groups. Further, the aromatic fragment series 77, 91, 115, 128 and 141 is of rather high intensity. Ions 113 and 149, considered contaminants in Figures 1-11, are completely absent,

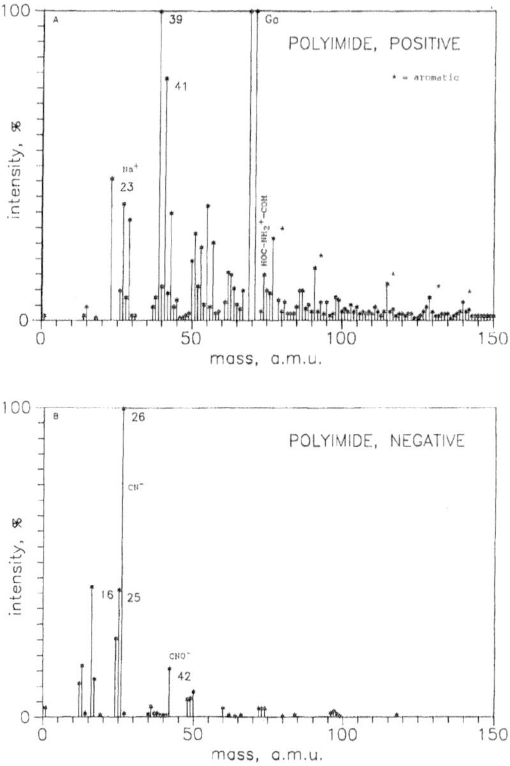

Figure 12. Positive and negative SIMS spectra of PMDA-ODA polyimide spin-coated and cured on a silicon wafer; analysis by quadrupole SIMS using a rastered 5 keV, 500 pA Ga^+ beam; acquisition time 200 s; identical analytical conditions for samples of Figures 13-16.

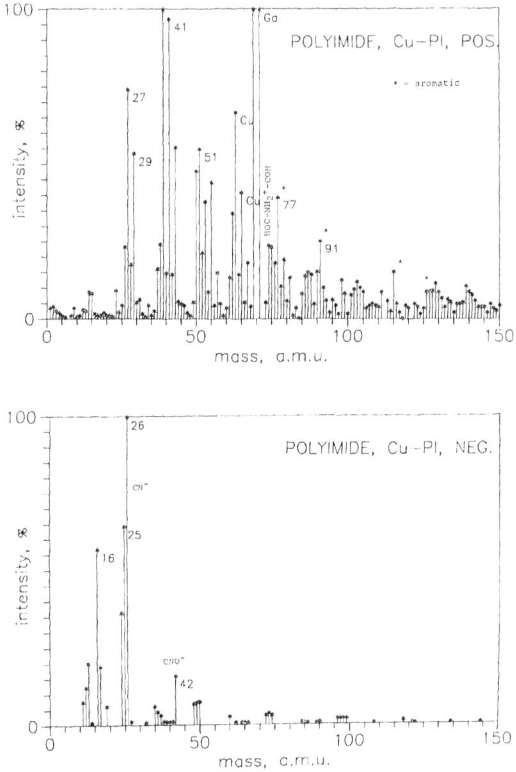

Figure 13. SIMS spectra of PI side of interface between PI of Figure 12 and 50 μm Cu film.

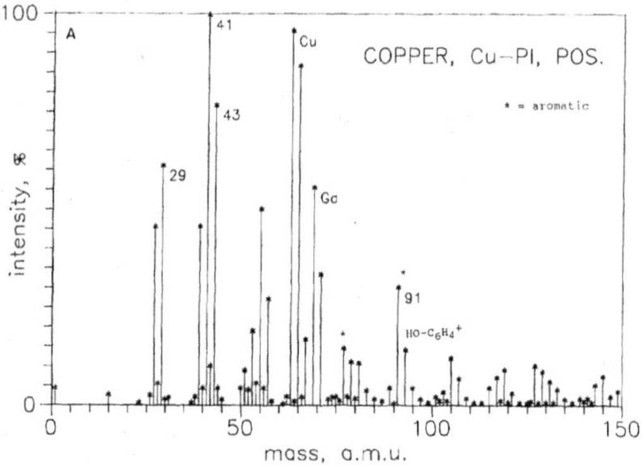

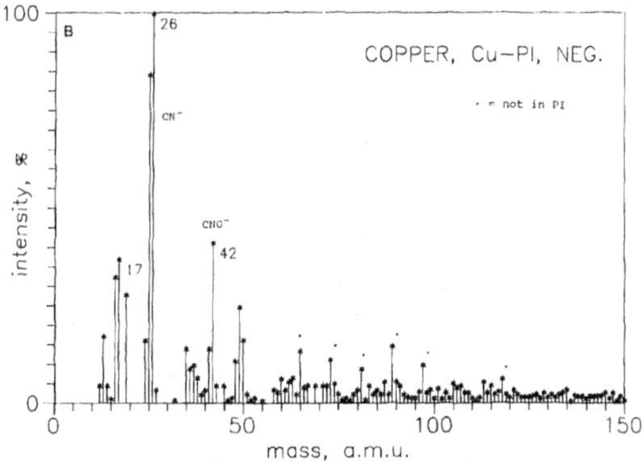

Figure 14. SIMS spectra of Cu side of interface of Figure 13.

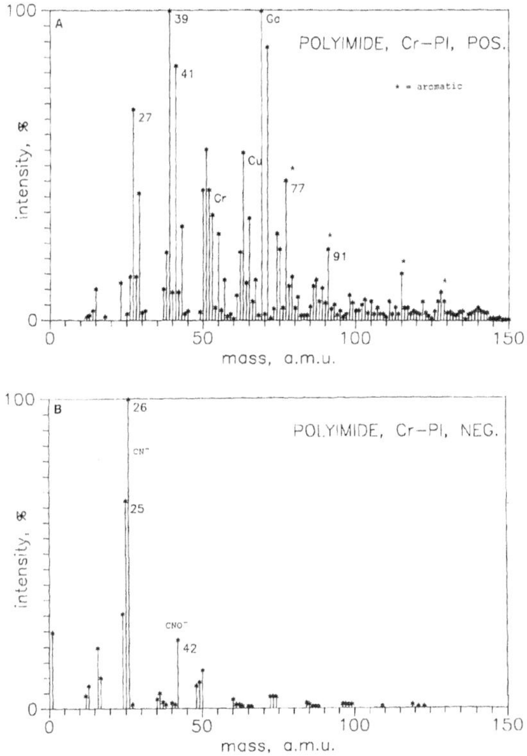

Figure 15. SIMS spectra of PI side of interface between PI of Figure 12 and 50 nm Cr film coated with 50 µm Cu layer.

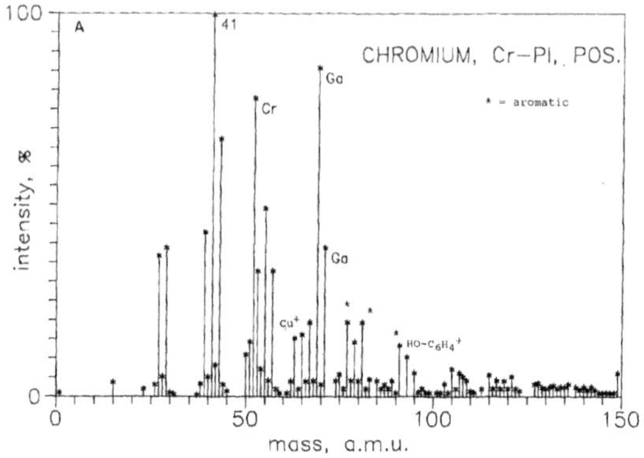

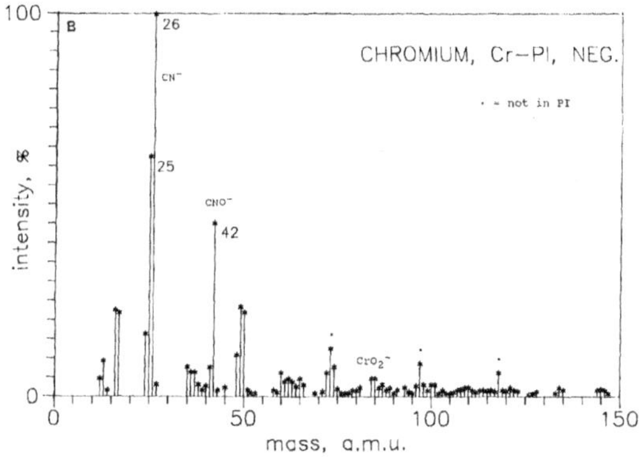

Figure 16. SIMS spectrum of Cr side of interface of Figure 15.

but Na and K are detected. Several N-containing fragments are observed in the positive and the negative spectra, e.g. 26 (CN^-) and 42 (CNO^-). A complication in the comparison of the SIMS spectra of polymers with the same or similar structures is that the surface orientation of the molecules may be different and it has been suggested that static SIMS is sensitive to such differences.[12]

A detailed comparison of the SIMS spectra of the interfaces and of the original PI (figures 12-16 and Table I) leads to the following observations:

1. Failure is in an organic layer for both metals, but this layer is extremely thin or incomplete as is shown by the high metal ion intensities (Figures 14 and 16);

2. The PI side of the Cu-PI interface resembles that of the original PI more than the PI side of the Cr-PI sample (Figures 12, 13 and 15);

3. Some specific aromatic fragments, e.g. the doublet 50/51, are higher in the PI sides of the two interfaces than in the PI control (Figures 12, 13 and 15);

4. The organic material remaining on the backsides of the metal films is different from the original PI andfrom the PI sides of the two interfaces; specifically, higher yields of CNO^- ions are observed, and the aromaticity and unsaturation of this material is less;

5. The metal sides of the two interfaces could be covered with a film of a metal-organic complex which is different for the two metals suggesting a different mode of attack for the two metals;

6. The yields of CN^- and of CNO^- ions from the Cr-PI interface are higher than from the Cu-PI interface (Figures 13 and 15); this suggests a cleavage of a C-C bond between an aromatic ring and a CNO group in the Cr-PI system;

7. In the PI sides of the two interfaces Cu or Cu and Cr are detected; upon annealing at 300°C more copper diffuses into the polymer;

8. After annealing the spectra from the Cu-PI interface have become practically identical to those of Cr-PI (Table I).

These results suggest a modification of the polyimide by the two metals. However, the observation that the spectrum of a polymer changes does not necessarily imply that the polymer has decomposed. If a complex is formed between the metal film and a polymer, it can be expected that the fragmentation pattern changes because this pattern is the result of the dissipation of excess energy that is deposited into the polymer by the momentum transfer of the impacting primary ion.[8] If the electron distribution in the polymer unit changes, for instance, because of the interaction of a metal ion with a carbonyl group, then such an effect will modify the fragmentation pattern because several modes of energy dissipation will now be modified. Unfortunately, such effects in the SIMS spectra have not yet been studied in any detail. However, the results show clearly that Cr has a greater effect on the SIMS spectrum of PI than Cu. It has a stronger interaction, i.e. it forms a stronger bond. After annealing, the spectra have become identical. This can be explained in the following simplified model. Both models interact with the -N-C=O groupings in the PI surface. In the case of Cr, the C-C bond between this group and the aromatic PMDA ring is broken and a covalent Cr-O-C-N- bond is formed. In the case of copper, although there is a similar interaction, the C-C bond is not yet broken. After annealing, however, the same complex has also been formed with Cu. In other words, copper needs an extra activation in the form of annealing in order to form the same complex that the more reactive metals form without annealing.

CONCLUSIONS

Clearly, more work in this area is needed before the SIMS spectra can be interpreted fully. This could be done with pure, well-characterized model systems and metal complexes. Nevertheless, SIMS already has an enormous potential for the study of polymer surfaces and modification of such surfaces, e.g. in the reaction with evaporated metals. This is of importance for the understanding of adhesion phenomena in metallized polymers. It has further been shown here that SIMS can detect and identify various contaminants and polymer structures that deviate from the bulk material. Especially commercial polyimide films have been shown to contain a wide variety of low-molecular weight contaminants in the outermost surface layer

ACKNOWLEDGMENT

This work was supported, in part, by a grant from IBM Corporation.

REFERENCES

1. N.J. Chou, D.W. Dong, J. Kim and A.C. Liu, J. Electrochem. Soc., 131, 2335 (1984).
2. N.J. Chou and C.H. Tang, J. Vac. Soc. Technol., A2, 751 (1984).
3. P.S. Ho, P.O. Hahn, J.W. Bartha, G.W. Rubloff, F.K. Legoues and B.D. Silverman, J. Vac. Sci. Technol., A3. 739 (1985).
4. Y.-H. Kim, G.F. Walker, J. Kim and J.M. Park, J. Adhesion Sci. Technol 1, 331 (1987).
5. M. Grunze, W. Unertl, S. Gnanarajan and J. French, MRS Symp. Proc., 108, 189 (1988).
6. S.C. Freilich and F.C. Ohuchi, Polymer, 1909 (1987).
7. J.W. Bartha, P.O. Hahn, F. Legoues and P.S. Ho, J. Vac. Sci. Technol., A3, 1390 (1985).
8. W.J. van Ooij and R.H.G. Brinkhuis, Surf. Interface Anal., 11, 430 (1988).
9. D. Briggs, Org. Mass Spectrom., 22, 91 (1987).
10. A. Brown and J.C. Vickerman, Surf. Interface Anal., 8, 75 (1986).
11. W.J. van Ooij, R.H.G. Brinkhuis and J.G. Newman, in: "Secondary Ion Mass Spectrometry, SIMS VI", A. Benninghoven, A.M. Huber and H.W. Wern editors, p. 671, John Wiley and Sons, New York, 1988.
12. W.J. van Ooij and R.S. Michael, in preparation for publication in Surf. Interface Anal.
13. W.J. van Ooij and R.H.G. Brinkhuis, in: "Secondary Ion Mass Spectrometry, SIMS VI", A. Benninghoven, A.M. Huber and H.W. Werner, editors, p. 635, John Wiley and Sons, New York, 1988.

XPS AND HREELS STUDY OF THE ALUMINUM / POLY (ETHYLENE TEREPHTHALATE) INTERFACE AND THE INFLUENCE OF THE POLYMER SURFACE PRETREATMENT ON ADHESION

Y. Novis, M. Chtaïb, J. Vohs, J.J. Pireaux, and R. Caudano
Facultés Universitaires Notre-Dame de la Paix - LISE - IRIS
rue de Bruxelles, 61, B - 5000 Namur (Belgium)

P. Lutgen and G. Feyder
Research Department - Du Pont de Nemours (Lux.) s.a.
L - 2984 CONTERN, (Grand -Duchy of Luxembourg)

Interfacial properties between evaporated metals and polyester films, specially adhesion, are important technological issues in several industrial fields, such as food packaging, audio and video tape and magnetic data storage. The presence of polar functionalities as well as the polymer surface structure are known to be important parameters which affect the metal adhesion on polymers. In this work, poly (ethylene terephthalate) (PET) has been surface modified by various techniques (UV lamp, corona) in order to change its surface structure and chemical composition. Creation and modification of surface chemical functional groups have been analysed by X-ray Photoelectron Spectroscopy (XPS) and a correlation is made with loss or gain of macroscopic adhesion, as measured by peel test . From our results, we also discuss what are the dominating factors that promote the adhesion of a polymer surface with an evaporated metal. Also, some preliminary results of a promising technique, High Resolution Electron Energy Loss Spectroscopy (HREELS), are presented. This vibrational spectroscopy has been used to monitor the in situ creation of an interface between evaporated aluminum and an unmodified PET substrate.

INTRODUCTION

The ever increasing consumption of composite materials in modern technology has made interface properties more and more important for the quality of many products. Polyester films are now used as composite materials coated with a thin in vacuo evaporated or sputtered metal layer (Al, Co-Cr, Co- Ni) in various applications such as food packaging, magnetic recording, capacitors, floppy disks. Interfacial properties such as adhesion, moisture resistance, etc are of key importance for the quality and reliability of the final product, which should withstand various environmental conditions (temperature, humidity, mechanical stress).

When a pretreatment process is required on the polyester substrate before metallization, e.g. for adhesion improvement, the most suitable treatment may be different according to the type of final product. The choice of a proper pretreatment process will be guided by several considerations such as the cost of the process, its durability (aging), its stability in a "difficult environment" (corrosives, moisture, temperature, ...). The last two criteria do indeed require a deep understanding of the physical and chemical modifications produced during a given pretreatment process.

This paper presents results obtained on surface modification of biaxially stretched PET films. Surface pretreatments (UV lamp and Corona) of the said films have been carried out in air and nitrogen. The surface modifications have been analysed by XPS and were correlated with peel test.

It is generally accepted[1,2,3] that a distinction should be made between "practical adhesion", a value measured by a macroscopic test (e.g. the force in a pull or a peel test) and the actual strength of the microscopic interfacial forces between two materials.

In this paper, when speaking of adhesion, we refer to the "practical adhesion" as measured by a 180° peel test, and not to the actual interfacial forces between the polymer and the metal layer. These forces, such as chemical bonding, polar and dispersion forces,electrostatic interaction, hydrogen bonding, etc are not the only governing factors in the "practical adhesion" of a thin metal film with a polymer. This would be the case only for the adhesion between two flat, defect-free surfaces, but it is not applicable to ill-defined materials such as polymers where structural factors play a dominant role: mechanical interlocking and interdiffusion may increase the adhesion of a polymer surface, while a weak boundary layer (WBL) in the polymer may decrease the adhesion of a strongly reactive polymer surface. This is typically illustrated by a temperature-dependent study performed on the adhesion of copper on polyimide [4] where, after annealing to 200°C, copper diffusion in the polymer increases strongly the adhesion measured by peel test. A correlation thus cannot be directly established between adhesion and the interfacial forces. Though interfacial forces are needed to obtain cohesion of the interface, the actual adhesion is also governed by "structural factors" of the interface : WBL, interdiffusion, etc ...

From XPS measurment of surface atomic concentration, temperature stability, penetration depth and effect on peel test adhesion, we try to picture the effect of the treatment (reticulation, chain scission, etc) and the cause of adhesion failure between evaporated metal layer and PET films.

One important issue is to know whether an increase of polar functions in the polymer surface is automatically linked with an increase of adhesion. PET being already polar, adhesion increase will not automatically result from treatments giving more polar groups, but from treatments giving a positive structural change to the polymer surface.

HREELS has been recently started with the aim of better understanding the type of interfacial link between metal and polymer. Though it may not be directly correlated with "practical adhesion" of metal on polymer, we see it as a complementary and fundamental approach in the study of metal-polymer interface.

EXPERIMENTAL

Different types of surface modifications have been carried out on semicrystalline, biaxially stretched 12 μm thick PET films (Mylar® from Du Pont). These commercial films contain no additives except 0.2 to 1 % by weight of a mineral and chemically inert additive of 1 - 5 μm grain size. UV lamp treatments were made in air and nitrogen atmospheres by the continuous UV irradiation (λ=185 nm and above) from a BHK mercury vapour grid lamp. Corona discharge was performed in an industrial system where the 12 μm film was rolled on a drum at a speed of 0.2 m/s. A 20 kHz discharge was appplied by 4 flat electrodes for a total duration of 0.5 second. The distance between the drum and the electrodes was 3 mm. Different atmospheres could be introduced in the processing chamber through blowers directed towards the treated surface.

XPS analysis was performed with a Hewlett-Packard 5950-A ESCA spectrometer using a monochromatized Al K_α radiation source, an hemispherical analyser and a low energy electron flood gun for charge compensation. Peel tests for adhesion measurement were standardized as follows : each set of metallization was performed in the same bell jar with the same setup, on the same day, preferably in the same run. The jar was turbo-pumped to the 10^{-6} Torr range. The metal was hot pressed (105° C, 10^5 Nm^{-2}) with an adhesive EAA copolymer strip, 25 mm wide. The peeling was performed on an Instron tensile tester with a peel angle of 180° and a rate of 5 cm/minute. High Resolution Electron Energy Loss Spectrometry (HREELS) analysis was performed on a SEDRA machine (Riber) using two 180° hemispherical electrostatic deflectors, one as monochromator of the incident electron beam and one as analyser. The incident beam had an energy of 7 eV and a take-off angle of 45°, and the analyser was in the specular direction (45°). The best resolution of this spectrometer has been measured as 2.5 meV on crystalline samples and around 8.5 meV on polymer samples. For this set of experiments, *in situ* Ultra High Vacuum ($\simeq 1*10^{-9}$ Torr) evaporation of aluminum was performed by incremental steps (smaller than the monolayer) using a Knudsen cell.

RESULTS AND DISCUSSION

U.V. Lamp Exposure

Treatment with mercury lamp did not bring adhesion improvement of subsequently evaporated metal layer. However, this type of irradiation, conducted in two different atmospheres, namely air and nitrogen, yielded qualitatively different results.

Peel test performed on samples irradiated in air showed a rapid and strong decrease from 180 g/25 mm for untreated sample to around 20 g/25 mm after 2.5 and 5 min; while in nitrogen atmosphere, irradiation caused adhesion to remain around 70 g/25 mm after 200 minutes exposure. In order to understand this in terms of surface polarity modification, we performed XPS analysis on the modified surfaces. C_{1s}, O_{1s} and N_{1s} core level XPS spectra of the modified and unmodified samples were recorded and their area ratio was calculated.

Given the fact that polar functional groups involve C-O (or C=O) and C-N (or C=N or C≡N) links, a quantitative evaluation of the increase or decrease of the polymer's surface polarity was possible by observing the evolution of O_{1s} and N_{1s} XPS signals (Figure 1 and 2 resp.) . Air treatment shows a slight increase in oxygen content (10 %) and creation of nitrogen functional groups ($\simeq$ 1 %) after 5 minutes. This means a small increase in surface polar groups which is correlated with a sharp decrease in adhesion. For nitrogen treatment, there is a small decrease in oxygen content, which is not totally compensated, in terms of polar groups, by the inclusion of nitrogen. However, there is only a small adhesion drop in this case.

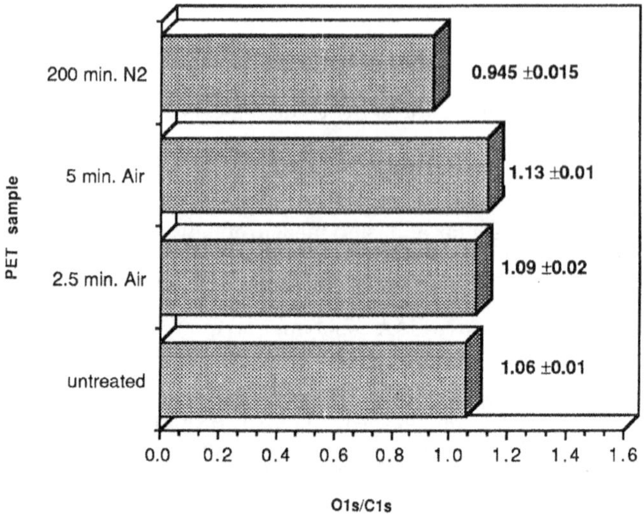

Figure 1 . XPS (O 1s/C 1s) ratio for U.V. treated PET films.

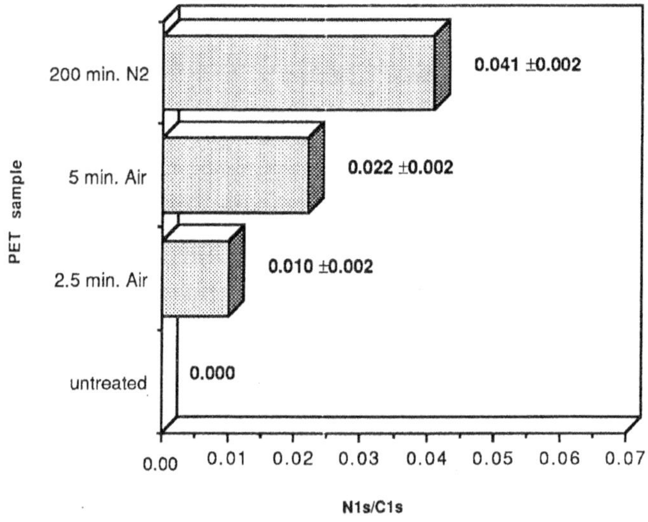

Figure 2 . XPS (N 1s/C 1s) ratio for U.V. treated PET films.

These results can be explained in terms of photolytic effect of UV treatment on PET surface, which breaks the linear structure of the polymer backbone. In the presence of oxygen radicals from the air, these fragments are oxidized and low molecular weight products are created, resulting in a sharp adhesion loss. The newly incorporated oxygen functions are known to be unstable hydroperoxy groups[6].

In the case of nitrogen, the same backbone breaking occurs; however, the low reactivity of N_2, which is only slightly ionised by the UV lamp, enables recombination processes to take place inside the polymer. Moreover, it is known that when nitrogen is included, its trivalent state enables reticulation to take place, yielding a tridimensional network at the surface [5,7]. Here, the relative stability of adhesion seems to be the result of a balance between the UV-induced degradation and the crosslinking effect of nitrogen.

Corona Treatment

Peel tests have been performed on corona treated samples, (in air and N_2 atmosphere). From Table I, we see an adhesion increase in both cases, more significant with nitrogen.

Table I. Peel Strength of Evaporated Aluminum on Corona Treated PET Films.

Untreated :	100 g/25mm
Air treated :	300 to 550 g/25mm
N_2 treated :	400 to 1000 g/25mm

XPS results. Corona treatment in air shows that there is a strong incorporation of oxygen (≃ 30 %). Oxygen reactivity is such that, with our discharge parameters, nitrogen was prevented from reacting with the polymer, at least in XPS detectable amount [Table II]. For N_2 treatment, a smaller (≃ 12 %) increase of oxygen is combined with nitrogen incorporation [Table II].In both cases, some atmospheric humidity fixes on the polymer surface when it is removed from the corona treater, and this contributes to the increase of the oxygen signal.

Table II. XPS Peak Ratio for Air and N2 Corona Treated PET.

	O_{1s}/C_{1s}	N_{1s}/C_{1s}
Air Corona	1.37	-
N_2 Corona	1.18	0.112 (6.2 % atom. comp.)
Untreated sample	1.06	-

In order to test the stability of the treatment, we performed in vacuo heating of both types of samples and monitored the changes in the surface composition after annealing at different temperatures (10 minutes). The change in oxygen concentration in air corona treated samples, shown in Figure 3, can be interpreted as follows :

* below 150° C, loss of unstable peroxide groups by thermal activation with evaporation or diffusion into the bulk of oxygen-rich low molecular weight fragments is observed
* above 150° C (Tg ≃ 70° C for PET), the kinetics of molecular motion is sufficient to have a restructuring of the surface with fresh polymer coming from the bulk[8].

In the case of N_2 corona, oxygen and nitrogen functions are also lost below 180° C (Figure 3 & 4) for the same reasons, but there is no similar restructuring of the material and intermixing with the bulk. Even at 210° C, the polymer surface does not come back to a composition close to the unmodified PET. This is due to the crosslinking effect of nitrogen atoms which creates a stable tridimensional molecular network at the surface. This network obviously causes better adhesion of the polymer surface with metal.

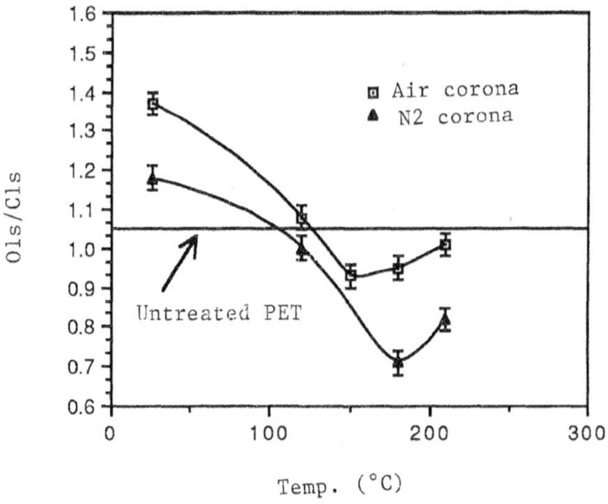

Figure 3 . XPS (O 1s/C 1s) ratio for Air and N_2 corona treated PET films.

Amouroux and co-workers [5] point out that in a corona treatment, reactive species diffuse in PET as deep as several microns. On the other hand, it has been shown more recently [9,10] that lack of adhesion of PET films is caused by a weak boundary layer (WBL) which is peeled away with the metal during adhesion tests. This WBL could be approximately 1 μm thick or thinner. An explanation for successful improvement of adherability in the corona treatment as opposed to the UV treatment would be the deep penetration of reactive species in corona treatment[5], while UV radiation (λ = 184 nm) penetrates only a few hundred Ångströms [6] and cannot anchor the WLB to the polymer bulk. This is also consistent with the observation reported by Briggs et al [11] which shows that, after washing of an air-corona treated film, all the new functions detected by XPS had disappeared while the adhesion properties of the film after washing remained as good as before. Though the authors gave an explanation of this in terms of molecular reorientation, it seems more likely here that the main adhesion reinforcement effect is not dependent on the polarity of the extreme surface but on the way the WLB is grafted to the bulk of the polymer.

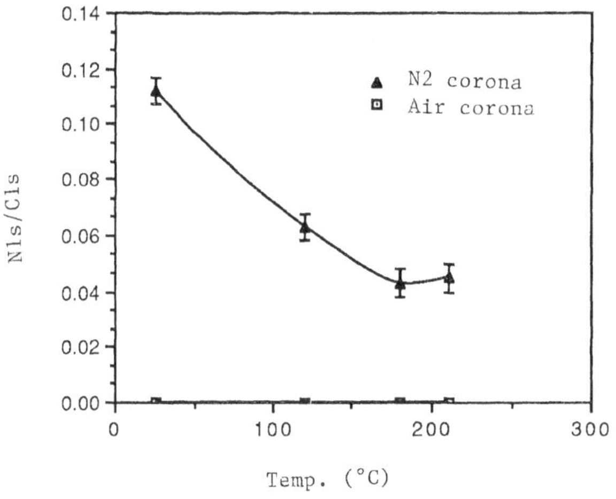

Figure 4 . XPS (N 1s/C 1s) ratio for Air and N_2 corona treated PET films.

HREELS Measurement

Though we believe that the key to metal-PET adhesion lies in the understanding of phenomena occurring in a ≃ 1 μm thick layer of the polymer surface, we also initiated a very surface sensitive (<100 Å) spectroscopic approach (HREELS) to study the interface between the two materials.

Two reasons have guided us in this direction :

1) HREELS offers an alternative way to distinguish surface functions[12] : as an example, aromatic and aliphatic carbon, which cannot be distinguished by XPS, can be easily differentiated by HREELS.

2) We also plan to analyse by HREELS both UV and corona modified PET samples, which could give interesting results compared to XPS, where the newly created chemical functional groups overlap in the C1s spectrum with the already existing oxygenated functionalities of PET[13].

An HREELS spectrum of a thin PET layer (< 200 Å), spin coated from a trifluoroacetic acid solution on a gold coated glass substrate, is shown in Figure 5. This polymer layer was washed in acetone, to remove traces of solvent, and was checked by ESCA as being pure PET. The spectrum shows as many as 17 peaks or shoulders. It has been shown elsewhere[12] that such spectrum can be interpreted by comparison with Infrared (IR) spectra: the oscillating dipole created by the impinging electron on the surface produces the same type of electromagnetic interaction as a photon, except for the quantum selection rules (1 meV electron energy loss is equivalent to a photon of wavenumber 8.066 cm^{-1}). Comparing with several bulk PET IR spectra[14] (Figure 6) a fairly good agreement is found. Most of the peaks (# 1 to 8, # 12 to 14, # 17) can be assigned (see Table III). # 9 to 11 and 15,16 seem to be produced by hydrogen bonded water at the surface[10]. The dominant peaks are :

- # 4 (≃ 1100 cm^{-1}) OH stretching
- # 5 (≃ 1260 cm^{-1}) C-O-C stretching
- # 8 (≃ 1720 cm^{-1}) C=O stretching
- # 12 to 14 (≃ 3000 cm^{-1}) three component C-H stretching structure.

Aluminum was vacuum deposited on this sample in a sub-monoatomic layer regime in order to determine where Al atoms were reacting on the polymer surface. Figure 7 shows the changes in the HREELS spectrum at different Al coverages. As soon as 3 x 10^{14} atoms of Al/cm^2 are deposited, O-C (# 4) and O-C-O (# 5) vibrations are almost completely dampened.Rather surprisingly, C = O vibration dampens more slowly.A new band appears at 820 cm^{-1}, accounting probably for a O-Al-O vibration[15].

After depositing a large amount of aluminum (≃ 6.3 10^{15} atoms/cm^2), the only visible peaks are C-H stretching (around 3000 cm^{-1}), C-H bending (≃ 1450 cm^{-1}) and a new interfacial peak around 820 cm^{-1}.

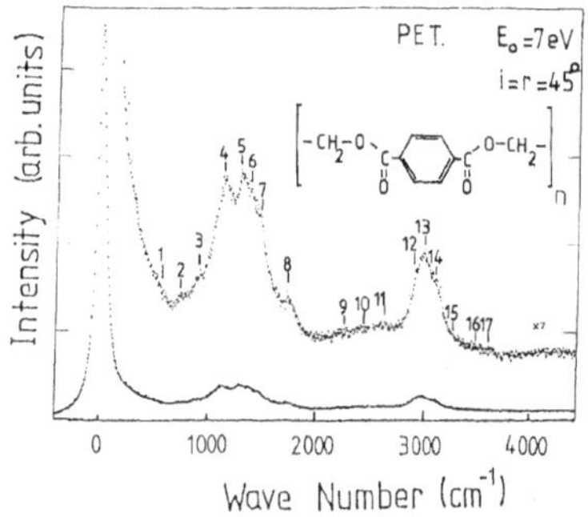

Figure 5 . HREELS spectrum of a clean PET film.

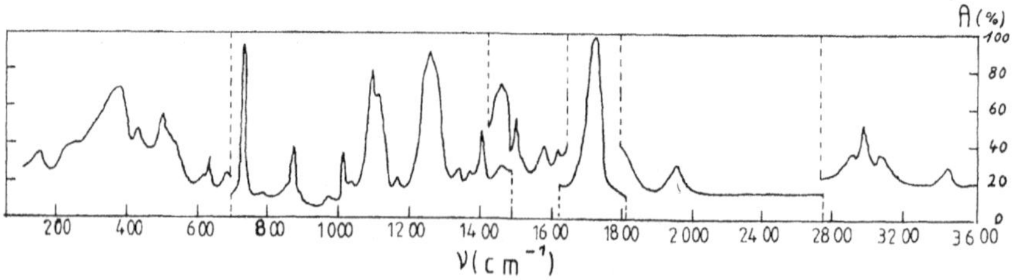

Figure 6 . Infrared spectrum of bulk PET.

Table III. Comparison of Bulk IR and HREELS Spectra of PET.

Line #	IR assignment	IR frequency (cm-1)	IR intensity	HREELS frequency (cm-1)
1	γ_w (C = O)	502	m	550
2	ν_{11} (B_1 u)	730	s	725
3	ν_{17} B (B_1 u)	875	m	900
4	ν (O-C)	1120	s	1125
5	ν (C-O-C)	1263	s	1275
6	γ_w (CH2)	1343	m	1350
	γw (CH_2)	1370	w	same
7	ν_{13} (B_2 u)	1410	s	1450
	δ (CH_2)	1455	m	same
	δ (CH_2)	1473	w	same
8	ν (C = O)	1725	s	1725
12	ν_s (CH_2)	2890	w	2910
13	ν_a (CH_2)	2970	m	2990
14	ν_{20} (B_2 u)	3068	w	3075
17	ν (OH) end group	3560	w	3610

This preliminary HREELS study shows that :

1) after a coverage of as low as ≃ 3 Å of Al (12 E), all polar functional groups are linked with Aluminum atoms, at the interface.

2) surprisingly, O-C and O-C-O vibrations are dampened faster than C = O. This could be due to an inward orientation of the C = O group : the O-C-O link would then be oriented toward the surface, which is consistent with its strong vibration intensity (# 4 and # 5, Figure 5) compared to C = O (# 8, Figure 5) in the bare PET.

3) after a coverage of 10 Å (42 E), only the C-H (nonpolar) segments of the molecules are still free to vibrate, i.e. not disturbed by the metal atoms.

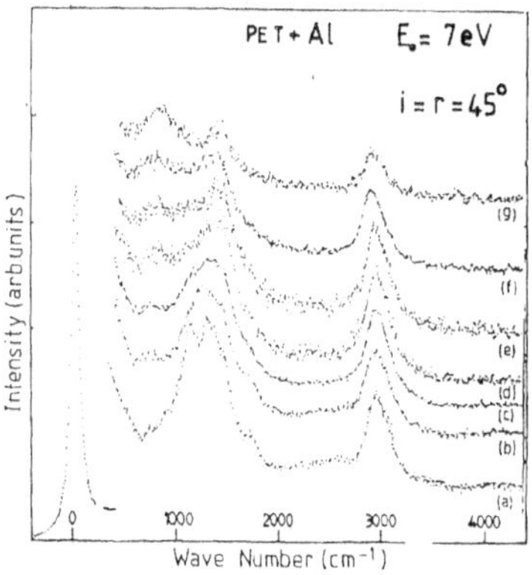

Figure 7 . General view of Al-PET interface by HREELS at different Al coverages

(a) clean PET

(b) to (g): increments of 2, 4, 6, 9, 12, 42 E respectively

(E = 1.5 * 10^{14} atoms /cm^2 of Al or 0.25 Ångströms approximately).

CONCLUSION

We have tried by different approaches to understand the adhesion mechanisms involved at the aluminum-PET interface.

From XPS and peel tests measurement, we come to the conclusion that the dominating factor for the lack of adhesion in biaxially stretched PET films is the existence of a ≃ 1 μm weak boundary layer (WBL).

Treating and grafting of the surface with nitrogen through UV exposure could not go deep enough in the material to improve its adherability; while grafting by corona treatment, which goes deeper, brought out substantial improvements.

From our HREELS results, it is observed that, at the interface, aluminum atoms react readily with the polar oxygenated functional groups, leaving the nonpolar ones unreacted.

ACKNOWLEDGMENTS

We thank Dr. S. Lazare (University of Bordeaux I, France) for providing U.V. lamp treated samples and Mr. R. Elmore (Du Pont de Nemours, Circleville, USA) for providing corona treated samples.

Mr. R. Bertrand and Mr. F. Mousset (Du Pont de Nemours, Luxembourg) are acknowledged for performing peel tests.

This work was supported by EEC grant n° RI 1B - 0178 (BRITE program).

REFERENCES

[1] K.L. Mittal, Polymer Sci. Eng. 17, 467 (1977).

[2] K.L. Mittal, in " Adhesion Measurement of Thin Films, Thick Films and Bulk Coatings", K.L. Mittal, editor,pp 5 - 17, ASTM, Philadelphia 1978.

[3] D.L. Allara, F.M. Fowkes, J. Noolandi, G.W. Rubloff and M.V. Tirrell ; Mater. Sci. Eng. 83, 213 (1986).

[4] P.S. Ho, P.O. Hahn, J.W. Bartha, G. W. Rubloff, F.K. Le Goues and B.D. Silverman, J. Vac. Sci. and Techn., A3, 739 (1985).

[5] J. Amouroux, M. Goldman and F. Revoil, J. Polym. Sci., Chem. Ed., 19, 1373 (1982).

[6] R. Srinivasan and S. Lazare , Polymer, 26, 1297 (1985).

[7] A.T. Bell and J.R. Hollahan , Eds., " Techniques and Applications of Plasma Chemistry", Wiley-Interscience, New York, 1974.

[8] See e.g. : C.J. Heffelfinger and E.L. Lippert , J. Appl. Polym. Sci., 15, 2699 (1971)

or : Tongyin Yu, Haishan Bu, Jianhua Chen, Jibang Mei and Jialun Hu,,Makromol. Chem. , 187, 2697 (1986)

[9] D.J. Brown, A.H. Windle, D.G. Gilbert and P.W.R. Beaumont ; J. Mater. Sci., 21, 314 (1986).

[10] S. Lazare and P.Lutgen, private communication (1988)

[11] D. Briggs, D.C. Rance, C.R. Kendall and A.R. Blythe , Polymer, 21, 895 (1980).

[12] J.J. Pireaux, M. Vermeersch, N. Degosserie, C. Gregoire, Y. Novis, M. Chtaïb and R. Caudano, Proceed. Workshop on Interface Phenomena, Halifax (Canada) August 23 - 27, 1988 (Springer - Verlag) in press.

[13] D. Briggs , Polymer, 25, 1379 (1984).

[14] See e.g. : - A. Miyake , J. Polym. Sci., 38, 479 (1959).

- L. D'Esposito and J.L. Koenig , J. Polym. Sci. : Phys. Ed., 14, 1731 (1976).

[15] P. Thiry, doctoral thesis, University of Namur (1984), unpublished.

INTERFACIAL CHEMISTRY OF METAL FILMS ON POLYMERS: DIFFUSION, OXIDATION, TRACE COMPONENTS STUDIED BY XPS, ISS, AND SIMS

Gene R. Sparrow*, and Larry Homstad[t]

*Advanced R & D Inc.; [t]American Thin Films Inc.

245 E. 6th St., St. Paul, MN 55101

The advantages of flexible circuit boards (PC) made from sputtered metals on polymer substrates are becoming very obvious and demand for such products is dramatically increasing. In order to optimize the production and performance of such films it is essential to understand the chemical processes involved. The chemistry of these thin films, which consist of tens to hundreds of Angstroms of metals on various polymers,exhibits large variations from the outer surface to the interface. This chemistry is largely affected by the nature of processing and treatment and generally includes three distinct areas of interest: The outer surface; the central or bulk material, and the interfacial area. This chemistry is amenable to analysis by only a few modern analytical tools including XPS, ISS, and SIMS. The requirements for ion sputtering metals onto polymers and the equipment used are discussed below along with detailed analytical information obtained about the interfacial diffusion, oxidation, and trace components monitored throughout some of the resulting films.

INTRODUCTION

The switch from rigid printed electronic circuit boards to flexible electronic boards has been prompted by several advantages of flexible boards. This has also been followed by increasing interest in adhesiveless or two-layer flexible material. Two-layer boards have several advantages.

Elimination of adhesives and adhesive outgassing.
Reduced board flammability.
Elimination of drill smearing.
Lighter weight - less material(both Cu and polymer).
Great flexibility.
Improved dimensional stability.
Better line quality - finer line widths and denser packing.
High-speed production utilizing "roll" processing.
Clean, continuous manufacturing.

Figure 1 illustrates typical flexible PC boards, one utilizing adhesives to laminate copper to Kapton; the other is a sputter-coated Cu film.[1]

Before addressing the chemical nature of sputtered metal systems on polymers, it should be noted that there is almost no similarity between metals sputtered onto polymers and polymers coated onto metals. This should be especially noted when comparing flexible PC boards made by sputter coating metal onto polymers and those using the adhesive laminati of polymer onto copper foil techniques. Very typical comparison would be that of microwave popcorn bags to coated beverage cans. The major difference which dramatically affects polymers coated or laminated to metals is the existence of an extensive oxide layer on the metals before the lamination or coating occurs. In most industries where polymers are coated on metals, it is well known that the process must be directed towa bonding to a metal oxide - not a metal. Such systems generally require special conversion coatings or coupling agents to act as a bridge between the metal oxide and the polymer. In sputter coated polymers or metalliz polymers the kinetics and reactions of concern are primarily those with metals; whereas in laminated or polymer coated metals the metal oxide interactions are major.

The ultimate appearance and performance of metallized polymers is greatly affected by the thickness and chemistry of the particular thin film. Even subtle changes in deposition conditions can greatly affect th chemistry. Three obvious features of these films include the nature of th outer 5 to 10 Angstroms, the nature of the bulk or major film itself, and especially the chemistry of the interface. Although the chemistry of the three layers can be controlled to some extent by processing, some aspects are inherent in this chemistry and processing.

One of these is the occurrence of at least some diffusion of metal a polymer at the interface. This not only depends on the original chemistr of the polymer surface, but also on the specific conditions used to prepa the polymer for deposition and the initial deposition conditions. The sputtering variables critical to the production of uniform thin films are

1. Base Pressure
2. Vacuum constituents (Residual Gases)

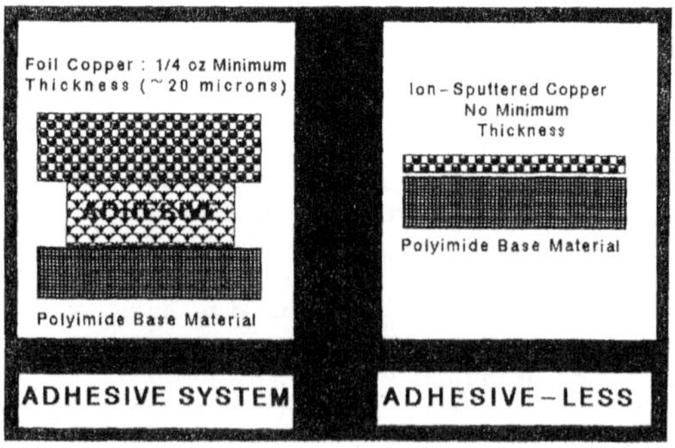

ADHESIVE-LESS FLEXIBLE PC BOARDS

Figure 1. Diagram showing adhesive laminated Cu film and sputter coated films for flexible printed circuit boards.

3. Angle of incidence
4. Plasma density
5. Mass flow of sputtering gases
6. Thickness of the film

All of the above are monitored and controlled by a computer in the system used for preparing the samples used in these studies. The second naturally occurring problem is that of oxidation of the metal at the outer surface and its subsequent contamination by foreign materials. Such oxidation and contamination may or may not also occur at the interface. Schiller et al.[3,4] have discussed some of these instrumental effects of DC sputter coating.

A third and very difficult problem involves the level of contaminants incorporated into the bulk of the thin metal. For active metals such as Al and Ti this can be very extensive oxidation which is very dependent on the rate of deposition and the nature of the vacuum in sputter or vapor deposition systems. In electrochemical depositions, it can be very dependent on the specific conditions and the amount of water, -OH, and trace components incorporated during depositions.

A fourth inherent problem in the deposition process is the presence of elastic stress in the thin film. This stress can be either compressive or tensile and can cause actual delamination or poor bonding between the metal and the polymer. Stress should not affect the chemistry of the interface although it may affect the bond.[2] However, conditions and variables which are still of interest include deposition rates, the kinetics and conditions of reaction at the interface, long-range effects on the interface and the effect of film thickness on interfacial chemistry.

We have utilized the three extremely sensitive techniques, XPS (X-ray Photoelectron Spectroscopy) or ESCA (Electron Spectroscopy for Chemical Analysis), ISS (Ion Scattering Spectroscopy), and SIMS (Secondary Ion Mass Spectroscopy), to investigate the detailed chemistry of several systems including Al, Ti, and Cu films. ISS has been extremely useful in quantitatively monitoring the detailed changes occurring in the outer few Angstroms of the surface as well as in the bulk and at the interface, and ESCA or XPS has been key at monitoring actual changes in bonding chemistry of individual elements, especially C and O throughout the thin film. SIMS has been unique in monitoring the nature of trace contaminants and their changes throughout the entire films as well as providing unusual information concerning variations indicative of changes in the nature of bonding within the bulk metal film itself.

Most of the work presented was accomplished using sputter-coated metals on either polyester or polyimide webs 10 to 20 inches in width. The majority of this work involved deposition of Al, Ti, or Cu metals. The analysis was conducted using a KRATOS X-SAM 800 combined Scanning Auger Microprobe (SAM) / ESCA (XPS) system. Concentration depth profiles of the various elements detected were obtained by slowly sputtering the surface using a low energy of Ar ions across an area about 20 mm^2. The distribution of oxygen and carbon in different forms was also monitored quantitatively using software deconvolution of the XPS peaks obtained using high energy resolution to acquire the data. The distribution of oxygen at the interface was of primary interest. SIMS data were obtained using a KRATOS Model 535 BX combined ISS / SIMS instrument. Trace levels of Na, K, Cl, and other impurities were monitored by SIMS.

Although most systems were found to be susceptible to heating, which broadened the interface, very thin films of Al were found to be the most severely affected. This was evident by extensive incorporation of oxygen into the entire thickness of the film. Cu was least affected. However, it exhibited the greatest complexity as far as contaminants.

Illustrations obtained from these three analytical techniques are presented and include profiles showing the variation of oxygen from the surface to the interface, the effects of heating and other processing variations on the nature of the interfacial boundary, and where various trace contaminants tend to segregate due to processing.

EXPERIMENTAL AND INSTRUMENTAL

Figure 2 illustrates the vacuum sputtering apparatus used to produce the metal coated polymers discussed below. It consists of a large cryogenically pumped vacuum chamber capable of obtaining vacuum pressures in the low 10^{-7} Torr range. It includes two 20" D chill drums to permit coating polymer substrates which are not stable at higher temperatures. Coating can be done on both sides in a roll-to-roll operation at deposition rates from atoms / minute to nearly 1 μm per minute and pretreatment of the surface can be done using reactive gases. Besides roll-to-roll coating, flat surfaces up to 30 " in diameter can be coated. Up to five different metals can be coated successively.

The two major concerns in the development of such films include:

1. Continuous manufacture of high resolution, flexible printed circuits with fine line resolution.

2. Manufacture of clean, well bonded metallized polymer films.

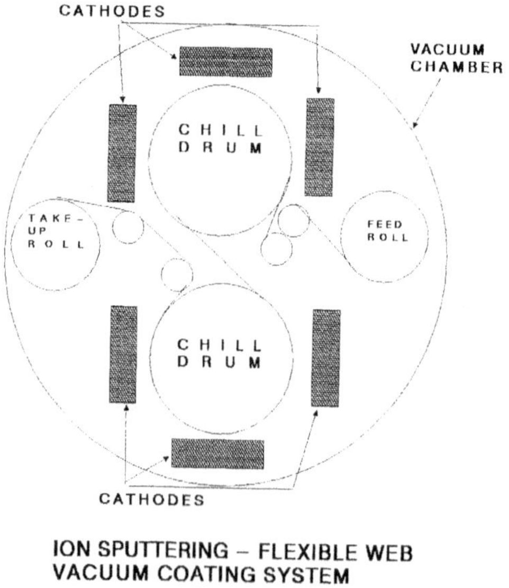

Figure 2. Diagram of Ion Sputtering, Roll-Coating System. Three cathodes located at each of two 20" chill drums permit coating of three different metals on both sides of a roll.

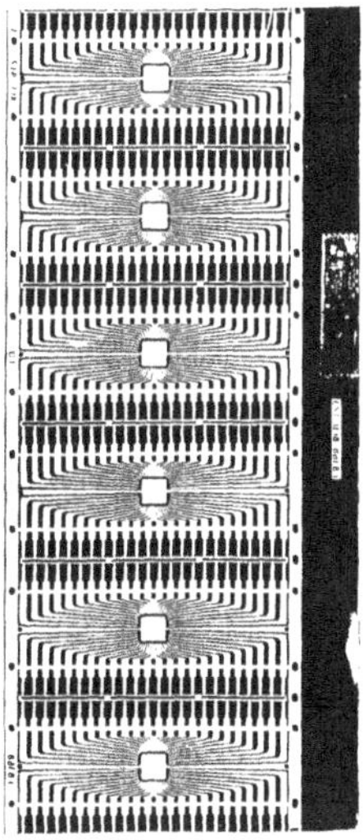

Figure 3. Photocopy of a portion of an actual flexible PC film made from Cr sputtered onto Mylar.

Figure 3 is a copy of an actual PC film made from Cr on Mylar using a positive dry film resist. Figures 4A, 4B, and 4C are SEM photographs obtained from another flexible PC showing line width of less than 2 μm. Previous work[5] using x-ray diffraction was used to investigate some effects of sputter chamber conditions on film integrity.

Several films manufactured under different conditions were subjected to surface analysis to determine the nature of the chemistry on the surface, at the interface and throughout the metal film. Although Ti, Al, Cr, and Cu films were studied, only typical results obtained from analysis of the Cu films will be illustrated. The films studied were from about 50 to 2000 Angstroms thick and with considerable variation in processing variables. Practical application of ISS toward studies of surface contamination has been described[6] and the role of surface chemistry in analyzing electronic packaging failures has been discussed.[7] There are also several other techniques which can be utilized for chemical analysis of these types of materials but the above three are considered the most applicable due to their extreme sensitivity to the surface. Other techniques, such as FTIR and EDX, often cannot provide reliable detection of information within the outer 10 to 50 Angstroms of a surface, or cannot detect the interface.

Samples approximately 1/2" x 1/2" were cut from the center of larger sheets of coated film for each experiment. Each surface was analyzed in detail before proceeding with depth profiles completely to the interface. These were obtained by continuously monitoring the spectrum while the metal thin film was sputtered using a mixed Ar^+, $^3He^+$ ion beam.

ISS and SIMS Instrumental

The sample holder containing the samples for analysis was placed in the insertion chamber of the instrument, evacuated to about 4×10^{-7} Torr

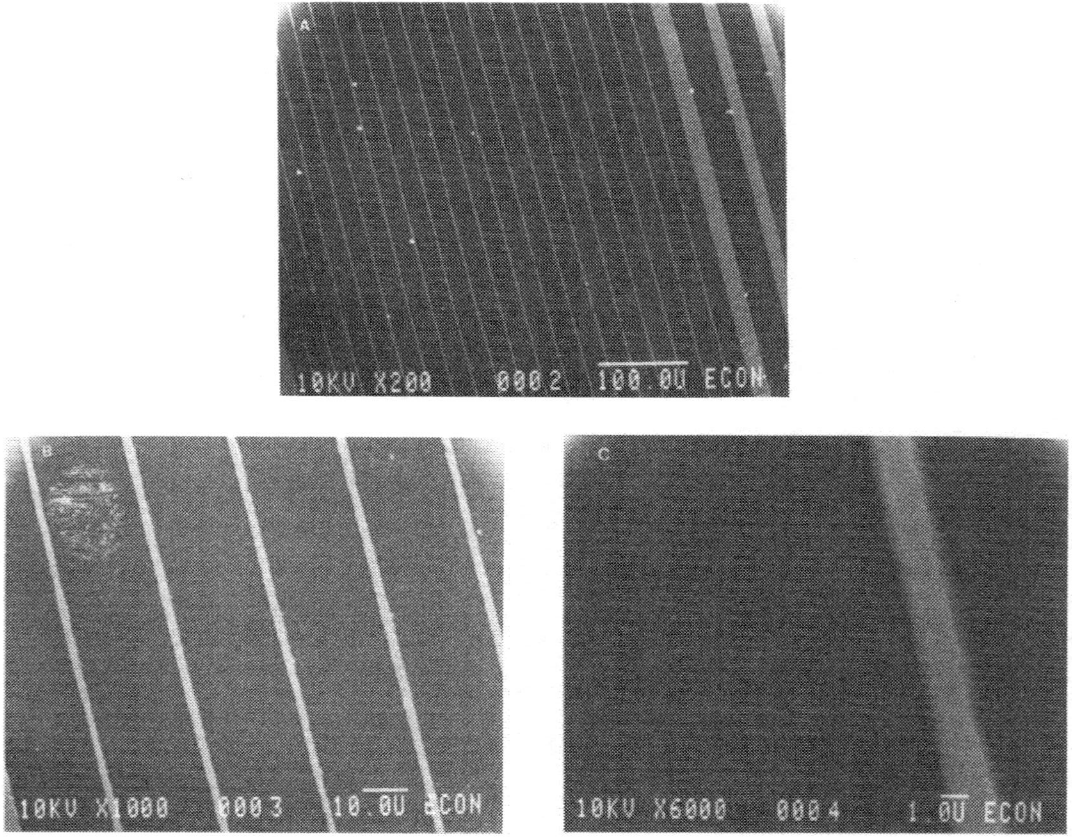

Figure 4A,B,C. SEM photographs of Cr lines on a flexible PC film. Line width of this sputter coated Cr is less than using a turbomolecular pump, and inserted through a vacuum isolation valve into the analysis vacuum chamber. The analysis vacuum chamber was re-evacuated to about 6 x 10^{-8} Torr prior to starting analysis. Prior to sample insertion, the base pressure was 2 x 10^{-8} Torr.

All samples for ISS and SIMS analysis were analyzed under identical conditions using a KRATOS/ 3M Model 535 BX combined ISS/SIMS System equipped with a Balzers 110 l/sec turbomolecular pump on the sample insertion chamber. All data were obtained automatically using a PC Computer Data Acquisition System--available from Advanced R&D.

For ISS analysis ,an ion beam composed of a mixture of Ar^+ and $^3He^+$ and operated at 1 keV and about 200 nA was used. The ion beam, about 300 microns diameter (FWHM),was rastered across an area about 3mm x 1 1/2mm to minimize crater edge effect. Further reduction of crater edge effects was obtained by electronically gating the signal so as to accept signal only when the ion beam was transversing the center 50% of the rastered area. The signal was measured using a continuous dynode electron multiplier operated in the saturated pulse counting mode.

In surface analysis, it is essential that all measurements include the three-dimensional aspects of a material, i.e., depth. It is difficult to attach a high degree of meaning to a single surface measurement that does not offer an indication of concentration with respect to depth. In all of the following analyses, this has been done.

Detailed ISS Depth Profiles of the concentration gradients within the outer 40-50 Angstroms were first obtained. A series of about 10-20 ISS spectra are obtained at successively deeper depths into the surface as the ion beam slowly sputters away outer atomic layers. All of these spectra are recorded on a single sheet for convenience - Figure 5A. They are recorded in two ways: one in which the intensity of the peaks in all spectra relate to an absolute signal, and the other in which each spectrum is normalized so that its maximum peak is 100. In general, these depth profiles are very reproducible and a simple convenient technique by which the overall surface integrity of a sample may be visually observed. They show basically how extensive surface contamination is, how deep it may extend, low level impurities, and trace, high mass impurities. Two ISS spectra from each depth profile, one near the surface and one at about 20 A, are expanded on a normal scale to permit more detailed observations and peak assignments - Figure 5B. Often hydrogen can be semiquantitatively detected indirectly as a broad sputtered ion peak at low energy.

A simple ISS spectrum has several features which should be noted. Except for C, N, and to some extent O, ISS detection sensitivity increases uniformly with atomic mass and only varies by a factor of about 10-20. Sensitivity for C and N is low, but in general about 1% absolute. This corresponds to about 1 atom per 100 on the outer atomic layer of a sample. Peak shape is normally broad, even for very weak peaks, and mass resolution is poor, only about 3-5% relative. Since ISS does not exhibit extreme matrix effects, like SIMS, the signal is a direct, accurate measurement of concentration. Due to the extreme surface sensitivity of ISS, detailed high resolution concentration depth profiles can be obtained and O/metal ratios can be accurately measured.

One unusual feature of ISS is that the signal generally depends only on the density of surface atoms of each particular element. Thus, the depth (sputtering time) it takes for the signal of a particular element to reach maximum intensity usually indicates the thickness of contaminants covering that element. Also, the relative intensity of that element in the

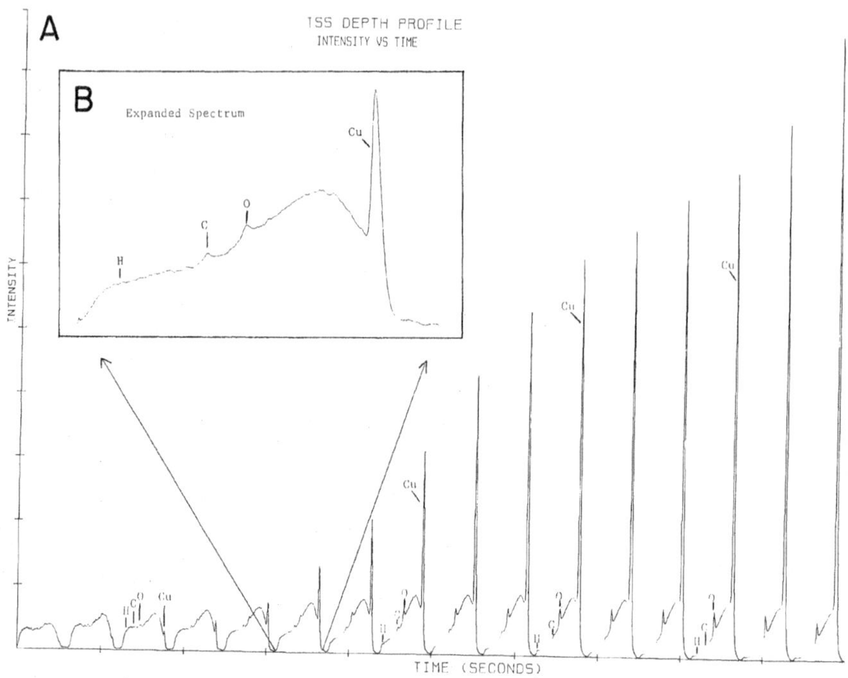

Figure 5. A. Raw data from 16 spectra acquired from outer surface analys of Cu film by ISS using $^3He^+$ at 2000 eV. B. Expansion unit of one typical spectrum showing peak identifications.

first few ISS scans (as a ratio of the maximum signal achieved) will indicate roughly the % of surface not contaminated or covered.

Surface Sensitive SIMS spectra are obtained using $^3He^+$ at 2000 eV to correspond with the same sputtering conditions used for ISS. These provi high detection sensitivity for trace elements, resolve elements not separated by ISS, and provide some indications of surface bonding or chemistry. In general, $^3He^+$ does not sputter high mass particles, hence, peaks are normally not observed much beyond mass 100 to 150. The same sputtering rate as estimated for ISS occurs during these analyses, i.e., about 3 to 5 A/minute. These conditions are very favorable for detecting low level amounts of low molecular weight organic fragment ions such as CH_3, C_2H_5, and others. In general, the quadrupole mass resolution is decreased for these scans in order to maximize sensitivity.

One undesirable and potentially confusing characteristic of SIMS involves the extreme dependency of the signal on the presence of surface oxide or O_2. During the first few minutes of sputtering, the signal intensity often decreases. When sputtering through a metal oxide into th base metal, intensity can decrease two to three orders of magnitude. For this reason, the first SIMS spectrum can often be much more intense than subsequent spectra, even when subsequent spectra are signal averaged over longer period of time. Fortunately, the relative intensities of most of the other elements and peaks are also affected in nearly the same way; hence, for these studies, their calculated concentrations are good approximations. This is not true for some metal substrates.

$^{40}Ar^+$ SIMS also provides some indications of surface chemical bonding or chemistry, although it sputters at a much greater rate. For example, peaks are commonly observed at m/e 43, 59, and 73 in silicone or siloxane polymers, but not in SiO_2. Their assignments are usually $SiCH_3$, $OSiCH_3$, and CH_2OSiCH_3, respectively, and thus strongly indicate Si bonded to C, possibly as a methyl unit. Similarly, it is common to observe peaks due to C, CH, CH_2, CH_3, C_2, C_2H, C_2H_2, and MC_2H which vary in relative intensity ratios depending upon the type of organic compound present and the extent of sputtering.

ESCA Instrumental

All samples for XPS or ESCA analysis were analyzed with a Kratos X-SAM 800 combined Scanning Auger Microprobe (SAM)/ ESCA (Electron Spectroscopy for Chemical Analysis - also termed XPS - X-ray Photoelectron Spectroscopy) equipped with a turbomolecular pumped sample introduction system and DEC microPDP-11 computer data acquisition system with a Tektronix color terminal and color printer. System base pressure is normally maintained below $2x10^{-10}$ Torr using a 300 l/sec ion pump and Ti sublimators to minimize contributions from vacuum contaminants. The sample is excited by a dual anode x-ray source with a choice of Al or Mg anodes. The electrons emitted from the specimen are focussed via an electron transfer lens (ACIL Lens) into a sophisticated Hemispherical Energy Analyzer and detected by a multichannel electron multiplier. The limited acceptance angle and focal depth of the lens allows a spot size of 2 to 3 mm in diameter on the surface to be analyzed.

Samples were fastened to stainless steel sample holders and introduced to the ultra-high vacuum analysis chamber through the introduction chamber. The samples remained in the introduction chamber while they were pumped down to approximately $2x10^{-7}$ Torr for typically 20 minutes prior to insertion into the analysis chamber through a vacuum isolation valve.

All data were acquired and stored on magnetic disk using an integrated computer system. ESCA spectra were acquired by recording the total electron current vs kinetic energy. All data were replayed and manipulated for detailed analysis, comparison, and quantification and subsequent plotting on the color printer.

High resolution analysis of smaller regions of particular interest were obtained by specifying certain regions (elements) for analysis instead of recording the entire spectrum. In ESCA, these regions were subjected to extensive computer deconvolution to separate complex elemental peaks into their various components, i.e., the C peak into C=O, CH_2, C-O-C, etc. For these experiments only C, N, O, Cu, and/or Cr were monitored.

For ESCA, the escape depth of the photoelectrons, typically 5-50 A, limits the depth analyzed to the top few surface layers of the sample. Briggs[8,9] discusses the principles and application of ESCA (XPS) quite thoroughly.

In order to profile into the sample, the instrument is equipped with an ion gun which allows the surface of the sample to be sputtered. By alternately sputtering and collecting ESCA spectra, the variation of elemental composition with respect to depth can be obtained. The computer system then allows the generation of informative displays of the variation of peak shape and intensity versus depth. For these experiments, the ion gun was operated at 4 keV using $^{40}Ar^+$ at about 2 μA across a sample area about 5mm x 5mm. The electron extraction lens (ACIL lens) was operated at high magnification during the depth profiles to restrict the ESCA signal to

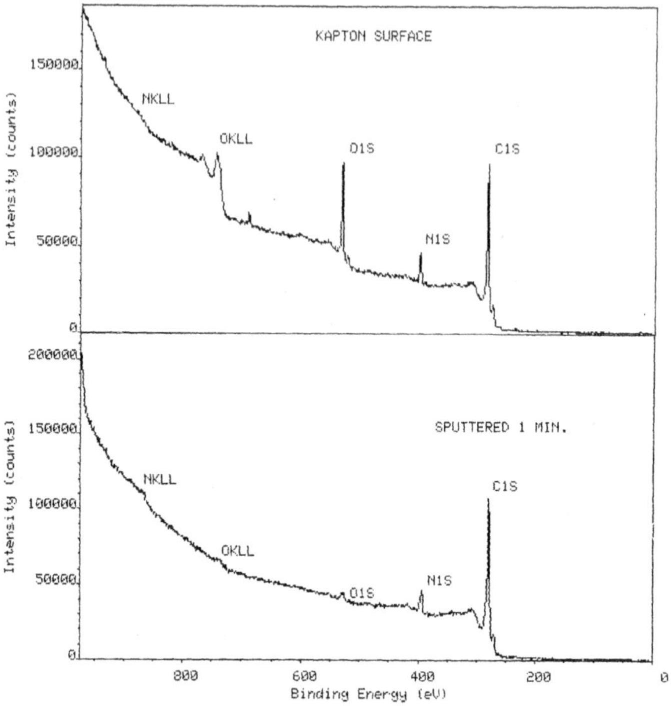

Figure 6. ESCA Spectra of Kapton Surface and after sputtering one minute.

only the center 0.5 x 0.5 mm area of the sputtered crater.

RESULTS AND DISCUSSION

One of the major problems involving successful metallization of polymers is the preparation of the substrate polymer before coating begins We have discussed preparation effects on the cleaning of Kapton® elsewhere[10], whereas van Ooij[11,12] has done extensive SIMS work on characterizing the surface bonding of polymers. In order to illustrate th difference between typical Kapton® as received, and its chemistry only a fe Angstroms below the surface, Figure 6 shows ESCA spectra before sputtering and after sputtering. Note that the original surface is substantially oxidized compared to the surface after sputtering. In addition to illustrating the level of oxidation and contamination of the surface, this analysis of Kapton without any metal layer helped establish the changes which are induced by the sputtering mechanism itself.

Figure 7 illustrates the variation of the C1s peak at different depth of analysis through a typical Cu film. Note that the C does not disappear completely even within the major portion of the Cu film. This type of diffusion into the Cu can vary as does the oxygen. This incorporation of C into the bulk of a thin metal film can also occur if the vacuum within the sputter coater is badly contaminated with volatile organic material. ESCA data subjected to quantitative analysis are displayed in Figures 8 an 9 which show the atomic concentration depth profiles for two different films. Note one film is especially high in oxygen, whereas the other show

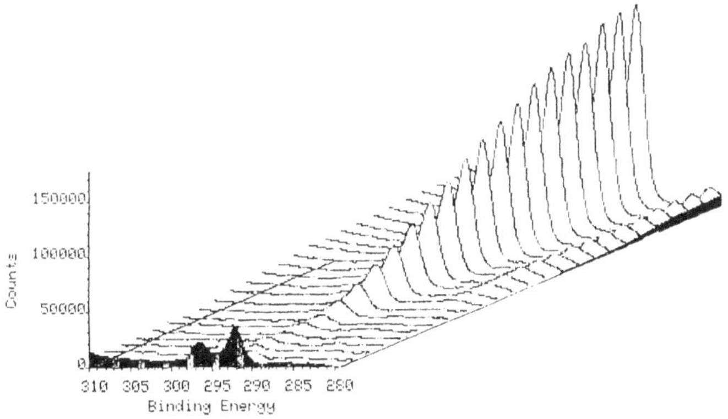

Figure 7. ESCA depth profile showing C1s peak through depth profile of Cu film on Kapton.

a fairly sharp interface. This latter film is more typical of a clean, sputtered thin film. Yet careful inspection of the peak shapes of the C, O, and Cu at the interface yields some interesting information concerning the nature of the bonding chemistry which occurs there.

Figure 10 shows a comparison of the C1s peak shape within normal sputtered Kapton (past the interface) and at the onset of the interface. The additional shoulder on the left side of the normal C-H bonding is attributable to carbon bonded to either N or O but not as a double bonded group such as COO. Figure 11 shows the Cu 2p bonding structure at the surface and within the bulk of the film. Cu is split into a doublet by bonding to oxygen and normally displays an additional doublet when Cu is

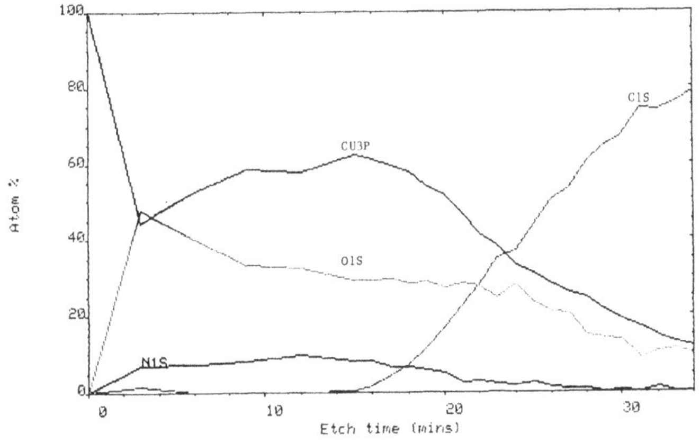

Figure 8. Concentration depth profile of sputtered Cu film on Kapton as obtained by ESCA. Note incorporation of high levels of C and O into metal film.

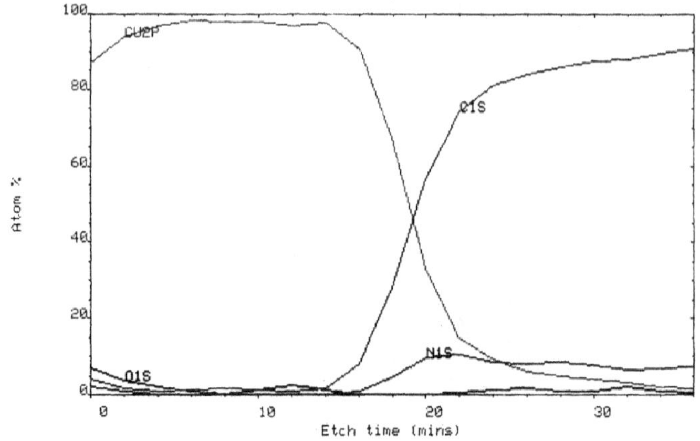

Figure 9. ESCA concentration depth profile of sputtered Cu film on Kapton. Note low C and O throughout film.

present in its divalent state. Careful inspection of the Cu peak shapes throughout the interface indicates that the second set of peaks due to the divalent Cu state is not very obvious. This is strong indication that the bonding at the interface is either a lower order, or that the thickness of any divalent bonding stage is too thin to provide significant contribution to the ESCA signal which encompasses about 40 to 50 Angstroms.

Similar profiles were obtained by ISS and SIMS. Figure 12 illustrates the ISS depth profile. It can be noted from this profile that the N to C ratio is excessively high at the onset of the interface. This is strong indication that there is a preferential concentration of the N at the interface and that it is most likely very involved in the actual metal bonding of these systems. Significant amounts of oxygen are not observed indicating that bonding through formation of Cu-O-Kapton bonds probably does not occur unless the Kapton surface is contaminated by excessive

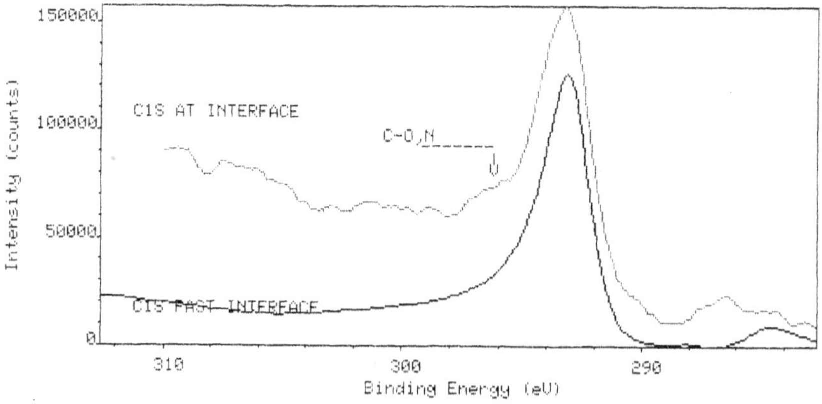

Figure 10. Comparison of Cls bonding structure at the Cu/Kapton interface and within the Kapton bulk.

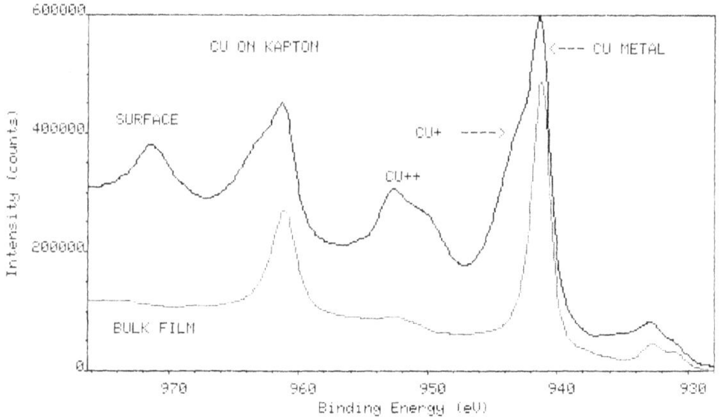

Figure 11. ESCA Cu2p peak at the surface and within the bulk of sputtered Cu film on Kapton.

oxidation before sputter coating. Van Ooij has studied Cu and Cr on Kapton interfaces by delaminating films and using SIMS.[13]

Our SIMS analysis did not indicate any obvious segregation of impurities at the interface although significant amounts of Na, K, and Ca were detected throughout the films, Figure 13.

CONCLUSION

An ion sputtering system capable of sputter coating three metals on both front and back sides of most types of polymer rolls has been described. Flexible printed circuit boards with line width less than 2 µm have been produced. Extensive analytical testing of these thin sputtered metal films on Kapton and Mylar has been performed using surface analytical instruments includng ESCA (XPS), ISS, and SIMS. These results have provided information useful and valuable toward understanding the nature of

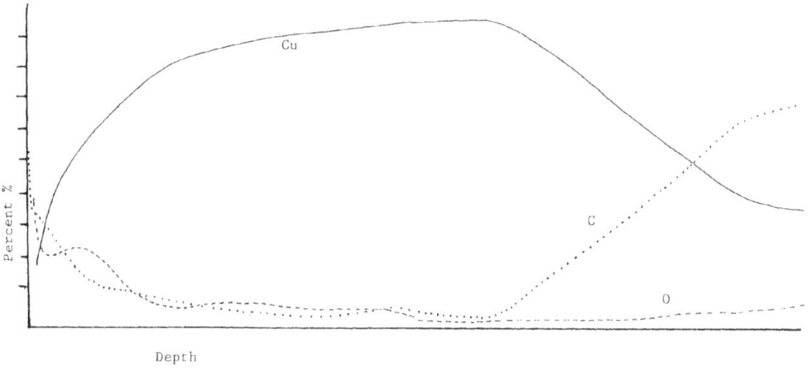

Figure 12. ISS depth profile of sputtered Cu film on Kapton.

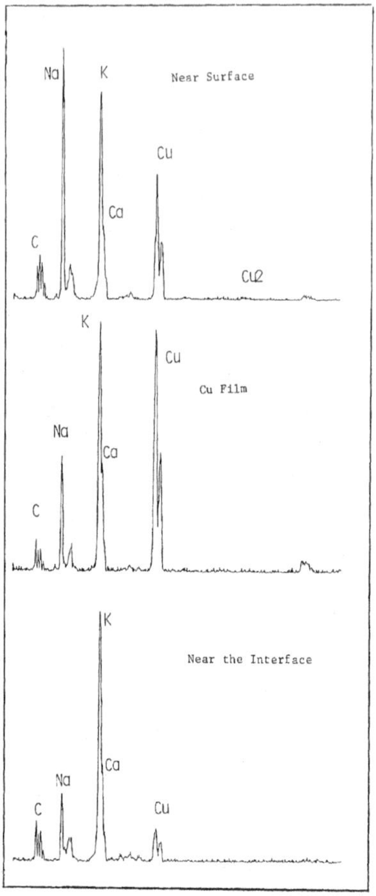

Figure 13. Typical SIMS spectra obtained from the near surface and near the interface of sputtered Cu thin film on Kapton.

original oxidation and contamination of the films, the chemical integrity of the metal film, and the nature of bonding at the interface.

REFERENCES

1. DOD, ANSI/IPC-FC-241C. "Flexible Metal-clad Dielectrics for Use in Fabrication of Flexible Printed Wiring".

2. E. Klokholm, Delamination and fracture of thin films, IBM J. Res. De 31,5 (September 1987).

3. S. Schiller, G. Beister, and W. Sieber, Reactive D.C. high-rate sputtering -- Deposition rate, stoichiometry and features of TiO_2 and TiN films with respect to the target anode. Paper presented at the International Conference on Metallurgical Coatings, San Diego, CA, April 1983.

4. S. Schiller, U. Heisig, C. Kordorfer, G. Beister, J. Peschke, K. Steinfelder, and J. Strumpfel, Reactive DC high-rate sputtering as production technology. Paper presented at the International Conference on Metallurgical Coatings, San Diego, CA, March 1987.

5. L. Homstad, "X-ray Diffraction Measurements of Single Layer Flexible Media". Unpublished internal report, Magnetic Peripherals, Inc., November 1984.

6. G. R. Sparrow, Applications of ISS/SIMS in characterizing thin layers (~10nm) of surface contaminants, in "Surface Contamination: Genesis, Detection and Control," K.L. Mittal, editor, Vol. 2, pp. 635-654, Plenum Press, New York, 1979.

7. G. R. Sparrow, The role of surface chemistry in electronic packaging failure analysis, paper presented at the Twelfth Annual International Symposium for Testing and Failure Analysis, Los Angeles, CA, October 1986.

8. D. Briggs, "Handbook of X-ray and Ultraviolet Photoelectron Spectroscopy", Heyden, London ,1978.

9. D. Briggs and M. P. Seah, editors, "Practical Surface Analysis by Auger and X-ray Photoelectron Spectroscopy". John Wiley and Sons, London 1983.

10. G. R. Sparrow and E. Mishmash, Surface analysis of polymers and glass by combined ISS/SIMS, paper presented at the 1977 Pittsburgh Conference on Analytical and Applied Chemistry, Cleveland, OH, March 1977.

11. W. J. van Ooij, Investigation of metal-polymer interfaces, paper presented at NSF Workshop on Adhesion Science and Technology, Stanford Science Camp, Lake Tahoe, CA, October 14-16, 1987.

12. W. J. van Ooij, and R. H. G. Brinkhuis, Interpretation of the fragmentation patterns in static SIMS analysis of polymers. Part I - Simple aliphatic hydrocarbons, Surface Interface Anal. 11, 430-440 (1988).

13. W. J. van Ooij, R. H. G. Brinkhuis, and J.M. Park, Adhesion of copper and chromium films to polyimide studied by static SIMS of the interface, Private Communication, 1988.

PART III. SURFACE MODIFICATION AND ADHESION ASPECTS

CHARACTERIZATION OF SURFACE MODIFICATIONS

DURING METALLIZATION OF POLYETHERIMIDE

Michael C. Burrell, Bradley R. Karas, Donald F. Foust,
William V. Dumas, Edward J. Lamby, Willard T. Grubb, and John J. Chera

GE Corporate Research and Development
Schenectady, New York 12301

The chemical and compositional modifications of a polyetherimide surface during metallization processes have been monitored using x-ray photoelectron spectroscopy (XPS). The following aspects will be discussed: 1) detection of contaminants on the polymer surface and the effectiveness of various cleaning agents; 2) determination of the degree of imide hydrolysis; 3) quantification of the coverages of metallization catalysts using Rutherford backscattering spectrometry (RBS) as a calibration method; 4) measurement of the changes in catalyst composition (colloidal Pd/Sn) during sensitization and acceleration steps; and 5) verification of the cause of leakage current between isolated conductor runs.

INTRODUCTION

Molded circuit boards and interconnects have attracted considerable attention in the electronics industry.[1-5] The method of molding a device, and then metallizing and patterning the plastic, has created new opportunities previously unobtainable with conventional laminated glass/epoxy resins. New devices such as three-dimensional circuits and combinations of circuitry and connectors are now feasible. With such new opportunities come new challenges with the properties of the thermoplastic, techniques for molding the device, metallization of the plastic, and definition of the conductor traces. This paper discusses some of the chemical modifications of a polyetherimide surface which occur during circuit board manufacture. The use of surface analysis methods to characterize plastic surfaces at different stages of an electroless metallization sequence has aided both in process development and in the solution of production problems. Figure 1 is a schematic diagram of a typical process scheme. The process steps at which these surface modifications will be discussed are indicated by asterisks. X-ray photoelectron spectroscopy is used for characterizing changes in the polymer surface chemistry prior to deposition of the metal layers, and after the metal layers have been patterned (e.g., between runs). Several of the problems addressed are unique to molded circuit boards (versus those prepared by glass/epoxy lamination techniques), and to electroless metallization schemes which involve activation of the surface by exposure to catalyst solutions.

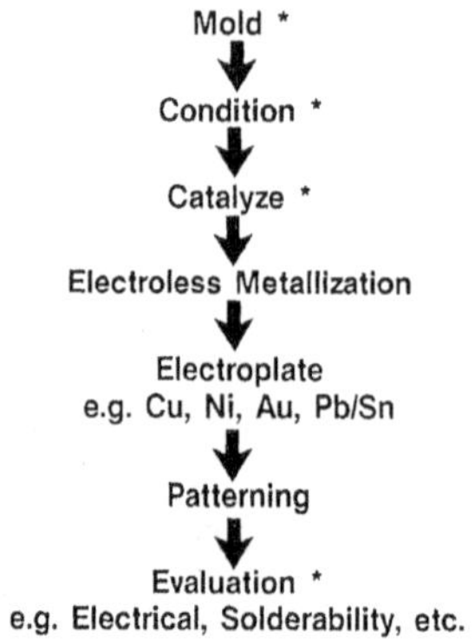

Figure 1. Schematic of molded plastic circuit board metallization sequence.

EXPERIMENTAL

Substrates utilized in this study were molded from Ultem® 1000 (unfilled), or from Ultem 2312 (30% milled glass-reinforced) polyetherimide (Figure 2). Both are products of the General Electric Co. Cleaning solutions consisted of 2 vol.% All® detergent (Proctor and Gamble), 2 vol.% Shipley Acid Cleaner 1118 (Shipley Chemical Co., Newton, MA) at 50 °C, or different grades of Freon (Ashland Co.). Thin films of Ultem used in the hydrolysis study were prepared by spin coating a 5 wt.% solution of Ultem 1000 in methylene chloride onto silicon wafer substrates. The resultant films were exposed to 1 N KOH in absolute ethanol (15 to 120 s) and rinsed with ethanol or DI water. All samples were air dried prior to XPS analysis.

Some of the surfaces containing adsorbed metallization catalysts were prepared specifically for XPS analysis. Palladium (II) solutions were prepared in toluene. The Ultem plaques were soaked for 5 min, water rinsed, and air dried. In some cases, an Enthone 402 electroless copper bath (Enthone Inc., West Haven, CT) was employed at room temperature for 5 s. Samples treated with colloidal Pd/Sn were prepared as follows: 1) 1 min in Shipley Cataprep 404, 270 g/l, 22 °C; 2) 3 min in Shipley Cataposit 44 1.5 vol.% in Cataprep 404 270 g/l, 45 °C; 3) 2-min flowing water rinse; 4) 3 min in Shipley Accelerator 19, 16 vol.%, 22 °C; and 5) 2-min flowing water rinse.

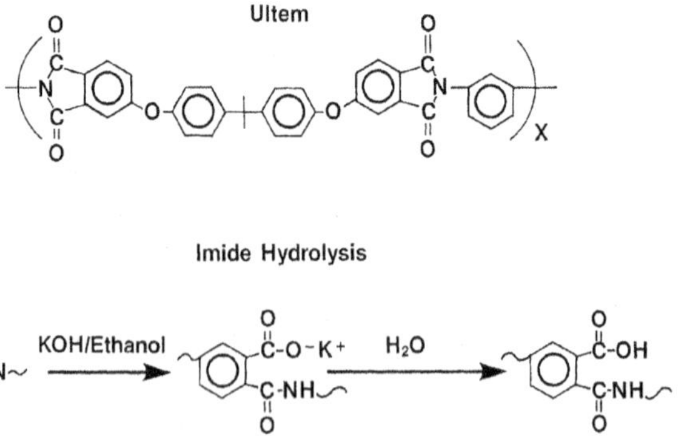

Figure 2. Ultem® polyetherimide structure, and hydrolysis reaction.

The sequence employed to make actual circuit boards is shown in Figure 1. After molding, the substrates were treated with a proprietary General Electric Co. conditioning process. This treatment leaves the Ultem substrate undamaged and receptive to catalysis and electroless copper deposition. Catalysis was achieved using the Shipley system described above, preceded by a 5-min immersion in Shipley 1175A cleaner/conditioner at 65 °C, 2.5 vol.%, with a water rinse. The Shipley Cuposit 250 electroless copper system imparted conduction to the otherwise insulating Ultem polyetherimide. The copper thickness was then built up to ~25 μm electrolytically from an acid copper bath, e.g., Lea Ronal Copper Gleam PCM+ (Lea Ronal,Inc. Freeport, N.Y.), or Electrochemicals PC-667 (Electrochemicals Inc., Youngstown OH.).

An etch mask was then applied and the undesired copper removed using either Ultra Etch 50 (MacDermid Inc., Waterbury, CT) or Shipley Hydro-Etch 536. The former is an ammoniacal etchant while the latter is a peroxide-sulfuric acid system. A General Electric Co. proprietary process was utilized to remove residual catalyst.

XPS analysis was conducted on a Surface Science Instruments SSX-100 spectrometer. In this instrument, the monochromatic AlKα x-ray source is a focused to a 150- to 1000-μm spot which defines the analysis area. A pass energy of 150 eV was used for survey spectra, or 50 eV for high-resolution scans of individual lines. Surface compositions were computed from spectral peak areas using calculated sensitivity factors. These are derived as the product of the tabulated ionization cross section,[6] and a term proportional to $KE^{0.7}$ to account for the energy dependence of the photoelectron sampling depth. Sample charging was controlled by an electron flood gun, and spectral energies were referenced to the C 1s line at 284.6 eV.

RESULTS AND DISCUSSION

Initial Surface Characterization

The capability of a conditioning process, e.g., swell and etch or chemical modification of the polymer surface, to successfully prepare a substrate for metallization, is dependent upon the ability to wet and interact with the polymer surface. The presence of contaminants such as grease or fingerprints from handling or mold release agents (typically silicones) can adversely affect the metallization process or the metal-polymer bond. The efficiency of various cleaners in removing silicones from the surface was monitored by XPS, and the results are summarized in Table I. The two most effective cleaners were Freon TMC and Freon TA. However, Freon TMC contains methylene chloride which is incompatible with polyetherimide.

TABLE I. XPS Surface Composition, Atom %, for Different Substrate Cleaning Processes.

Cleaning Process	C	N	O	Si
None	71.9	2.1	17.2	8.8
All® detergent	81.5	3.1	13.4	2.0
Shipley Acid Cleaner 1118	81.1	3.1	11.1	4.5
Freon TF	78.6	2.3	12.3	5.2
Freon TMS	84.9	3.2	10.3	1.3
Freon TMC	86.0	3.0	9.7	0.4
Freon TA	83.2	3.9	11.5	0.4
Ultem theoretical	82.3	4.4	13.3	–

Imide Hydrolysis

Exposure to various solutions used in the metallization sequence may chemically alter the structure of the polymer surface. For the polyetherimide, hydrolysis of the imide group is likely to occur during contact with alkaline plating solutions, according to the reaction sequence illustrated in Figure 2. Treatment of a cast-film Ultem surface with 1 N KOH in ethanol, followed by rinsing with either ethanol or water, resulted in the surface compositions summarized in Table II. Potassium is detected when the surface is rinsed with ethanol, but is absent when water is employed. The potassium salt formed in the imide cleavage reaction is easily converted to the acid upon rinsing with water. Figure 3 shows the accompanying changes in the core level photoelectron spectra. The C 1s line from the untreated surface shows three major components: (C1) 284.6 eV, due to carbons bound only to other carbons and/or hydrogen; (C2) 286.0 eV, due to carbons singly bound to O or N; and (C3) 288.3 eV, from the carbonyl in the imide groups. A fourth weak component (C1*) at ca. 292 eV is attributable to a shake-up satellite in the aromatic groups. The O 1s line exhibits two components from doubly bound oxygen (O1) and singly bound (O2) species at 531.1 and 533.2 eV, respectively. The N 1s line consists of a single component at 400.2 eV consistent with the imide moiety. The relative intensities of the C 1s and O 1s components agree with the structural formula of the Ultem repeat unit. Treatment of the surface with KOH/ethanol gives rise to the presence of potassium peaks in the spectrum, and changes in the polymer spectra. The C 1s line is broadened in the carbonyl region due to the presence of a new component at 287.4 eV. This new component is consistent with the amide functionality formed by the opening of the imide ring.[7] An equal number of $O{=}C\text{-}O^{-}K^{+}$ groups are formed, appearing at 288.3 eV (coincident with remaining unreacted imide, if any). Broadening and shifting of the signal maxima are observed in the O 1s spectrum, due to formation of new unresolvable oxygen environments. The N 1s peak appears at 399.7 eV, consistent with its conversion into amide. Rinsing the KOH-treated surface with distilled water removes the potassium and causes further changes in the polymer spectra. The presence of two carbonyl components in the C 1s line at 287.8 and 288.9 eV is due to amide and carboxylic acid groups, respectively. This is confirmed by the increased surface oxygen content (Table II), and the additional intensity in the O 1s line due to $O{=}C\text{-}\underline{O}H$.

The rate of imide hydrolysis was determined by measuring the K/N ratio with XPS for samples after various levels of KOH/ethanol treatment. For molded samples, the limiting K/N ratio of about 0.5 (50% reaction) was attained after 15-s immersion. The limiting K/N value for molded surfaces was somewhat lower than for solvent-cast films (Figure 3), possibly due to a surface orientation effect. The total potassium uptake determined by RBS was ca. 1×10^{15} atom/cm^2, indicating that the hydrolysis reaction was

TABLE II. XPS Surface Composition, Atom %, After Imide Hydrolysis Reaction.

Specimen	C	O	N	K
Ultem, Formulation	82.3	13.3	4.4	
Ultem, Measured	82.7	13.7	3.6	–
KOH/Ethanol Treated; Ethanol Rinse Only	77.9	15.7	3.4	3.0
KOH/Ethanol Treated; DI Water Rinsed	80.5	16.1	3.4	–
Hydrolyzed Ultem, Theoretical	77.9	17.7	4.4	–

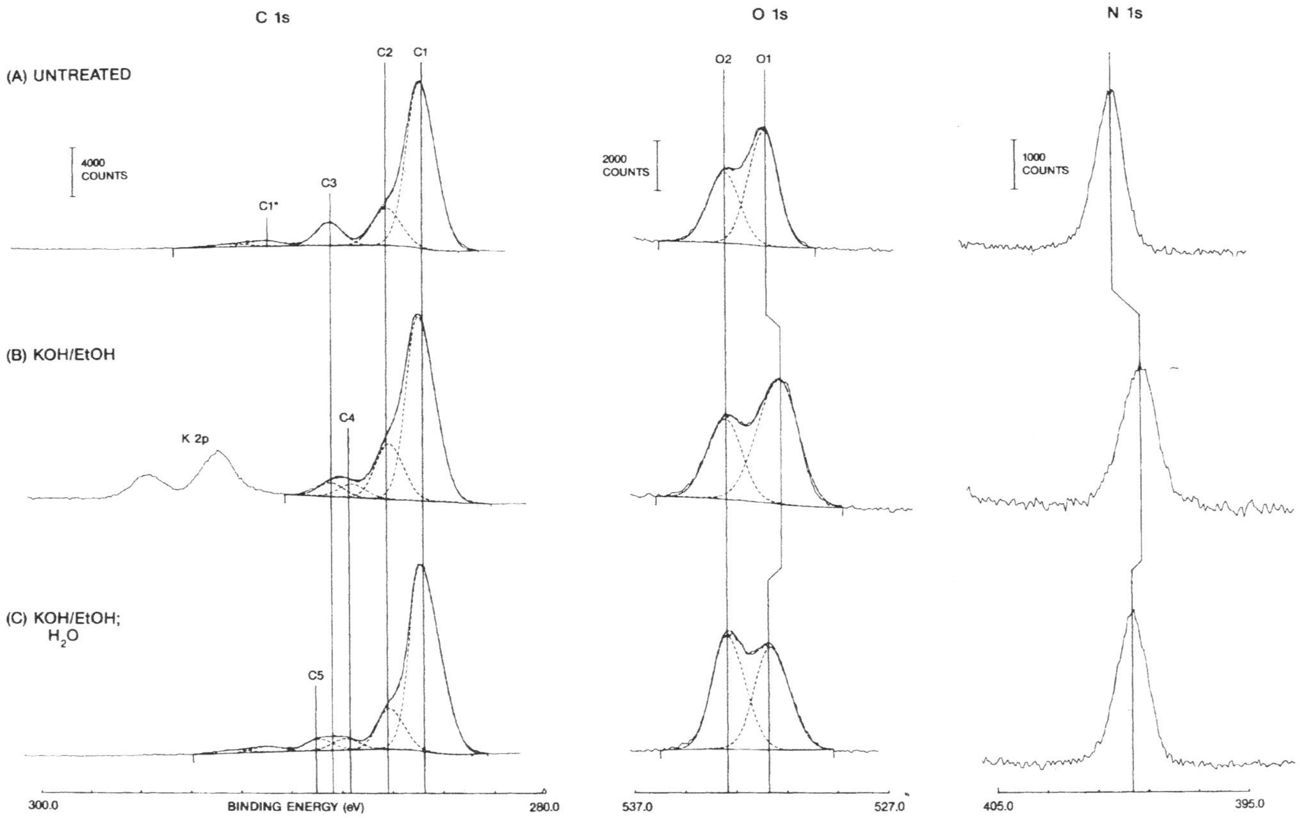

Figure 3. Core level photoemission peaks of Ultem and hydrolyzed surface.

confined to the outer 150 to 200 Å of the surface (assuming the degree of reaction determined by XPS).

If a hydrolyzed surface is subsequently metallized and heat treated, it is conceivable that re-imidization may occur and the released water could react with the metal deposit, resulting in a weak boundary layer of metal oxide or hydroxide. The results above show that the amount of water would be on the order of only one monolayer, which would be negligible compared to the amount of water absorbed in the bulk of the polymer (often about 1% by weight).

Adsorbed Metallization Catalysts

Polyetherimide surfaces activated with two different forms of palladium have been characterized. The first scheme involves treatment of the surface with Pd(+2) compounds, which is subsequently reduced to the active metallic form. The second scheme utilizes a commercial two-step sensitization/activation sequence which deposits colloidal Pd/Sn particles on the surface. Figure 4 compares the XPS Pd 3d lines for a surface treated with a Pd(+2) compound and one treated with colloidal Pd(0). The spectra from Pd foil and $PdCl_2$ are also shown for comparison, and illustrate the difference in binding energy which permits differentiation of Pd(0) and Pd(+2) species. Quantification of the surface Pd coverage from the XPS line intensities is complicated due to a lack of knowledge of the Pd distribution within the surface layer for the two types of treatments. For example, simple models assuming homogeneous distribution or an overlayer/substrate relation are not applicable in the case of the colloidal Pd particles which range in size and only partially cover the surface (see below). These absolute Pd coverages were determined directly by Rutherford backscattering spectrometry. A representative RBS spectrum for an Ultem surface treated with a palladium (2+) compound (Figure 5) shows that the Pd backscatter peak from the adsorbed surface layer is well resolved from the substrate signals. The width of the Pd peak indicates that the Pd is contained within a surface layer 200 Å or less.

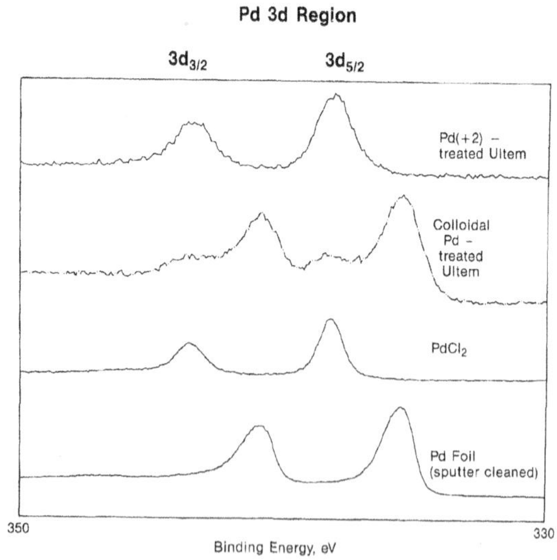

Figure 4. Pd 3d region for adsorbed Pd compounds, and Pd standards.

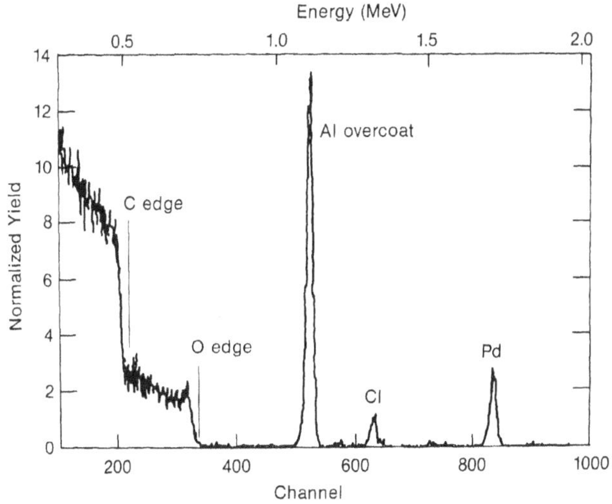

Figure 5. RBS spectrum of Pd(+2) adsorbed on an Ultem surface. Total Pd coverage is 4.5×10^{15} atom/cm^2.

Figure 6 is a plot of the Pd concentration from the XPS measurements versus the absolute coverages determined by RBS. The surface Pd content was varied by controlling the solution concentration and the exposure time. The XPS composition was computed assuming a homogeneous distribution of Pd within the XPS sampling depth (ca.

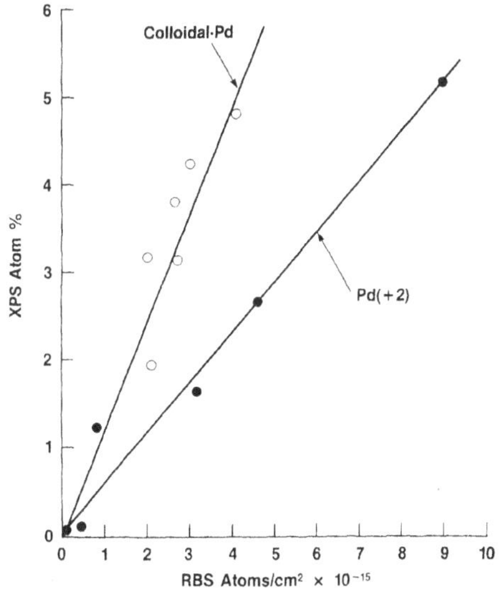

Figure 6. Plot of the surface Pd signal from XPS, versus the absolute coverage determined from RBS.

75 Å). The intensity of the XPS Pd signal is linearly related to the total Pd determined by RBS. However, a different slope for the calibration curves is observed for the two types of deposited Pd. This results from their different physical distribution within the surface layer. For a given amount of total Pd, the higher Pd signal observed by XPS for the colloidal Pd, compared to the amount detected for the Pd(+2), shows that the colloidal Pd/Sn is more confined to the surface. A TEM micrograph of the adsorbed colloidal Pd/Sn catalyst (Figure 7) reveals that the size of the particles ranges from 10 to 50 Å, and that they cover only a fraction of the surface.

The colloidal Pd/Sn catalyst is deposited onto the surface in a two-step process, and the chemical changes in the catalyst composition which occur have also been monitored by XPS and RBS.[8] Representative XPS spectra for surfaces treated with the sensitizer only, and then treatment with the accelerator solution, are shown in Figure 8. These spectra confirm the role of the accelerator in removing the excess Sn from the sensitized surface. The surface Pd/Sn ratio increases from about 0.2 to 3.6 during this sequence. (The total Pd/Sn ratio determined by RBS for these same samples increases from 0.4 to 3.6). High-resolution scans of the Pd 3d region (Figures 9a and 9b) show that the Pd is

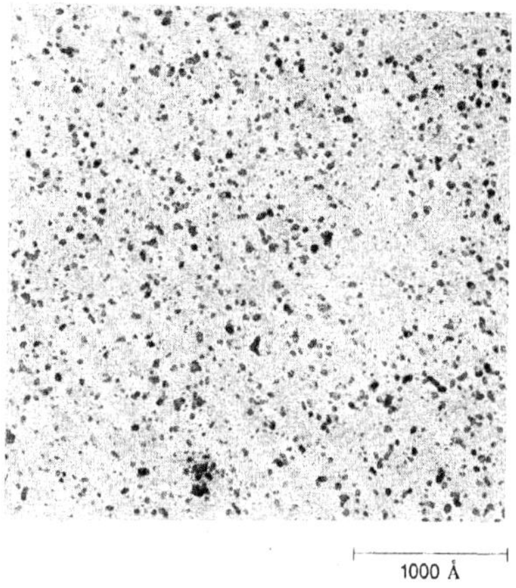

Figure 7. TEM micrograph of colloidal Pd/Sn catalyst on Ultem surface prior to metallization with copper.

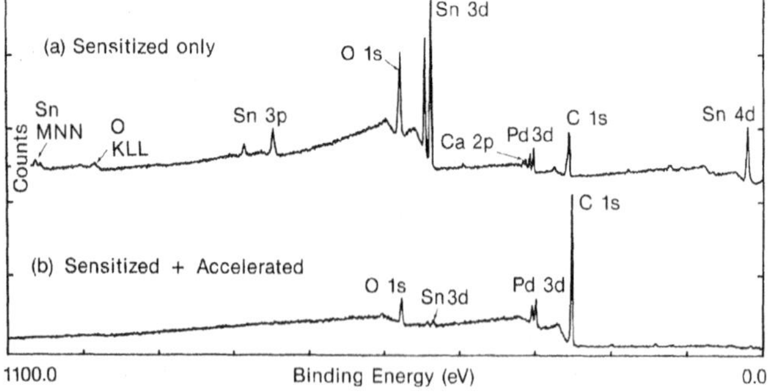

Figure 8. XPS survey spectra for Ultem surfaces treated with sensitizer and accelerator solutions.

present in a metallic form prior to and after treatment with the accelerator solution. The initial inactivity of the catalyst towards promoting electroless deposition is probably due to a layer of Sn compounds surrounding the active metallic catalyst. The Sn 3d line after sensitization (Figure 9c) shows a single component at 486.3 eV, consistent with stannic oxide or hydroxide. After acceleration (Figure 9d), the Sn signal is greatly reduced in magnitude, and displays an additional component attributed to metallic Sn. These results are consistent with previous studies[9–12] which indicate that the active metallic catalyst is a Pd/Sn alloy (Pd-rich).

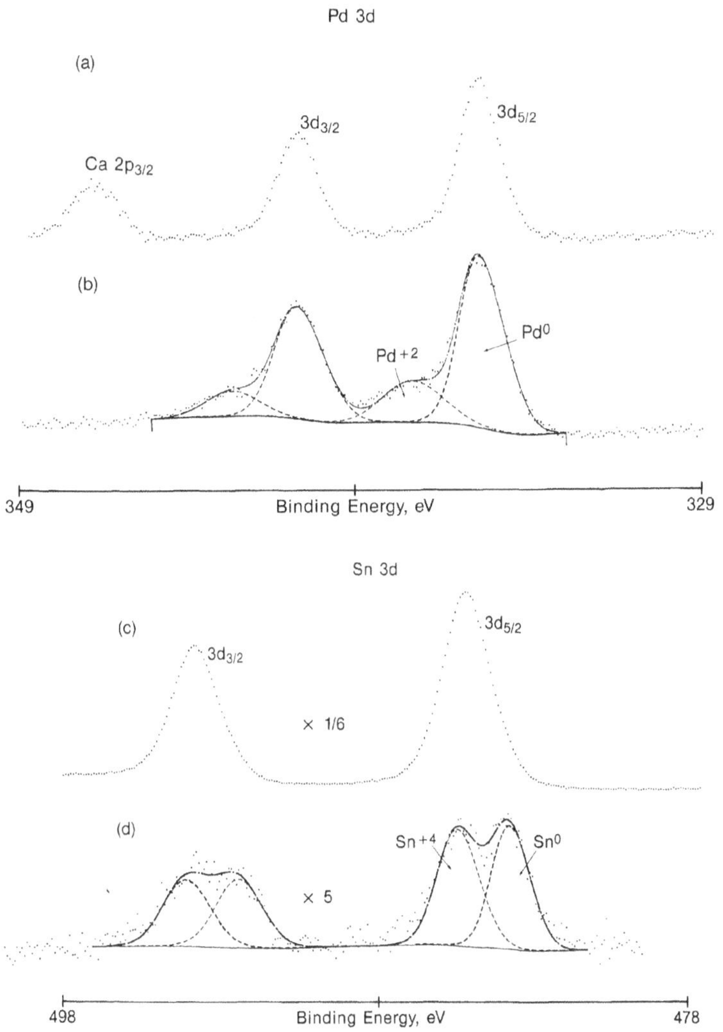

Figure 9. **High-resolution scans of Pd 3d region for surfaces treated with (a) sensitizer only, and (b) sensitizer + accelerator. The corresponding scans of the Sn 3d region are shown in spectra (c) and (d), respectively. From Reference 8, with permission of John Wiley and Sons, Ltd.**

The composition of the active catalyst deposited on various surfaces by this two-step process has been the subject of several studies. Osaka and coworkers[9] determined Pd/Sn ratios less than unity using XPS, in contrast to previous RBS measurements[10-11] which showed Pd/Sn = 3 to 7. They also reported a Pd $3d_{5/2}$ binding energy about 1 eV higher than pure Pd, which they attributed to Pd/Sn alloy formation. However, our results for catalyzed polyetherimide surfaces indicated Pd $3d_{5/2}$ binding energies identical to pure Pd, and Pd/Sn ratios of 3 to 7 which agreed with RBS results from the same surfaces.[8] Thus, the discrepancy between the previous XPS and RBS measurements is probably due to actual differences in the sample surfaces, rather than a reflection of the different sampling volume of the two techniques.

Since Pd(0) is required to catalyze electroless deposition, Pd(+2) compounds must be reduced to become active. However, in certain cases, surfaces treated with palladium(+2) compounds were active towards electroless deposition without an intervening reduction step. This occurs when the plating bath itself is capable of reducing the Pd(+2) to the active Pd(0). For example, after brief immersion (ca. 5 s) into an Enthone 402 electroless copper bath, XPS analysis clearly showed that the Pd(2+) had been reduced to Pd(0).

Failure Analysis

Surface analysis has also been used to examine surfaces of completed circuit boards which fail during final evaluation. One example is the occurrence of short circuits between copper runs on patterned circuit boards. Microscopic examination reveals no remaining copper in the etched areas, and this is confirmed by small-spot XPS analysis of the same areas. However, residual Pd and Sn (residual plating catalyst) are detected on the surface, which are not removed by the copper etchants typically employed during the patterning process. Table III shows typical values of the measured surface Pd and Sn concentrations, and the corresponding electrical resistance measured between isolated copper conductors (600-μm line/space). A new (GE proprietary) process was implemented to specifically remove the remaining Pd and Sn, without attack on the metal traces, and short circuits are no longer observed. Table IV presents the Ultem surface compositions following three of the key steps. As can be seen in Table IV, the first step oxidizes the residual Pd to the divalent state and slightly reduces its concentration. Subsequent steps solubilize the Pd(+2) ions and remove them from the surface.

CONCLUSIONS

Surface analytical studies of polyetherimide circuit boards at different stages of processing have been discussed. XPS measurements have aided in the identification of production problems such as surface contamination, which may affect the metal/polymer adhesion, and residual metallization catalyst, which is responsible for electrical leakage between runs. More fundamental studies have considered the chemical modification of the polyetherimide due to hydrolysis, and the amount and chemical form of different

TABLE III. Test Resistance versus Surface Composition, Atom %, Between Copper Runs on a Patterned Circuit Board.

Resistance	C	O	N	Pd	Sn
100 Ω	65.9	21.7	2.8	2.8	0.3
200 Ω	71.7	17.9	2.0	1.1	0.2
>20 MΩ	79.7	14.4	3.6	0.4	–

TABLE IV. Ultem Surface Composition, Atom %, Following Process Steps Used to Remove Residual Catalyst (After Copper Pattern/Etch).

Treatment	C	O	N	Pd	Sn	Others
None	45	39	–	2*	13	–
'Oxidizer'	52	35	1	1**	0.7	5
'Solubilizer'	70	21	2.5	–	0.4	4
'Cleaner'	78	16	3.1	–	–	–

* Pd $3d_{5/2}$ binding energy = 335.1, 337.2 eV (Pd^0 and Pd^{2+})
** Pd $3d_{5/2}$ binding energy = 336.6 eV (Pd^{2+})

metallization catalysts. These studies have led to a better understanding of the surface conditions during successful electroless metallization.

REFERENCES

1. T. Naegele, Electronics 63, (April 21,1986).
2. W. Jacobi and M. Kirsch, Printed Circuit Fabrication 9, 50 (1986).
3. M. O'Loughlin and S. Duensing, Electronic Manufacturing 33, 22 (1987).
4. British Plastics and Rubber 24 (February 1988).
5. D. C. Frisch, Material Engineering 34 (August 1988).
6. J. H. Scofield, J. Electron Spectrosc. Relat. Phenom. 8, 129 (1976).
7. H. J. Leary and D. S. Campbell, Surface Interface Analysis 1, 75 (1979).
8. M. C. Burrell, G. A. Smith and J. J. Chera, Surface Interface Analysis 11, 160 (1988).
9. T. Osaka, H. Takematsu and K. Nihei, J. Electrochem. Soc. 127, 1021 (1980).
10. L. G. Svendsen and G. Sorensen, in "Physicochemical Aspects of Polymer Surfaces," K.L.Mittal, editor, Vol. 2, p. 1213, Plenum Press, New York (1983).
11. R. L. Cohen and R. L. Meek, J. Colloid Interface Sci. 55, 156 (1976).
12. N. Feldstein, M. Schlesinger, N. E. Hedgecock and S. L. Chow, J. Electrochem. Soc. 121, 738 (1974).

EFFECT OF PLASMA TREATMENT AND ION-PLATING ON ADHESION OF METALLIZED POLYIMIDE AND POLY(ETHYLENE TEREPHTHALATE)

Katsuhiko Nakamae, Satoshi Tanigawa and Tsunetaka Matsumoto

Department of Industrial Chemistry, Faculty of Engineering
Kobe University
Rokkodai, Nada, Kobe 657 Japan

The adhesion of thin metal films on polymeric substrates is a very important factor in any application of metallized plastics. The effect of plastic surface treatment by O_2 plasma and R.F. ion-plating on adhesion was investigated. Polyimide [PI] and poly(ethylene terephthalate) [PET] were used. The following results were obtained: 1) When the polymeric surface was pretreated with high power O_2 plasma (the radio frequency (R.F.) power : 100W), the peel strength at the Fe/PI interface abruptly increased with increase in the treatment time. When it was pretreated with lower power O_2 plasma (the R.F. power : 15W), its peel strength increased at earlier treatment time and then kept constant. ESCA spectra of the polymeric surface treated with O_2 plasma were analyzed, and the surface density of functional groups on the polymeric surface was determined. The production of the functional groups and their surface density were correlated with the adhesion of thin metal films on polymeric substrate. 2) When the ion-plating technique was used to metallize plastics, the peel strength at the polymer/metal interface increased with increase in the R.F. power and the accelerating voltage. The deposited metal formed chemical bonding with the polymeric substrate through oxygen atoms at the interface as shown by ESCA or FT-IR results.

INTRODUCTION

The metallized plastics find applications in many fields, for example, they are used in the electronics and information storage industry as magnetic recording media for high density[1] electro-conductive films[2] and flexible printed circuits. The thin metal films are prepared on polymeric substrates by physical and chemical methods, such as vacuum deposition, sputtering and electrolytic deposition.

The metallized plastics require many properties depending on their applications. Particularly, the adhesion of thin metal films on polymeric substrates is the most important property in any application[3]. By the vacuum deposition method, thin metal films can be obtained easily. Since the adhesion of metallized plastics prepared by the vacuum deposition method is not so good, so it is important as to how this adhesion can be improved[4].

We have studied the adhesion of metallized plastics from a chemical viewpoint of the polymeric surface. Various polymers and the surface modified polymers by chemical and physical techniques (alkaline treatment, corona discharge treatment, etc.) were used. The adhesion of the thin metal films on polymeric substrates correlated with the surface energy calculated from contact angles on polymeric surfaces, especially the polar component of surface energy[5].

In this study, first the plasma treatment which is selective and is confined only to surface, was used as surface treatment. The plasma treatment produces functional groups on the surface of a polymeric substrate[4,6]. So the relation between the functional group density on polymeric surfaces produced by plasma treatment and the adhesion of the thin metal film on such surfaces was investigated.

Secondly, the thin metal films were prepared by the ion-plating technique. The effects of radio frequency (R.F.) power and accelerating voltage during the deposition on the adhesion at the polymer/metal interface were investigated. Ion-plating is the technique in which the metal vapor possesses high kinetic energy[7,8]. So there should be chemical or physical interaction at the interface formed by the ion-plating technique.

EXPERIMENTAL

Materials

Poly(ethylene terephthalate) (PET) and polyimide films were used as polymeric substrates. Polyimide was poly[N-(oxy-1,4,-phenylene)-N'-(1,4-phenylene)pyromellitimide[produced by Du Pont Co. Ltd. (Kapton). Cobalt and iron ingots used for depositing thin films had a purity of 99.8% each.

Plasma Treatment, Vacuum Deposition and Ion-plating

Figure 1 shows the apparatus used for plasma treatment, vacuum deposition and ion-plating. In this apparatus, it is possible to carry out the metal deposition following the plasma treatment without exposure to air[9].

Plasma treatment was carried out at 1.3Pa (10^{-2}torr) evacuated by a rotary pump and an oil diffusion pump. The plasma was generated by the application of radio frequency (R.F.) electric field of 13.56MHz to the inner R.F. coil. The treating gas was introduced into the system by using a variable leak valve.

After the plasma treatment, the systems was evacuated to a pressure of 6×10^{-4}Pa(5×10^{-6}torr), and the vacuum deposition was carried out with electric beam (beam:voltage 4kV, emission current 150mA). The deposition rate was about 1.0-1.5 nm/s, and the thickness of the deposited thin metal film was 100 nm. The thickness of the deposted metal film was measured by optical interference microscope (λ =273.05 nm) and scanning electron microscope.

In the ion-plating, the vacuum deposition was carried out under the R.F. electric field (0-100W) and the accelerating voltage (0-1000V).

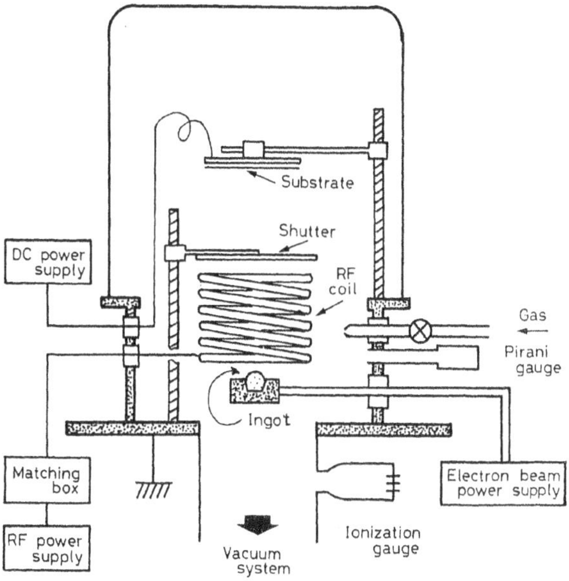

Fig. 1. Schematic diagram of the apparatus for plasma treatment, vacuum deposition, and ion-plating.

Measurement of Adhesion at polymer/metal interface

Adhesion of the deposited thin metal films on polymeric substrates was measured by the 90° peel test using a tensile tester (Shimazu Autograph IS-100) at the tensile rate of 200 mm/min and 20°C. The test specimen were bonded to the steel plate by placing the metal deposited face in contact with epoxy adhesive. Fig. 2 shows the schematic of the peel test.

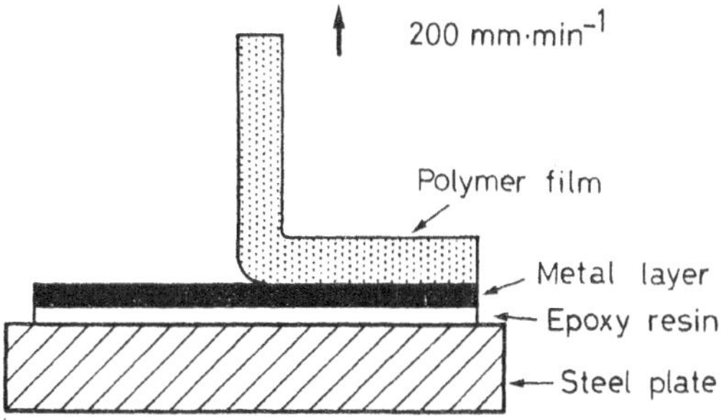

Fig. 2. Schematic representation of the peel test.

Characterization of Polymer Surfaces

Contact angles of water on the polymeric surfaces were measured by the droplet method. The contact angles were determined from advancing and receding contact angles[5,10-12], because the surface roughness of the polymers may be altered by the surface treatment. If the surface is rough, the advancing contact angle is distinct from the receding contact angle. The advancing contact angle (θ_a) was measured while the droplet was getting bigger, and receding contact angle (θ_r) was measured while the droplet was getting smaller. These contact angles were calculated from equation (1): h, height of droplet; x, diameter of droplet.

$$\theta_a, \theta_r = \begin{cases} 2\tan^{-1}(2h/x) & [\ \theta \leq 90^o\ ] \\ 90^o + \cos^{-1}\{x.h/(h^2+(x/2)^2\} & [\theta > 90^o] \end{cases} \qquad \ldots (1)$$

The contact angle was calculated from equation (2) with θ_a and θ_r.

$$\theta = \cos^{-1}\{(\cos\ \theta_a + \cos\ \theta_r)/2\} \qquad \ldots(2)$$

FTIR-ATR spectra were recorded with Nicolet model 7199: Internal reflection element (IRE): KRS-5; Integration number: 500. Polymeric film with metal of thickness 5 nm was measured.

C1s, O1s, N1s and F1s ESCA spectra of polymeric surfaces were recorded by XSAM800 ESCA system (KRATOS): X-ray: Mgk$_\alpha$ 12.5kV, 12mA (150W), 3×10^{-7} Pa (2×10^{-9} torr). Curve fitting of the spectra and the F/C ratio of surface atomic concentration were calculated with DS800(DEC) systems. Calibration of this apparatus was done by using a C-C bond with the aromatic ring of 285 eV. Polymeric surfaces were chemically modified before ESCA measurement. This modification was carried out according to the method of Everhart et al[13]. Fluorine atoms are detected more sensitively by ESCA measurement.

Reaction of TFAA : The 1x2 cm polymeric samples were immersed in the solution containing 1 ml trifluoroacetic-anhydride (TFAA) (0.4M) and 1 ml of pyridine in 15 ml of benzene for 1.5 hours at 25 ^oC. Non-reacted TFAA was removed by washing in pure benzene.

Reaction of PFBA : The 1x2 cm polymeric samples were immersed in the solution containing 300 μl of pentafluorobenzaldehyde (PFBA) in 15 ml of hexane for 2 hours at 30 ^oC. Non-reacted PFBA was removed by washing in pure hexane.

RESULTS AND DISCUSSION

Polymeric Substrate Treated by O_2 Plasma

The effect of O_2 plasma treatment of polyimide (PI) surface on the adhesion of the vacuum deposited thin iron film was investigated. Fig. 3 shows the relationship between the 90° peel strength for the Fe/PI system and the exposure time of O_2 plasma at the PI surface as pretreatment before metal deposition. When the R.F. power during plasma treatment was 100W, the peel strength increased remarkably with the increase in the exposure time of O_2 plasma. The cohesive fracture of polyimide in the Fe/PI system was macroscopically observed when the O_2 plasma was applied for 300 seconds. At the R.F. power 15W, the peel strength increased at short exposure time; however, it became constant at longer exposure times.

In general, the polymeric surfaces are oxidized and several hydrophilic functional groups are generated through the plasma treatment[6]. We have studied the adhesion of the deposited thin metal films on polymeric surfaces. It was shown that the adhesion correlated primarily with the surface energy of the polymeric substrate[5]. The effect of the plasma treatment on the adhesion was investigated on the basis of the chemical interaction.

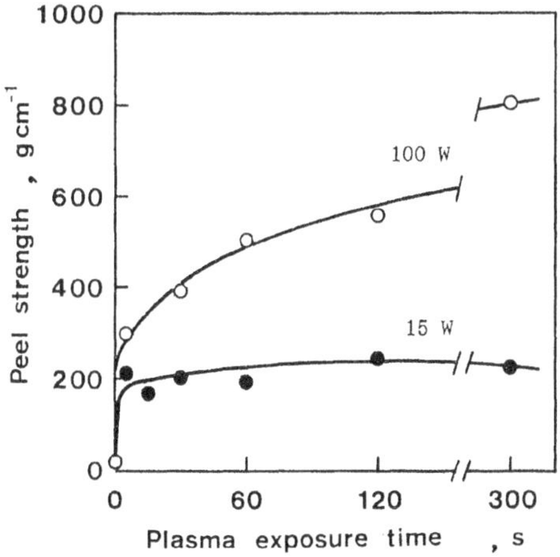

Fig. 3. Effect of O_2 plasma exposure time of polyimide on the adhesion of thin vacuum deposited iron film. R.F. power: O, 100W; ●, 15W.

The wettability of PI surface treated with O_2 plasma was examined using water. Fig. 4 shows the relationship between the contact angle of water on the PI surface and the exposure time of O_2 plasma. The contact angle of PI surface decreased in a short treatment time (5-15s) irrespective of R.F. power. Such decrease of the contact angle is caused by the introduction of hydrophilic groups onto the PI surface through the breakage of polymer chain by O_2 plasma. At longer exposure times, the contact angle decreased further at R.F. power 15W; however, it tended to increase at 100W of R.F. power. The PI surface treated with O_2 plasma at higher power could be much rougher than that treated at lower R.F. power, as the difference in the advancing contact angle and the receding contact angle increased with the increase of the treatment power for O_2 plasma. But these difference could not be observed by scanning electron microscope. From these results, for PI treated with O_2 plasma, no correlation was observed between the adhesion and the wettability of water on the polymeric surface.

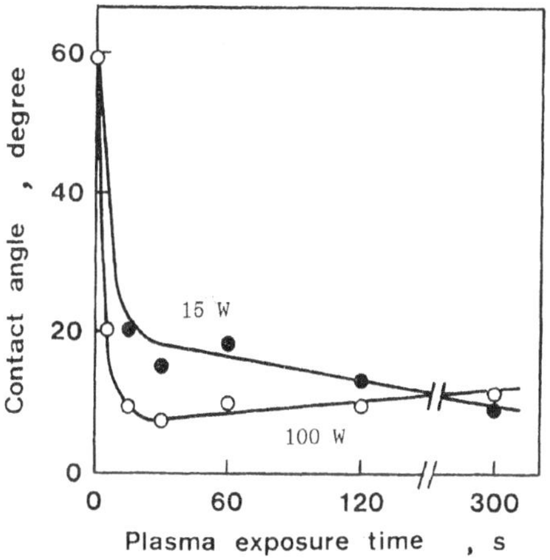

Fig. 4. Effect of O_2 plasma exposure time on the contact angle of water on polyimide R.F. power: O, 100W; ●, 15W.

One would think that the adhesion depends on the interaction between the functional groups at the polymeric surface and the thin metal film. The surface property of a polymer can be evaluated qualitatively from the contact angle measurements. ESCA is superior technique for quantitative analysis of the polymeric surface property. The functional groups density at the PI surface treated by O_2 plasma was determined by ESCA.

Figure 5 shows the results of the curve resolution of Cls spectra at the PI surface treated with O_2 plasma. The spectrum calculated from the chemical structure is shown with dotted line. As a result of the plasma treatment, the peaks assigned to the carbonyl, ether group or the carbon bonded with nitrogen decreased, and the peaks assigned to the peroxide, hydroxyl or amino groups newly appeared.

The curve fitting technique for ESCA is a useful method for the analysis of chemical structure at a material surface. Since the polymer is an insulater, it is difficult that the functional group at a polymeric surface can be identified with only the curve fitting technique. Because the chemical shifts of the Cls spectra by the functional groups were small and these spectra had broad peaks.

The chemical modification is one of the best methods for the identification of functional groups at a polymeric surface. TFAA was used as the reagent to detect peroxide and hydroxyl groups, and PFBA was used to detect amino group[13-16]. Poly(vinyl alcohol) film reacted with TFAA, and poly(acrylic amide) film reacted with PFBA. Cls ESCA spectra of these films were recorded. Hydroxide group reacted with TFAA and the amino group reacted with PFBA. The efficiency of reaction was over 95%.

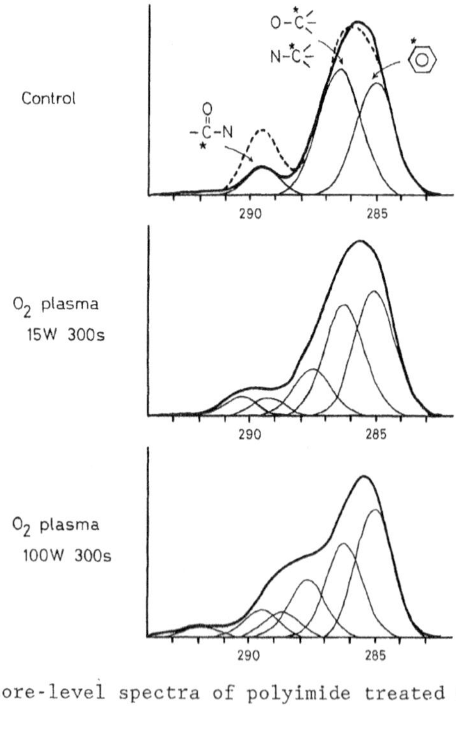

Fig. 5. Cls core-level spectra of polyimide treated with O_2 plasma.

Figure 6 show the effect of the O_2 plasma treatment time on the ratio of the atomic concentrations of fluorine and carbon, F/C, as calculated from the ESCA spectra. The F/C represents the amount of reagent reacted with the functional group at the polymeric surface. The F/C increased abruptly with the increase in the O_2 plasma treatment time at high R.F. power. At low R.F. power, the F/C increased first with the increase in the treatment time, however, it became constant afterward. It is clear that the functional groups, such as peroxide, hydroxyl, amino were produced at the PI surface by the O_2 plasma. The change in the density of the hydrophilic groups at the polymeric surface which was not detected by the contact angle technique was recognized by ESCA.

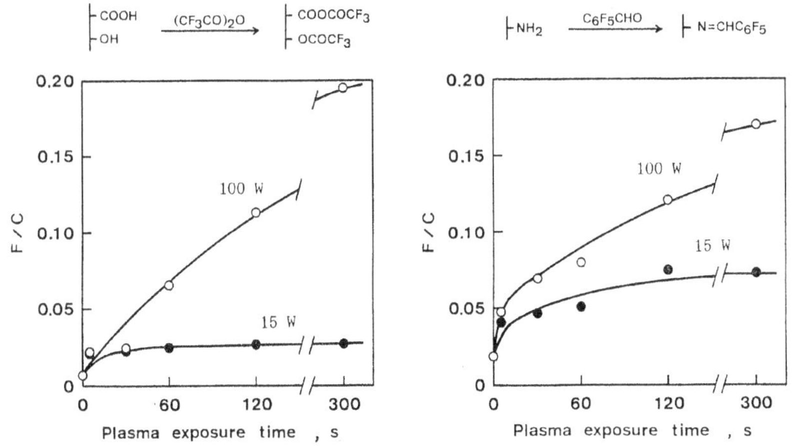

Fig. 6. Relationship between F/C ratio of atomic number at the PI surface after derivatization determined with ESCA and the plasma exposure time. R.F. power: ○, 100W; ●, 15W.

Figure 7 shows the relationship between the peel strength for the Fe/PI system and the F/C measured by ESCA. The F/C ratio was calculated from the sum of the F/C obtained from TFAA modification and PFBA modification. The adhesion correlated primarily with the F/C. So it is shown that the adhesion of vacuum deposited thin metal film on a polymeric substrate is strongly influenced by the density of hydrophilic groups at the polymeric surface. This may depend on the chemical bonding between the deposited metal oxide and the functional groups at the polymeric surface.

Ion-plating

Figure 8 shows the effect of the accelerating voltage during ion-plating technique on the 90° peel strength of thin film of cobalt deposited on untreated PET film. Irrespective of the R.F. power during deposition, the peel strength increased with the increase in the accelerating voltage. Cohesive fracture of PET film was observed for the sample with high peel strength. The adhesion increased slightly with the increase in the accelerating voltage without the R.F. power. In this case, the metal vapor was ionized by using an electron beam as a heating source.

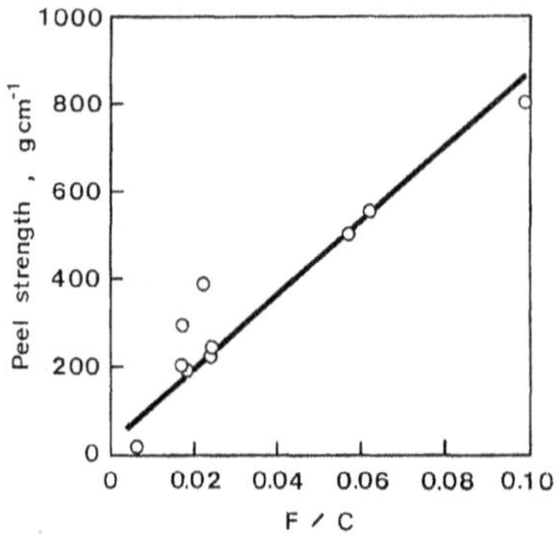

Fig. 7. Relationship between the peel strength and F/C ratio of atomic number at the PI surface after derivatization.

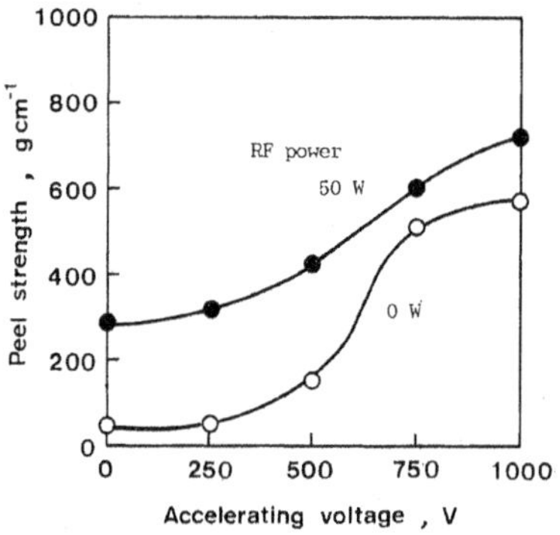

Fig. 8. Relationship between the peel strength of cobalt ion-plated PET film and the accelerating voltage. O , Supplying only accelerating voltage; ●, Supplying R.F. power (50W) and accelerating voltage.

Figure 9 shows the effect of the R.F. power during ion-plating on the 90° peel strength of Co/PET system. Irrespective of the accelerating voltage, the peel strength increased with the increase in the R.F. power. From this result, it is shown that the adhesion improves when the metal vapor is ionized positively only. Such superior adhesion obtained by ion-plating depends not only on physical effect such as the ionized metal diffused into the PET film, but also on the chemical bonding between the ionized metal and the PET film.

The chemical structure of the Co/PET interface prepared by ion-plating (R.F. power : 50W, accelerating voltage : 1000V) was analyzed by FTIR and ESCA.

Figure 10 shows the FTIR-ATR spectra of ion plated cobalt on PET film and the vacuum deposited cobalt on PET film. In the spectra of the ion-

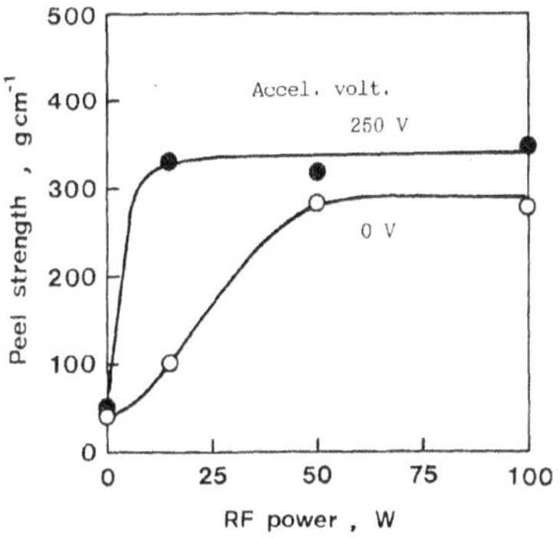

Fig. 9. Relationship between the peel strength of cobalt ion-plated PET film and the R.F. power.

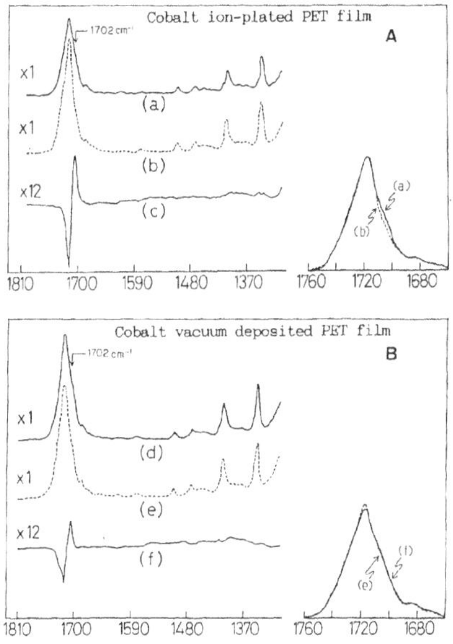

Fig. 10. FTIR-ATR spectra. (a) and (d) cobalt deposited PET film (the thickness of cobalt layer, 0.1 μm), (b) and (e) raw PET film, (c) and (f) subtracted spectra. [A]: Cobalt ion-plated PET film, [B]: Cobalt vacuum deposited PET film.

-plated cobalt on PET film, the peak at $1717cm^{-1}$ assigned to carbonyl group decreased, and the peak at $1702cm^{-1}$ appeared. The cobalt atoms at the interface interact strongly with the carbonyl groups at the PET surface through the ion-plating. One would think that such a chemical interaction at the interface should improve adhesion.

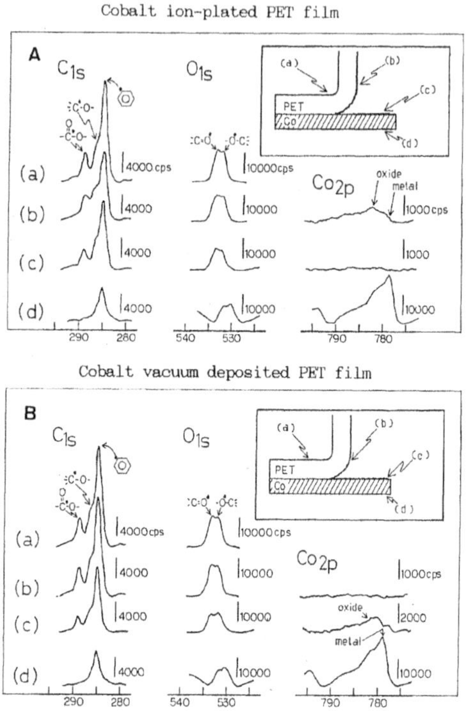

Fig. 11. ESCA spectra. (a) raw PET film, (b) PET film surface of peeled interface. (c) Cobalt layer surface of peeled interface, and (d) Ar^+ etched cobalt surface. [A]: Cobalt ion-plated PET film, [B]: Cobalt vacuum deposited PET film.

Figure 11 shows the ESCA spectra of PET and cobalt film surface after peeling. Because of good adhesion by the ion-plating, the cohesive fracture in PET film was observed. The Co2p ESCA spectra were recorded through the cobalt particles diffuse into the PET film at the peeled interface. Also the thin cobalt film fractured at the interface. The Co2p spectra at the PET surface after peeling of ion-plated cobalt had mainly the peak at 781 eV, that is, of cobalt oxide. (The peak for pure cobalt was at 779 eV, when the thin cobalt film was etched by argon ion-beam). The cobalt deposited by ion-plating at Co/PET interface formed metal oxide. Therefore, the adhesion was improved because of strong chemical interaction between the carbonyl and other functional groups at the polymeric surface and the metal oxide of the deposited metal film.

CONCLUSION

The effect of plasma pre-treatment of polymer surface on the adhesion of vacuum deposited Fe thin film was investigated. The peel strength at the Fe/PI interface increased with increase in the surface density of functional groups produced at the polymeric surface by the O_2 plasma pre-treatment. The adhesion of ion-plated Co thin film on PET was also investigated. The peel strength at the Co/PET interface increased with increase in R.F. power and the accelerating voltage during the deposition. The chemical bonding at the Co/PET interface during the ion-plating was observed by FTIR and ESCA.

REFERNCES

1. S. Iwasaki, IEEE Trans Magn., MAG-16, 71 (1980).
2. F. Soeda, K. Hayashi and A. Ishitani, J. Electron Spectrosc., 27, 205 (1982).
3. K.L. Mittal, J. Vac. Sci. Technol., 13, 19 (1976).
4. T. Tsukada and N. Hosokawa, ibid, 16, 348 (1979).
5. K. Sumiya, T. Taii, K. Nakamae and T. Matsumoto, J. Adhesion Soc. Japan, 18, 345 (1982).
6. J.M. Burkstrand, J. Vac. Sci. Technol., 16, 1072 (1979).
7. K. Nakame, T. Miyata, T. Taii, K. Yamaguchi, K. Sumiya and T. Matsumoto, J. Adhesion Soc. Japan, 22, 299 (1986).
8. K. Nakamae, K. Yamaguchi and T. Matsumoto, ibid, 22, 347 (1986).
9. Y. Murayama, M. Matsumoto and K. Kashiwagi, Oyo-Butsuri, 43, 687 (1974).
10. F.M. Fowkes, J. Phys. Chem., 66, 328 (1962); ibid, 67, 2538 (1963).
11. D.K. Owens and R.C. Wendt, J. Appl., Polym. Sci., 13, 1741 (1961).
12. T. Kitazaki and T. Hata, J. Adhesion Soc. Japan, 8, 131 (1972).
13. D.S. Everhart and C.N. Reilley, Anal. Chem., 53, 665 (1981).
14. D.S. Everhart and C.N. Reilley, Surface Interface Anal., 3, 126 (1981).
15. L.J. Gerenser, J.F. Elman, M.G. Mason and J.M. Pochan, Polymer, 26, 1162 (1985); J.M. Pochan, L.J. Gerenser and J.F. Elman, Polymer, 27, 1058 (1986).

ADHESION OF CHROMIUM, NICKEL, AND COPPER TO POLYMERS CONTAINING BASIC FUNCTIONAL GROUPS

S. R. Cain, L. J. Matienzo, and F. Emmi

IBM Systems Technology Division
1701 North Street
Endicott, NY 13760

Surface acidity of chromium, nickel, and copper and their oxides was investigated by performing extended Hückel type band calculations on these materials. Interactions with basic functional groups encountered in polymers were investigated by studying the adsorption chemistry of ammonia on these metal (metal oxide) surfaces. Results of the calculations were analyzed in terms of the density of states, the total energy, and overlap populations.

Results indicated that nickel and chromium oxides are much more acidic (in the Lewis sense) than the clean metals, owing to the decrease in metal orbital energy with increased oxidation state. Copper oxide, while more acidic than clean copper, was found to be no more acidic than clean chromium or clean nickel. Considering the metal oxides and the clean metals as two separate groups, the acidity was found to decrease going from chromium- to nickel- to copper-based substrates. Absolute ordering of acidity for all the systems studied is,

$$Cr_2O_3 > NiO > Cr \approx Ni \approx Cu_2O > Cu$$

The chemical component of adhesion is expected to follow the same trend, where the polymer of interest contains mainly basic functionalities (e.g. amino groups).

INTRODUCTION

Interfacial bonding of dissimilar materials is of crucial concern in a number of industries, including printed circuit manufacture. Metals which are used often in making printed circuits include chromium, nickel, and copper. Due to its high conductivity, copper is the metal of choice for circuit lines, but frequently does not bind well to polymers. Hence, a thin layer of chromium (or other suitable metal) is used as an adhesion promoter. Chemical bonding, mechanical interlocking, electrostatic attraction, and diffusion phenomena all contribute to adhesion[1]. This paper deals exclusively with the chemical component of adhesion.

Since acid-base chemistry is pertinent to description of interfacial bonding[2,3], evaluating the surface acidity of a substrate to which a polymer may be bound is of interest. For example, such comparisons have been carried out and summarized for metal oxides in aqueous media[4], but these are limited to the Bronsted-Lowry acid-base concept. Recently, quantum chemical calculations have been used in describing surface acidity in terms of the Lewis acid-base concept[5]. This approach is promising for evaluating acid-base interactions in non-aqueous systems (e.g. where a thin layer of metal is vacuum deposited onto a polymer), and was used in this study.

In studying the chemical component of adhesion, proper choice of model systems is essential. Since polymers commonly used in printed circuits contain functional groups with electron lone pairs which are available for bonding, a suitable model for these functional groups is some "generic" base: ammonia was used in this treatment. Knowledge of the metal surface chemistry is also important. An active metal surface (e.g. chromium) is very likely to be oxidized to some degree owing to small concentrations of oxygen in the deposition chamber[6]. For example, chromium has been shown[7] to oxidize readily at 30°C in an oxygen atmosphere as low as 10^{-7} Torr. Thus, one must understand the interfacial chemistry of the metal oxide as well as that of the clean metal.

Although small molecular adsorbates in the gas phase, typical of those used in high vacuum adsorption studies, possess high mobility not characteristic of functional groups in polymers, high vacuum studies provide useful insights into the interactions which govern chemical bonding. Adsorption and decomposition chemistry of ammonia has been the subject of several studies[8-16]. In an early study of ammonia adsorption on chromium oxides[8], infra-red spectra indicated that ammonia binds via electron lone pair donation to Lewis acid centers, chromium in the 3, 4, and 6 oxidatio states. In a later study of Cr_2O_3-MoO_3 systems, surface acidity was relat ed to the electronegativity of the chromium atoms[9]. Ammonia forms severa fragmented species on nickel even at temperatures as low as 190°K[10-13]. Dissociation to nitrogen and hydrogen is complete at high temperature on nickel[13] and chromium[14]. Even on copper and copper oxide, ammonia readily fragments into hydrogen and NH_2 species, but does not molecularly adsorb[1] However, dissociation may be due to photon or electron radiation used for in-situ analyses[16]. High vacuum adsorption studies are difficult to execute, and the results often are confusing; but from these studies, it seems clear that molecular ammonia adsorbs onto chromium and nickel based substrates, but only weakly (if at all) to copper based substrates.

Previous theoretical studies[17] provided insight as to appropriate geometries to consider. Due to anti-bonding interactions between the meta and the electron lone pair in ammonia-like adsorbates, the adsorbate attaches directly atop a surface atom rather than in a bridging position[17a] This effect is marked for transition metals with the 3D band completely filled (e.g. copper), but rather small for metals with the 3D band only partially occupied (e.g. chromium). For purposes of one-to-one comparisons, ammonia was assumed to attack directly atop a surface metal atom. Though simple binding schemes tend to exclude the 4S and 4P orbitals[18], these orbitals have been shown to play a dominant role in the surface adsorption[17b], and have been included in our bonding scheme.

Two factors govern acidity of a molecule or atom: 1) the energy of the empty orbitals, and 2) the number of empty orbitals available for bonding with a Lewis base. An empty orbital is acidic if it is low in energy. (For example, the H^+ ion has an empty orbital at -13.6 eV, and very acidic.) The two criteria mentioned above may be expressed in term of the density of states (DOS) and the Fermi energy (ε). An acidic surface will have ε at low energy, and a high DOS above ε. This is another way of saying that the strongest acids have many empty orbitals at low energy.

This paper deals with the chemical component of adhesion of polymers to metal and metal oxide surfaces, where ammonia has been used to model basic (in the Lewis sense) functionalities on the polymer. Though the decomposition reactions discussed above play a role in the interfacial chemistry, they are not considered here in the interest of clarity. This may be justified (at least partially) on the basis of the limited mobility of the "adsorbate" when incorporated into a polymer. Functional group decomposition is an interesting topic for later studies.

TECHNIQUE FOR PERFORMING BAND CALCULATIONS AND ANALYSES

Band calculations were carried out on model systems using NEWBAND1[19], which does tight binding calculations at the extended Hückel level[20,21]. Though extended Hückel type calculations do not yield very reliable results on an absolute basis, the calculations are useful in determining trends, if the output is analyzed thoroughly. Results were interpreted using density of states (DOS) analyses from which the Fermi level (ε) as well as the total energy and overlap populations were determined. Details are given[5,22,23] elsewhere. Atomic parameters used in the extended Hückel calculations are given in Table I. The difference between metal orbital energies in the oxides versus the clean metals arises from positive charging of the metal upon oxidation[24]. In this treatment, only chemical bond-

Table I. Orbital Parameters

Orbital		Exponent (Å^{-1})	H_{ii} (eV)
H	1S	1.3	-13.6
N	2S	1.95	-26.0
	2P	1.95	-13.4
O	2S	2.275	-32.3
	2P	2.275	-14.8
*Cr	4S	1.7	-6.6 [-8.7]
	4P	1.7	-3.5 [-5.2]
	3D	4.95 (0.506)	-7.8 [-11.2]
		1.80 (0.675)	
*Ni	4S	2.1	-7.5 [-9.2]
	4P	2.1	-3.9 [-5.2]
	3D	5.75 (0.568)	-10.9 [-13.5]
		2.00 (0.629)	
*Cu	4S	2.2	-7.7 [-11.4]
	4P	2.2	-3.9 [-5.2]
	3D	5.95 (0.593)	-11.8 [-14.0]
		2.30 (0.574)	

* For the metals, two Slater type functions were used for each 3D orbital. The coefficient of the function is given in parentheses by the corresponding exponent. Different sets of metal atom H_{ii}s were used in calculation of the clean metals and metal oxides. The H_{ii}s used for the metal oxides are given in brackets.

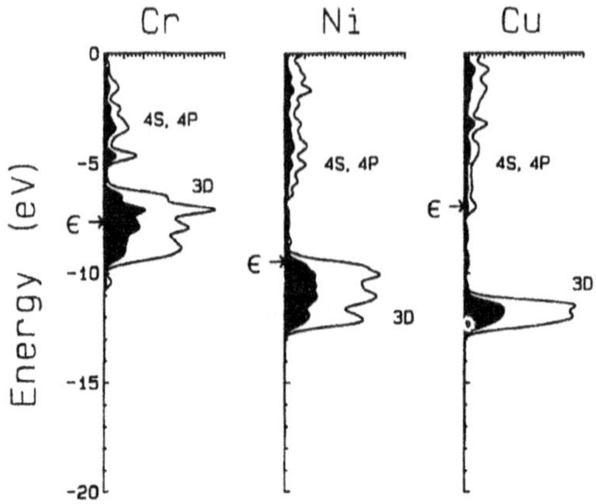

Figure 1. DOS for the clean metal slabs. The total DOS is indicated by the line, while the projected DOS for the surface atoms is indicated by the shaded area.

ing was addressed; no attempt was made to account for Van der Waals forces

Each system was modeled by a two-dimensional slab of material. The resulting two-dimensional reciprocal space was sampled with a set of nine wave vector points, chosen as prescribed by Monkhorst and Pack[25]. Components of the wave vectors were permuted through the sequence, $0.166\pi/a$, $0.50\pi/a$, $0.834\pi/a$; the Brillouin zone runs from $0-\pi/a$ in the "reduced zone" representation[26], where a is the size of the unit cell. The metal (100) crystallographic plane and the metal oxides were modeled by a three- or four-layered slab, which gives a reasonable accounting for the contribution from bulk material[27,28]. Experimentally determined geometries were used for the metals[29] and metal oxides[30]. No accounting was made for defects in the crystal.

In order to probe the acid-base interactions explicitly, ammonia was allowed to approach the surfaces directly above a surface metal atom. An idealized geometry of ammonia was assumed; N-H bond lengths of 1.01 Å and H-N-H bond angles of 109.47° were used. Only the metal-ammonia separation was varied; no other bond lengths or angles were relaxed. Subsequent bonding analyses were performed with the ammonia molecule fixed 2 Å above the surface.

RESULTS AND DISCUSSION

Clean Metal Surfaces and Ammonia Adsorption on the Clean Metals

Understanding the acid-base nature of clean metal surfaces is a convenient starting point in this study. DOS plots for metallic chromium, nickel, and copper are given in Figure 1. Qualitative features in the DOS plots are observed by valence band XPS[31-33]. The DOS for surface atoms follows closely the total DOS. Each has a prominent band derived from the 3D orbitals. Above the 3D band are the broad 4S and 4P bands. As shown in the figure, ε of chromium is near the middle of the 3D band, ε of nickel is near the top of the 3D band, and ε of copper is well above the 3D band. Therefore chromium has the largest number of empty orbitals with which a basic adsorbate can interact, followed by nickel, followed by copper. Because of incomplete screening[34], the 3D orbitals decrease in energy

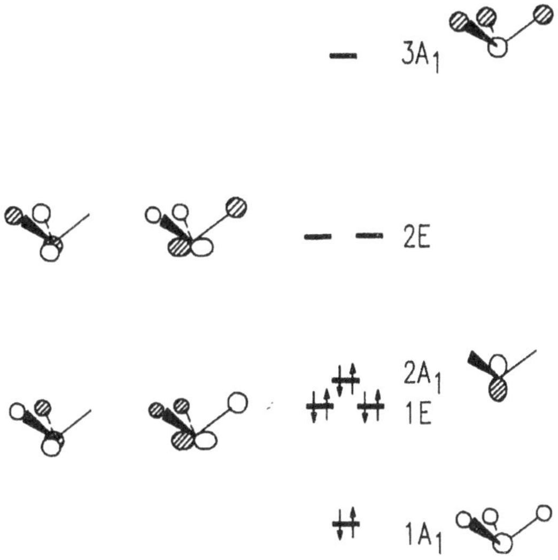

Figure 2. Valence orbitals of ammonia and their symmetry labels.

along the chromium to copper series. This decrease accounts for ε of nickel being lower than ε of chromium. In this sense, nickel is a stronger acid than chromium, but, as stated above, chromium has more vacant orbitals with which to interact with a base. Copper, with one more electron per atom than nickel, fills the 3D band and partially fills the 4S and 4P bands. The Fermi level of copper is higher than that of either nickel or chromium, and the DOS immediately above ε in copper is rather small. Thus copper is weakly acidic by virtue of both the high ε, and the small number of vacant orbitals which can "accept" electron lone pairs. Surface acidity for the clean metals should follow the trend,

$$Cr \approx Ni > Cu$$

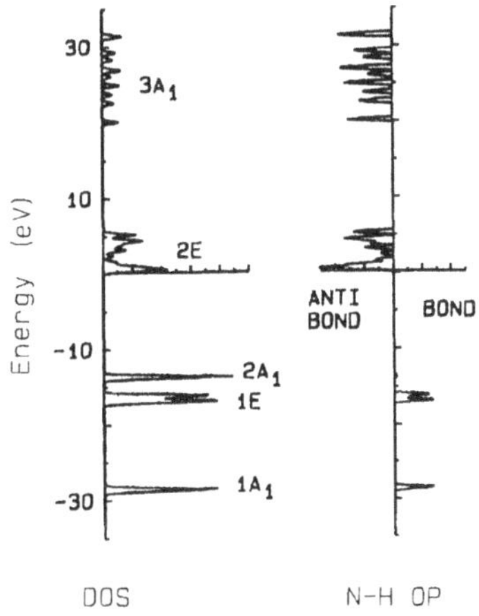

Figure 3. Total DOS for a monolayer of ammonia (left), and differential overlap population curve for the N-H bonds.

Surface interactions were probed by allowing ammonia, a model Lewis base, to approach the metal with the electron lone pair pointed directly at the metal surface. The valence orbitals of ammonia are sketched and ordered by increasing energy in Figure 2. Symmetry labels are numbered according to a scheme where only the valence orbitals are considered. Of these orbitals, the $2A_1$ (lone pair) dominates the metal-ammonia bonding. Other orbitals are not oriented for strong metal-ammonia interaction in the assumed configuration. To avoid excessive ammonia-ammonia interactions, a coverage of 1/2 was assumed. Total DOS and differential overlap population curves for a monolayer of ammonia are given in Figure 3. (In the monolayer, ammonia molecules were placed in a square lattice, with a separation of 3.6 Å.) The bands in the DOS plot are very narrow, save for the 2E and $3A_1$, which are unfilled, and too high in energy to figure in the ammonia chemistry. Narrow bands indicate small inter-ammonia interactions, hence these interactions did not complicate the analyses. The differential overlap population curve for N-H bonds shows the $1A_1$ and 1E bands to be strongly bonding, the 2E and $3A_1$ bands to be strongly antibonding, and the $2A_1$ (lone pair) band to be non-bonding. This is completely analogous to a single ammonia molecule.

Metal-ammonia adsorption strength may be inferred from the total energy versus metal-N separation curve. These curves are shown in Figure 4, which gives the energy relative to that for the metal-N separation equal to 4 Å. From these curves, ammonia may be expected to bind to both chromium and nickel, but not to copper, as suggested by the acidity trend given above. Since the energy minimum for the ammonia-chromium and ammonia-nickel systems occurred with a metal-nitrogen separation of about 2 Å, subsequent bonding analyses were performed using that separation.

A more detailed description of the bonding at the metal-ammonia interface is in order. Consider, as a representative example, the chromium-ammonia system. Figure 5 gives the total DOS plots for clean chromium, ammonia adsorbed on chromium, and the single monolayer of ammonia. As shown in the figure, the chromium-ammonia system retains much of the characteristics of the separate chromium and ammonia monolayer systems. The ammonia $2A_1$ band is pushed down by about 0.5 eV, and the bands derived from surface chromium atoms, the 4S, 4Pz, and $3Dz^2$, in particular, are

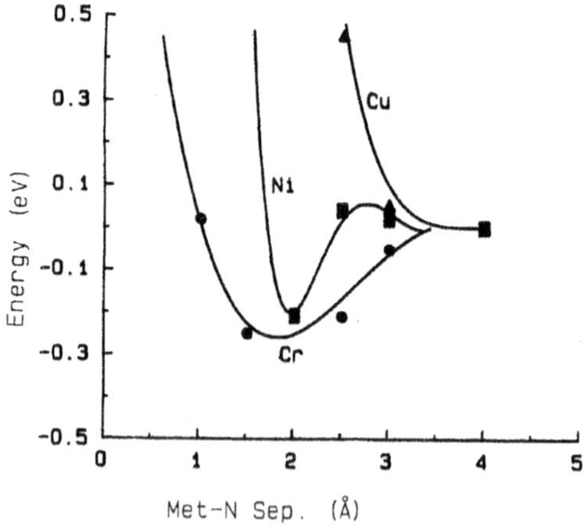

Figure 4. Total energy as a function of metal-N separation. The energy is taken relative to a metal-N separation of 4 Å. (The energies for Ni-N separations of 1.5 Å and 1.0 Å lie outside the energy range of the figure, hence those points are not shown.)

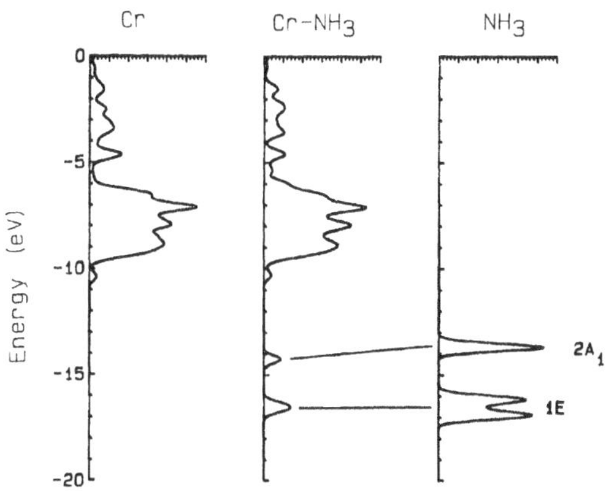

Figure 5. Total DOS plots for the chromium slab (left), chromium slab with ammonia adsorbed on one side (middle), and the non-adsorbed monolayer of ammonia (right). Peak heights of the 1E and $2A_1$ differ from the right and middle plots because different scales were used for clarity of presentation. Of the ammonia orbitals, only the $2A_1$ and 1E are in the energy range of interest.

pushed up in energy. This is analogous to interaction of ammonia with a single chromium atom; the simplified orbital diagram of which is shown in Figure 6. The ammonia lone pair orbital, pushed down in energy, mixes the chromium 4S, 4Pz, and $3Dz^2$ orbitals in a "bonding" manner; while the chromium orbitals, pushed up in energy, mix the ammonia lone pair orbital in an "anti-bonding" manner. In this scheme, the only anti-bonding component arises from the singly occupied chromium $3Dz^2$ orbital. The chromi-

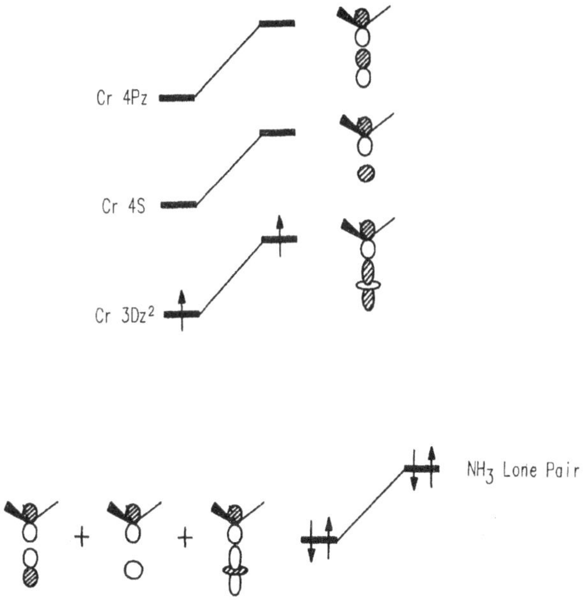

Figure 6. Important orbital interactions for a single chromium atom bound to an ammonia molecule.

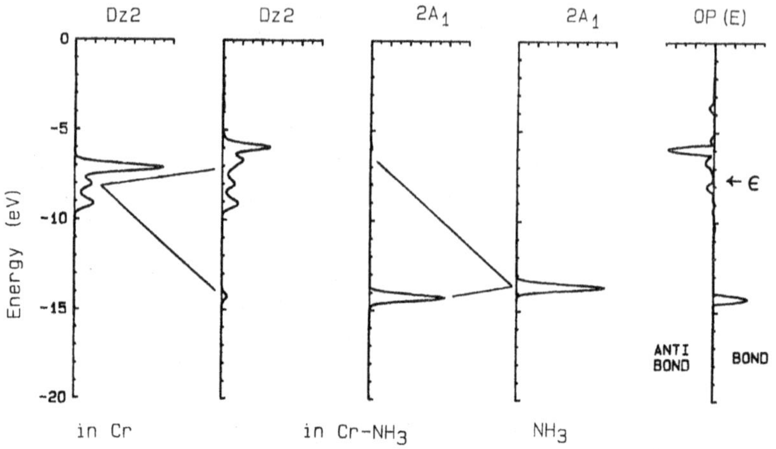

Figure 7. Projected DOS plots for a surface chromium $3Dz^2$ orbital in clean chromium (left) and with ammonia adsorbed on the surface (second from the left); for the $2A_1$ orbital of adsorbed ammon (third from the left) and the free ammonia monolayer (fourth from the left); and the $3Dz^2$-$2A_1$ differential overlap populat curve (right).

um-ammonia system becomes completely "bonding" if the $3Dz^2$ electron can transferred to some low-lying empty orbital.

The simple bonding scheme discussed above describes the interfacial chemistry of the chromium-ammonia system well. Consider, for example, t $3Dz^2$ orbital of a surface chromium atom interacting with the $2A_1$ (lone pair) orbital of adsorbed ammonia. Projected DOS plots for these orbita

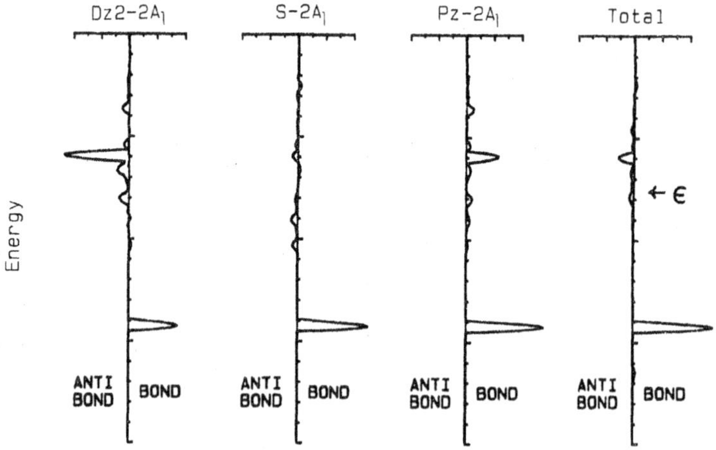

Figure 8. Differential overlap population curves for the electron lone pair of adsorbed ammonia interacting with chromium $3Dz^2$ (lef 4S (second from the left), 4Pz (third from the left) orbital and the net chromium-ammonia differential overlap population (right).

are given in Figure 7. Adsorption places the chromium $3Dz^2$ band at higher energy, and the ammonia $2A_1$ band at lower energy. (Owing to differences in peak heights, mixing of the $2A_1$ orbital into the $3Dz^2$ band does not show up in the projected DOS plots given in Figure 7, but analyses with an expanded scale showed that such mixing occurs.) Lines connecting peaks in the DOS plots are suggestive of a standard molecular orbital interaction diagram. The chromium $3Dz^2$-ammonia $2A_1$ differential overlap population curve is given in Figure 7 in order to complete the analogy to finite molecular orbital analysis. As shown in the curve, the lower band gives rise to a bonding type interaction, and the upper band gives rise to an anti-bonding type interaction. Note the position of the Fermi level; the antibonding band is essentially unoccupied. Electrons in the anti-bonding band have been transfered to other chromium 3D bands. Because ε splits the chromium 3D band, such electron transfer costs very little energy.

These concepts can be extended readily to describe interaction of the chromium 4S and 4Pz orbitals with the ammonia lone pair, but a detailed discussion is not given here. Instead, interactions are summarized by the differential overlap population curves in Figure 8. Peaks in the $3Dz^2$-$2A_1$ and 4S-$2A_1$ plots are consistent with the simple bonding and anti-bonding pair scheme suggested from the interaction diagram in Figure 6. The 4S-$2A_1$ anti-bonding band occurs at very high energy, and is outside the energy range shown in the plots. Both bands of the 4Pz-$2A_1$ interaction are bonding in character. The lower band arises from the typical bonding interaction. The higher energy band, coinciding with the $3Dz^2$ band, arises from a second order mixing[35] of the chromium 4Pz orbital into the $3Dz^2$-$2A_1$ band, and is bonding with respect to the chromium and ammonia molecule. Very high in energy (outside the range shown in the diagram) are bands derived from the anti-bonding component of the chromium 4Pz-ammonia $2A_1$ interaction.

Similar analyses may be applied to the ammonia-nickel and ammonia-copper systems. For example, differential overlap population curves for ammonia-on-copper are given in Figure 9. As in the chromium-ammonia case, the $3Dz^2$-$2A_1$ curve exhibits the simple bonding and anti-bonding pair of bands. However, ε of copper is well above the anti-bonding band; the $3Dz^2$-$2A_1$ interaction is net anti-bonding and repulsive. Because ε of copper (-7 eV) is above the copper 3D band (-12 eV), the 3D band is com-

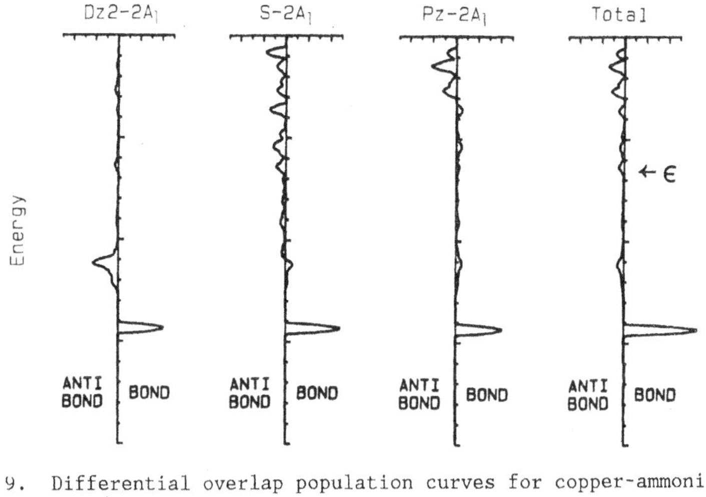

Figure 9. Differential overlap population curves for copper-ammonia interactions.

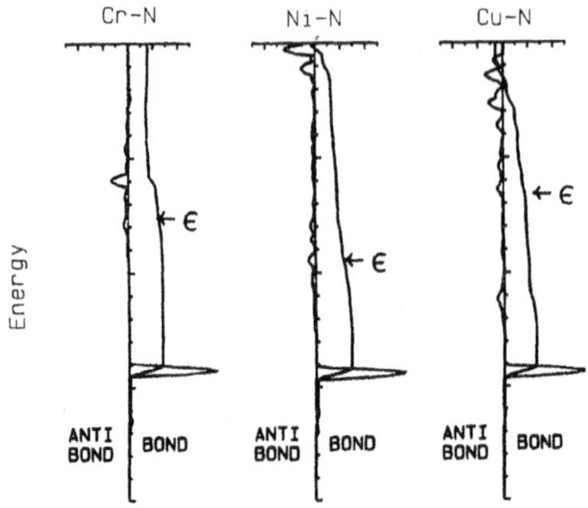

Figure 10. Differential and integrated overlap population curves for net metal-ammonia interactions.

pletely occupied. Electron transfer which occurs readily with chromium or nickel is not possible in the copper-ammonia system; the $3Dz^2$-ammonia antibonding orbital remains completely occupied. The $4S-2A_1$ and $4Pz-2A_1$ curves also exhibit the bonding and anti-bonding pair. (In the copper-ammonia system, the anti-bonding components of these curves lie at lower energy than in the chromium-ammonia system, and are within the energy range depicted in the plots.) Unlike chromium or nickel, copper binds to ammonia exclusively through the 4S and 4Pz orbitals.

Differential and integrated overlap population curves for the net metal-nitrogen interaction are given in Figure 10. As indicated in the figure, more of the metal-ammonia anti-bonding bands are filled as the electron count is increased. Overlap populations (differential overlap populations integrated up through the Fermi energy) for ammonia on the clean metals are given in Table IIA. As shown in the table, the metal 4S and 4Pz orbitals play an important role in binding ammonia to the surface If the metal 3D band is completely occupied, as in the case of copper, electrons cannot readily be transfered out of ammonia-$3Dz^2$ anti-bonding orbitals, and the metal 3D electrons repel the incoming base. The net trend in the overlap population is,

$$Cr \approx Ni >> Cu$$

Metal Oxides and Ammonia Adsorption on the Oxide Surfaces

Geometries used in the calculations on the metal oxides are shown in Figure 11. DOS plots for these oxides are given in Figure 12. As with the clean metals, DOS plots of the metal oxides resemble valence band photoelectron spectra[36-38]. Each plot exhibits a low energy peak (-33 eV) derived from the oxygen 2S orbital, and high energy peaks (-5 eV to 0 eV) derived from the metal 4S and 4P orbitals. The oxygen 2P band occurs at about -15 eV. Metal 3D bands range from -14 eV to -7 eV. As with the clean metals, the projected DOS of the surface atoms resembles the total DOS.

By comparing Figure 12 to Figure 1, note that a drop in the Fermi energy of roughly 3 eV accompanies oxidation of the metal. This decrease in ε implies an increase in the acidity provided that there is no change

Table II. Metal orbital-nitrogen overlap populations for ammonia adsorbed on the clean metals and on their oxides

A) Ammonia on the clean metals

metal orbital	chromium	nickel	copper
4S	0.088	0.107	0.071
4Pz	0.091	0.103	0.127
$3Dz^2$	0.059	0.030	-0.016*
total**	0.235	0.231	0.177

B) Ammonia on the metal oxides

metal orbital	Cr_2O_3	NiO	Cu_2O
4S	0.086	0.096	0.098
4Pz	0.124	0.108	0.177
$3Dz^2$	0.152	0.150	-0.021*
total**	0.352	0.334	0.248

* A negative overlap population indicates a very repulsive anti-bonding interaction.

** Includes interactions from the 1E orbitals.

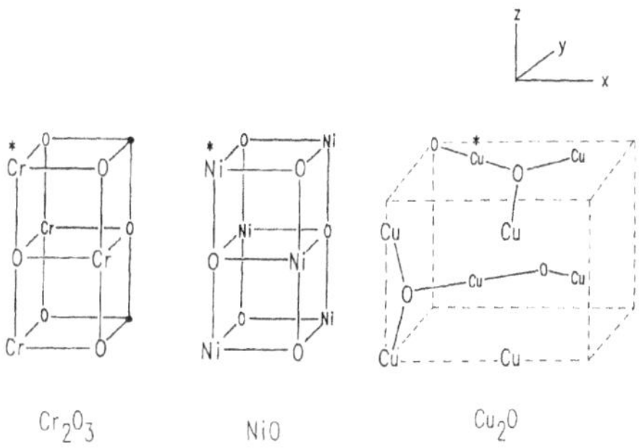

Figure 11. Unit cell geometry of the metal oxides. Asterisks indicate metal atoms attacked by ammonia in subsequent studies. The entire surface was generated by translations in the x-y plane.

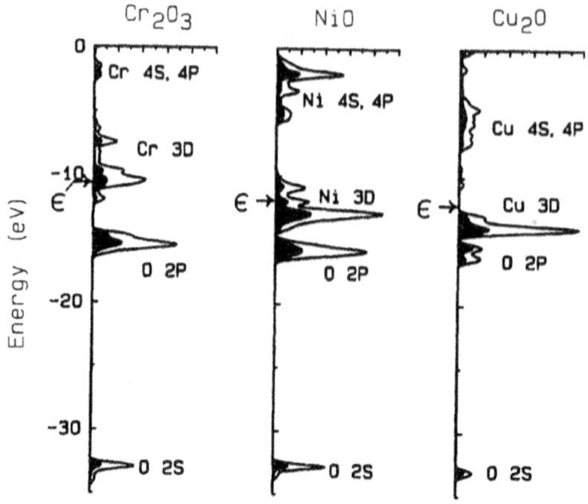

Figure 12. DOS for metal oxides. The total DOS is indicated by the line, while the projected DOS for surface atoms is indicated by the shaded area.

in the DOS immediately above ε. In a more general sense, the bands themselves occur at lower energy in the oxides than in the clean metals, and interact more strongly with base orbitals (e.g. the electron lone pair of ammonia). Thus repulsive as well as attractive interactions are stronger in the oxides. The Fermi level passes through prominent peaks in the DOS of Cr_2O_3 and NiO, as it did with the clean metals; Cr_2O_3 and NiO may be expected to be more acidic than clean chromium and clean nickel. The Fermi level of Cu_2O falls at the top of the copper 3D band; DOS immediately above ε is zero. Any acidic character results from the 4S and 4P orbitals, just as in unoxidized copper. From the DOS plots, surface acidity may be expected to follow the trend,

$$Cr_2O_3 > NiO >> Cu_2O$$

Surface acidity was explicitly probed by performing calculations on systems with ammonia adsorbed on the metal oxides. Ammonia was allowed to approach a surface metal atom, starting with a metal-nitrogen separation of 4 Å (cf. Figure 11 for geometries). Results are summarized in Figure 13, which shows the metal-ammonia interaction to be strongly attractive for adsorption on Cr_2O_3 and NiO, but only weakly attractive for adsorption on Cu_2O. These results follow the expected acidity trend. Since the energy passes through a minimum (for Cr_2O_3 and NiO) at a metal-nitrogen separation of roughly 2 Å, subsequent binding energy and bonding analyses

Table III. Binding energy for ammonia on the clean metals and their oxides (given in eV)

	chromium	nickel	copper
clean metal	0.21	0.21	-1.62*
metal oxide	1.07	0.71	0.24
at surface oxygen	-1.69*	-2.97*	-1.58*

* A negative binding energy indicates a repulsive interaction.

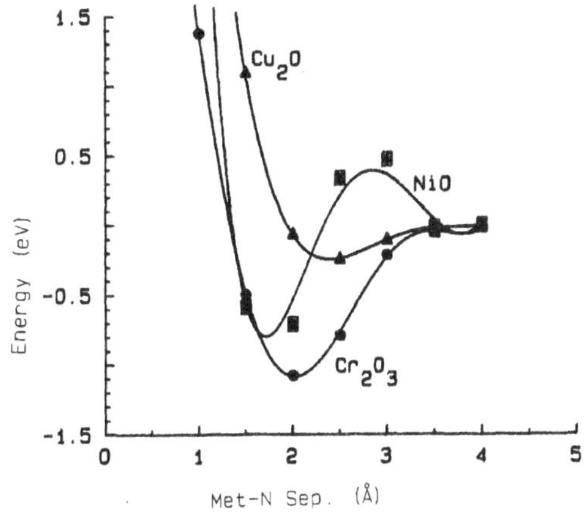

Figure 13. Energy versus metal-N separation curves for ammonia approaching a metal oxide surface. Energies are taken relative to that for ammonia 4 Å from the surface. (The energy for Ni-N separation of 1 Å lies outside the energy range of the figure, hence that point is not shown.)

were carried out assuming a metal-nitrogen separation of 2 Å.

Ammonia binding energies for the different systems, given in Table III, show that in addition to following the same acidity trend as the clean metals, the oxides bind ammonia more strongly than do the clean metals, in accord with the decrease of ε on oxidation. This result has been found in previous experimental studies for tungsten[39] and nickel[40].

Since the basic metal-ammonia bonding scheme is the same for the metal oxides as for the clean metals, a detailed discussion is not given here. The bonding analyses are summarized simply by the differential and integrated overlap population curves in Figure 14, and the total metal-N

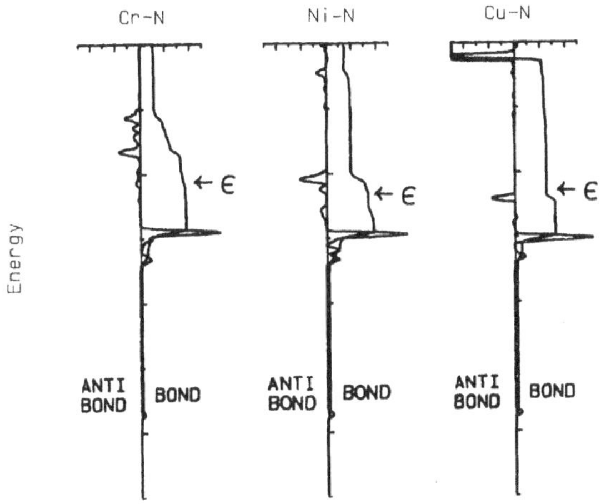

Figure 14. Metal-N differential and integrated overlap population curves for ammonia bonding to Cr_2O_3 (left), NiO (middle), and Cu_2O (right).

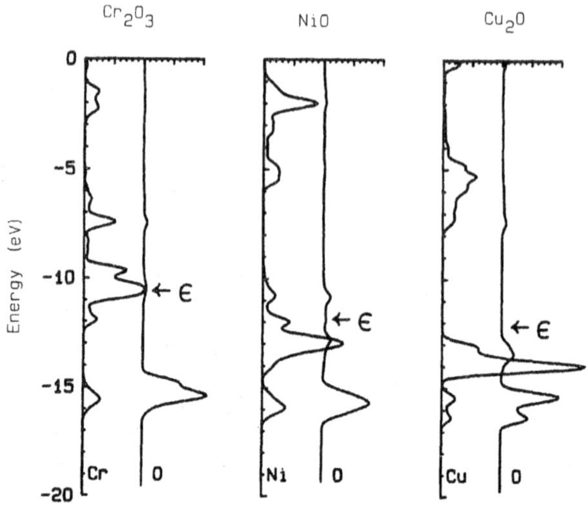

Figure 15. Projected DOS plots for surface metal and oxygen atoms in the metal oxides.

overlap populations given in Table IIB. The Fermi level in Cr_2O_3- and NiO-ammonia systems lies below the destablizing bands which arise from th anti-bonding metal $3Dz^2$-ammonia $2A_1$ interaction, while ε for the Cu_2O-ammonia system lies above these anti-bonding bands. This accounts for the significant difference in the metal-ammonia overlap population between th copper system and the chromium or nickel systems (cf. Table IIB). Furthe Cr_2O_3 and NiO systems show a substantially higher metal-ammonia overlap population than do clean chromium or clean nickel, a result of increased participation of the 3D orbitals in the bonding. The overlap population analysis substantiates the results based upon binding energies.

For completeness sake, attachment of ammonia to oxygen (via a nitrogen-oxygen "bond") was considered. The ammonia molecule was placed 2 Å above a surface oxygen atom. As indicated by the binding energies in Table III, this geometry is very unfavorable. Projected DOS plots of the oxides, Figure 15, indicate why this should be so. Empty bands are "loca ized" on the metal, not on oxygen; only the metal atoms can behave as acids. To put it simply, the oxygen electrons repel the ammonia electron

CONCLUSIONS

Surface acidity of chromium, nickel, and copper substrates has been probed. Based on the electronic structure (position of the Fermi level a the density of states), as well as the interaction with adsorbed ammonia, the acidity is expected to follow the trend,

$$Cr_2O_3 > NiO > Cr \approx Ni \approx Cu_2O > Cu$$

Though the 4S and 4P bands play a significant role in binding a base to the substrate, the marked difference between acidity of a chromium or a nickel surface and that of a copper surface is a consequence of the occu tion of metal 3D bands. The completely filled 3D band in copper (and Cu_2O) gives rise to a strong copper-ammonia anti-bonding interaction, which is not present in the chromium-ammonia and nickel-ammonia systems.

Surface acidity also is governed to a large extent by oxygen in the crystal. As a result of an increased oxidation state, the Fermi level drops, giving a more strongly acidic surface. It should be noted, however, that if the surface is sufficiently oxygen rich as to "cap" the metal atoms, ammonia will not bind via its electron lone pairs. In this case, binding may occur through weak hydrogen bonds.

In this study, ammonia has been used to simulate basic functional groups in a polymer. Thus, adhesion of the metals (metal oxides) to a polymer with basic functionalities is expected to follow the trend above.

REFERENCES

1. K.L. Mittal, Polymer Eng. Sci., 17, 467 (1977).

2. F.M. Fowkes, in "Physicochemical Aspects of Polymer Surfaces," K.L. Mittal, editor, vol. 2, p. 583, Plenum Press, New York, 1983.

3. J.C. Bolger and A.S. Michaels, in "Interface Conversion," P. Weiss and D. Cheevers, editors, p. 3, Elsevier, New York, 1968.

4. J.C. Bolger, in "Adhesion Aspects of Polymer Coatings," K.L. Mittal, editor, p. 3, Plenum Press, New York, 1983.

5. S.R. Cain and L.J. Matienzo, J. Adhesion Sci. Technol. 2, 395 (1988).

6. H. Brockmann, J. Adhesion, 22, 71 (1987).

7. J.S. Arlow, D.F. Mitchell, and M.J. Graham, J. Vac. Sci. Technol. A., 5, 573 (1987).

8. D.D. Eley, C.H. Rochester, and M.S. Scurrell, J. Chem. Soc. Faraday Trans. I, 69, 660 (1973).

9. A.A. Budneva and A.A. Davydov, React. Kinet. Catal. Lett., 11, 133 (1979).

10. K.K. Al-Shammeri and J.M. Saleh, J. Phys. Chem., 90, 2906 (1986).

11. I.C. Bassignana, K. Wagemann, J. Kuppers, and G. Ertl, Surface Sci., 175, 22 (1986).

12. M. Grunze, P.A. Dowben, and C.R. Brundle, Surface Sci., 128, 311 (1983).

13. A. Baiker and D. Monti, Ber. Bunsenges. Phys. Chem., 87, 602 (1983).

14. T. Miyano, K. Kamei, Y. Sakisaka, and M. Onchi, Surface Sci. Lett., 148, L645 (1984).

15. M.H. Matloob and M.W. Roberts, J. Chem. Soc. Faraday Trans. I., 73, 1393 (1977).

16. C. Klauber, M.D. Alvey, and J.T. Yates, Surface Sci, 154, 139 (1985).

17. Theoretical treatises of adsorption are extensive. For a brief sampling of the literature germane to this discussion, see,

    a. C. Zeng, Y. Apeloig, and R. Hoffmann, J. Am. Chem. Soc., 110, 749 (1988).

b. P. S. Bagus, K. Hermann, and C. W. Bauschlicher, J. Chem. Phys., 81, 1966 (1984).

18. Consider, for example, simple crystal field theory. See, F.A. Cotton and G. Wilkinson, "Advanced Inorganic Chemistry: A Comprehensive Text," 3rd ed., Wiley Interscience, New York, 1972.

19. NEWBAND1 is a program developed by the Hoffmann group at Cornell University. M.-H. Whangbo and R. Hoffmann, J. Am. Chem. Soc., 100, 6093 (1978).

20. R. Hoffmann and W.N. Lipscomb, J. Chem. Phys., 36, 2197 (1962).

21. R. Hoffmann, J. Chem. Phys., 39, 1397 (1963).

22. S.R. Cain, Chem. Phys. Lett., 143, 361 (1988).

23. S.R. Cain, L.J. Matienzo, and F. Emmi, J. Phys. Chem. Solids., 50, 87 (1989).

24. C.J. Ballhausen and H.B. Gray, "Molecular Orbital Theory," W.A. Benjamin, Inc., New York, 1965.

25. H.J. Monkhorst and J.D. Pack, Phys. Rev. B., 13, 5188 (1976).

26. For a brief discussion of Brillouin zones, see N.W. Ashcroft and N.D. Mermin, "Solid State Physics," Saunders, Philadelphia, 1976.

27. D.D. Coolbaugh, Ph.D. Thesis, State University of New York, Binghamton (1987).

28. J.-Y. Saillard and R. Hoffmann, J. Am. Chem. Soc., 106, 2006 (1984).

29. C. Kittel, "Introduction to Solid State Physics," 5th ed., Wiley and Sons, New York, 1976.

30. F.S. Galasso, "Structure and Properties of Inorganic Solids," Pergamon Press, New York, 1970.

31. for chromium photoelectron spectra see, J.A. Leiro and E.E. Minni, Philos. Mag. B, 49, L61 (1984).

32. for nickel photoelectron spectra see,

a. Y. Sakisaka, T. Komeda, M. Onchi, H. Kato, S. Masuda, and K. Yagi, Phys. Rev. Lett., 58, 733 (1987).

b. M.G. Thube, S.K. Kulkarni. D. Huerta, and A.S. Nigavekar, Phys. Rev. B, 34, 6874 (1986).

33. for copper photoelectron spectra see, P. Steiner, S. Hüfner, A.J. Freeman, and D.-S. Wang, Solid State Commun., 44, 619 (1982).

34. See for example,

a. J.C. Slater, Phys. Rev., 36, 51 (1930).

b. G. Burns, J. Chem. Phys., 41, 1521 (1964).

35. Mixing of orbitals in the molecular orbital formalism follows directly from second order perturbation theory. For more thorough discussions see,

a. T.H. Lowry and K.S. Richardson, "Mechanism and Theory in Organic Chemistry," pp. 538-567, Harper and Row, New York, 1976.

b. W.L. Jorgensen and L. Salem, "The Organic Chemist's Book of Orbitals," pp. 1-50, Academic Press, San Fransisco, 1973.

c. R.B. Woodward and R. Hoffmann, "The Conservation of Orbital Symmetry," Verlag Chemie, Germany, 1971.

d. S.R. Cain, R. Hoffmann, and E.R. Grant, J. Phys. Chem., 85, 4046 (1981).

36. for chromium oxide photoelectron spectra see, N. Beatham, A.F. Orchard, and G. Thornton, J. Phys. Chem. Solids, 42, 1051 (1981).

37. for nickel oxide photoelectron spectra see, S.-B. Lee, J.-H. Boo, and W.-S. Ahn, Bull. Korean Chem. Soc., 8, 358 (1987).

38. for copper oxide photoelectron spectra see,

a. D.D. Sarma, Ind. J. Chem. A, 19, 1046 (1980).

b. M. Scrocco, Chem. Phys. Lett., 63, 52 (1979).

39. M. Grunze, C. R. Brundle, and D. Tomanek, Surface Sci., 119, 133 (1982).

40. F.P. Nitzer and T.E. Madey, Surface Sci., 119, 422 (1982).

ADHESION AND DEFORMATION BEHAVIOR OF THIN METAL FILMS ON POLYIMIDE

Y.H. Jeng, F. Faupel, S.T. Chen and P.S. Ho

IBM Thomas J. Watson Research Center

Yorktown Heights, N.Y. 10598

Adhesion and interfacial failure of Cr ,Cu and Cu/Cr films on the PMDA-ODA polyimide substrates have been investigated using a stretch deformation technique. This method determines the adhesion energy by measuring the difference in the load vs. elongation curves between film/substrate and substrate structures. The effects due to the addition of a Cr interface layer between polyimide and Cu on the build-up of stress and the onset of failure have been investigated. This study shows that by varying the thickness of Cr interface one can alter the extent of ductile to brittle transition of the deformation behavior of the metal/polyimide layered structure.

INTRODUCTION

Metallized polyimide structures are widely used in microelectronic packaging and on the chip level because of the unique properties of polyimide such as low dielectric constant, high thermal stability, low processing temperature, and planarization capability. Adhesion is one of the major concerns since residual stresses generated due to thermal expansion mismatch during fabrication can introduce large deformations, which can lead to delamination failure at the metal/polyimide interface.

To form the metal/polymer structure, the choice of metal is copper because of its high conductivity. Previous experimental results[1-4] show that copper adheres weakly to PMDA-ODA polyimide in comparison to Cr, implying a weak chemical interaction at the molecular level. In order to improve the adhesion strength of copper to the PMDA-ODA polyimide, a thin Cr inter layer is often used. Cu and Cr have distinct ductile and brittle characteristics so they are good choices for studying the delamination behavior. Cu, being ductile, can effectively release stresses by plastic deformation before the fracture stress is reached. In comparison, Cr is brittle, so a high stress can be generated in the Cr/polyimide structure to cause interfacial delamination. In this study, we investigate the adhesion and deformation behavior of the Cu/polyimide structure as well as the effect due to the incorporation of a thin Cr interfacial layer.

Several experimental methods have been developed to measure the adhesion energy of thin films on ductile substrates which is defined as the minimum energy per unit area required to separate two surfaces. The standard tests, such as the peel and blister tests, are designed for measuring adhesion strength[5,6], but these yield results strongly dependent on the deformation behavior of the film and the substrate. To predict the actual performance of a film/substrate structure, a stretch-deformation method has been developed by us to investigate such deformation behavior in addition to measure the adhesion energy of metal/polymer structure[7]. This method is based on measuring the stress vs. strain dependence in combination with in-situ optical microscopy.

EXPERIMENTAL PROCEDURE

PMDA-ODA polyamic acid solution (DuPont 5878) was spun on oxidized Si wafers. The PMDA-ODA substrates were first baked for 30 min at 95 °C to remove the NMP solvent and then were processed through a curing cycle in nitrogen to complete the imidization process. The Cu and Cr overlayers were electron-beam evaporated at room temperature through a metal mask to define the geometry used in the stretch deformation test as shown in Fig. 1. A 2mm width of free standing metal/polyimide stripes were cut and peeled off from the wafer that was clamped at both ends. In the deformation tester, a stepping motor was used to produce the deformation and a piezoelectric tranducer was provided to measure the force experienced by the sample. This tensile test machine is computer-controlled and the progression of the delamination process can be continuously monitored.

RESULTS

Cu/Polyimide

The deformation behaviors observed for pure PMDA-ODA and a 4300Å Cu/ 5.3μm PMDA-ODA bilayer structure are shown in Fig.2. Within the initial 5% of elongation range, elastic and yield deformation of the Cu overlayer are observed. This is followed by a plastic deformation up to 20% straining. At this point cracks start to form in the Cu film. With further elongation (at ε_1 of about 30%) the Cu film starts to delaminate starting at the edges of the cracks. After the delamination, the deformation behavior of the sample converges to that of pure PMDA-ODA polymer substrate as expected.

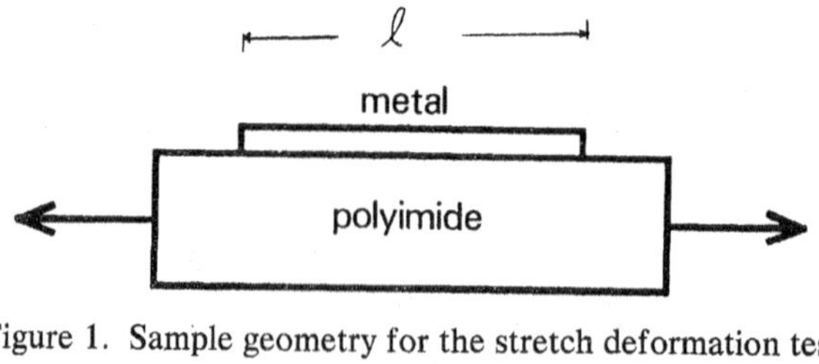

Figure 1. Sample geometry for the stretch deformation test.

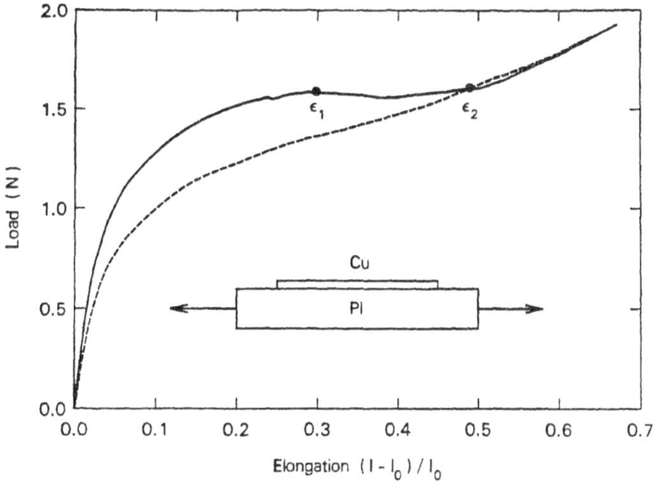

Figure 2. Load vs. elongation curve of
a) dashed line: 5.3μm PMDA-ODA.
b) solid line: 4300Å Cu/5.3μm PMDA-ODA bilayered structure.

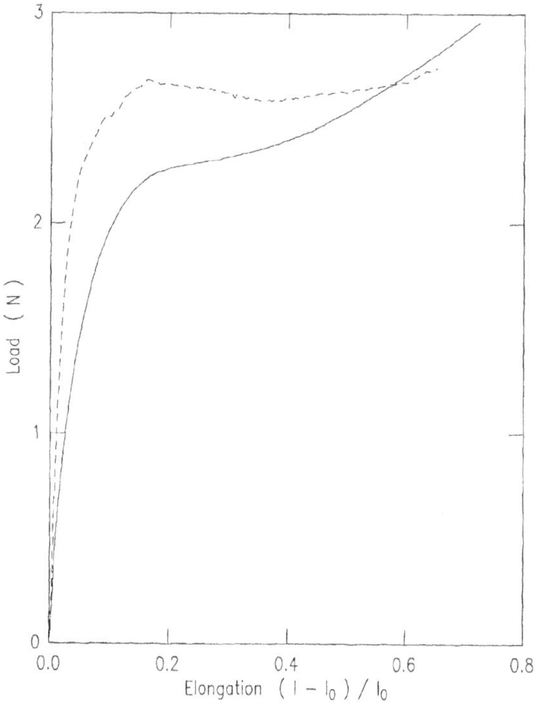

Figure 3. Load vs. elongation curve of
a) dashed line: 7.5μm PMDA-ODA.
b) solid line: 6000Å Cu/800Å Cr/8.5 μm PMDA-ODA trilayered structure.

The total work necessary to produce a given elongation can be directly obtained by integration of the load vs. elongation curve. In order to calculate the adhesion energy one has to consider the range between the starting and end point of delamination ε_1 and ε_2, respectively. In addition to the minimum adhesion energy required to separate the two surfaces, there are essentially three contributions to the total work. The work to strain the substrate (polyimide) from ε_1 to ε_2, the work for crack formation, and the work for plastic deformation. In the present work, the small energy contributions from the plastic and crack formation are neglected. On this basis, one can obtain the adhesion energy by integrating the load between Cu/PMDA-ODA and PMDA-ODA deformation curves from the starting point (ε_1) to the end point (ε_2) of delamination process, one obtains the adhesion energy as 9.6 ± 0.5 J/m^2. The details of the evaluation of the adhesion energy have been described previously.[8]

Cu/Cr/Polyimide Structure

Fig.3 shows a load vs. elongation curve of a trilayer structure with a 1100Å Cr interfacial layer between 6000Å Cu and 7μm PMDA-ODA. Here a high level of stress has already reached at a much lower strain than the Cu/ polyimide structure (Fig.2). The onset of crack formation and subsequent delamination have been shifted to lower strains as compared with the Cu/polyimide structure, starting at about of 3.5%. The effect due to the decrease in the thickness of the brittle interfacial Cr layer, from 1100 to 300Å, can be seen in Fig.4. By decreasing the Cr thickness to 300Å the onset of crack formation and delamination are shifted towards high strains, with ε_1 changing from 3.5% to 17%. Apparently a thinner Cr layer reduces considerably the sensitivity to film fracture and delamination. This trend is confirmed by additional studies using Cr layers of 500Å and 800Å. For these trilayer Cu/Cr/PMDA-ODA structures, the Cr/PMDA-ODA adhesion energy has been determined. The values were found to be 31 ± 6 J/m^2, 30 ± 8 J/m^2, 36 ± 8 J/m^2 and 46 ± 8 J/m^2 for Cr interfacial thickness of 1100Å, 800Å, 500Å and 300Å, respectively. Fig.5 plots the adhesion energy as a function of Cr thickness, showing that as the Cr thickness increases the adhesion energy decreases. In addition, the deformation behavior of the metal/polyimide structure becomes more brittle with increasing Cr thickness. Therefore, it appears that one can control the interfacial layer adhesion energy or its ductility by a proper combination of the overlayer and interfacial metal layer thickness.

DISCUSSION

The stretch deformation test has been applied to study deformation behavior and interfacial failure of Cr, Cu and Cr/Cu films on polyimide. In general, the stretch deformation test should be less affected by the energy dissipative processes than the peel test since this approach allows one to separate the different energy relief modes and to determine the adhesion energy without resorting to correction for the deformation energy. For the stretch deformation test the right choice of the substrate as well as the overlayer thickness is of vital importance. To ensure that strain energy is built up sufficiently per interface area during stretching to promote delamination, a certain film thickness combination is required. In addition, the

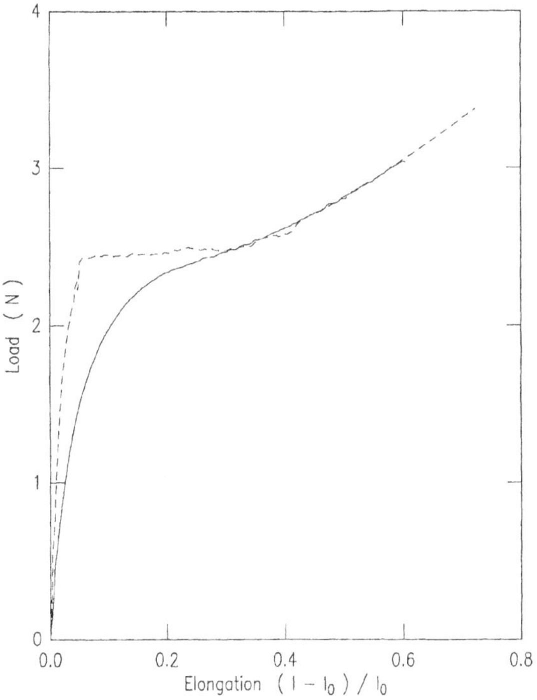

Figure 4. Load vs. elongation curve of
a) dashed line: 8.5μm PMDA-ODA.
b) solid line: 6000Å Cu/300Å Cr/8.5 μm PMDA-ODA trilayered structure.

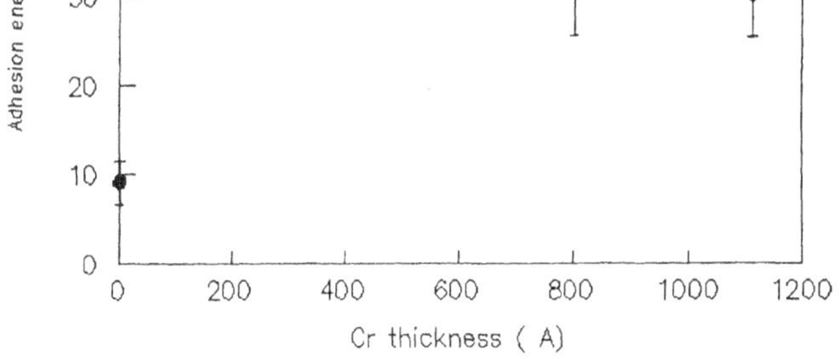

Figure 5. Adhesion energy vs. Cr interfacial thickness.

thickness of the substrate has to be properly chosen to support the additional stress introduced by the overlayer. Otherwise the thickness combination of the structure may not provide the proper energy partition in the layered structure to induce crack formation and a complete delamination of the interface.

Finite element analysis reveals that essentially tensile strain is introduced in the overlayer upon straining except close to the film edges or at the cracks, where a large local shear stress is dominant. Thick Cr films behave as a brittle material as expected, with stress relaxation primarily by crack formation and delamination. In contrast, Cu exhibits a ductile behavior with plastic deformation occurring via slip and twinning long before the fracture strain is reached. A Cr interface layer incorporated between Cu and polyimide can substantially enhance the built-up of stress upon stretching, and reduce the strain required for the onset of crack formation and delamination. As a result, the deformation behavior of the layered structure and the adhesion strength of the interface can be significantly changed. Indeed, the ductile deformation behavior of Cu/polyimide structure can be changed into a brittle behavior. The extent of the effect can be controlled by varying the Cr layer thickness.

The Cr thickness dependence of the adhesion energy and the deformation behavior can be qualitatively explained by considering the effect on crack formation due to the change in Cr thickness. As the layered structure is stretched, the Cr layer will first reach, before Cu does, a critical strain energy to initiate cracks, This critical energy varies about linearly with the thickness of the Cr layer, as the crack density in the Cr layer is expected to increase with increasing Cr thickness. Since the interfacial delamination commences at crack edges, a higher crack density will lead to an earlier delamination and requires a smaller strain energy , as observed in our experiments. Further experiments have been made to check the validity of this explanation and to observe the behavior of other metal layer combinations.

REFERENCES

1. R. Haight, R.C. White, B.D. Silverman, and P.S. Ho, J. Vac. Sci. Technol. A6(4) , 2188 (1988).
2. R.C. White, R. Haight, B.D. Silverman, and P.S. Ho, Appl. Phys. Lett. 51 , 481, (1987).
3. J.W. Bartha, P.O. Hahn, B.D. Silverman, and P.S. Ho, J. Vac. Sci. Technol. A3 ,1390, (1985).
4. N.J. Chou, and C.H. Tang, J. Vac. Sci. Technol. A2 , 751, (1984).
5. K.L. Mittal ed., "Adhesion Measurement of Thin Films, Thick Films and Bulk Coatings," Special Technical Publication, No.640, American Society For Testing and Materials, Philadelphia, 1978.
6. K.L. Mittal, J. Adhesion Sci. Technol. 1 , 247, (1987).
7. P.S. Ho, and F. Faupel, to be published.
8. F. Faupel, C.H. Yang, S.T. Chen and P.S. Ho, to be published.

ABOUT THE CONTRIBUTORS

KEVIN M. ANETSBERGER has been General Manager of Midwest Tungsten Service, Burr Ridge, IL since 1986. From 1985 to 1986 he served as Sales Manager for Midtech Systems, Inc. He attended the University of Illinois and the College of DuPage, majoring in Industrial Design and Plastics Technology, respectively.

T.G. BEAT is a senior engineering associate in the Vacuum Processes Group of the Materials Fabrication Division at Lawrence Livermore National Laboratory. He has 25 years of experience in vacuum deposion and vacuum engineering and is a frequent instructor for short courses on vacuum technology.

STEVE J. BEZUK has been since 1986 manager of the UNISYS GaAs Process Development Group and is now manager of the Advanced Materials Processing Group at UNISYS in Rancho Bernardo, CA. He joined Sperry Corp. (now UNISYS Corp.) CMOS process development group in 1983 and UNISYS's advanced silicon research group in 1984. Before that he was with RCA Sarnoff Research Center. He received his Ph.D. degree in Chemistry from the University of Minnesota in 1981. He has published papers on plasma diagnostics, $TaSiO_2$-polysilicon gate materials for CMOS devices, and laser photochemical and pyrolytic depositions of metals.

R.H.G. BRINKHUIS is currently a Ph.D. student in Polymer Chemistry at the University of Groningen in The Netherlands.

MICHAEL C. BURRELL is with the General Electric R&D Center in Schenectady, N.Y. where he joined in 1984 after receiving his Ph.D. in Analytical Chemistry from the University of Arizona. His research activities include the application of surface analysis in the areas of thin films, catalysis, adhesion and polymer surface chemistry.

STEPHEN R. CAIN is a staff engineer at the IBM Systems Technology Division, Endicott, NY. He pursued theoretical studies at Cornell University under Prof. Roald Hoffmann, and in 1982 earned the Ph.D. degree in Molecular Physics. Currently, his research efforts include molecular orbital studies of interfacial chemistry, plasma etching fundamentals, and statistical thermodynamics. He has published on a number of topics.

M. CAOLO has worked since joining Hewlett-Packard in 1981 in the Colorado Integrated Circuits Division in several manufacturing areas. She received her Ph.D. in Chemistry from Colorado State University (CSU) in 1980. Since 1985 she has been a Faculty Affiliate in the Department of Electrical Engineering at CSU. She has collaborated with Dr. Collins' group on soft e-beam patterning and teaches an introductory class in Microlithography.

R. CAUDANO is with Facultes Universitaires Notre-Dame de la Paix-LISE-IRIS, Namur, Belgium.

SHYNG-TONG CHEN is currently with IBM General Technology Division (Semiconductor Laboratory) as a staff engineer working in the area that uses metal/polyimide as conducting/insulating material. Prior to his current position, he was a postdoctoral fellow at IBM T.J. Watson Research Center studying metal/polyimide thermal mechanical properties. He received his Ph.D. in Physics from MIT in 1985.

TSE SHIH CHEN is currently pursuing doctoral studies in materials science at the University of Vermont. He received an MS degree in Materials Science and Engineering from Rochester Institute of Technology in 1988. His master's thesis research involved the dependences of resistivity and stress in sputtered copper films on polyimide on deposition conditions.

JOHN J. CHERA is a Chemist at General Electric Corporate R&D Center in Schenectady, NY. He earned his M.S. degree in Analytical Chemistry from Rensselaer Polytechnic Institute in 1988. He has been with GE for 10 years working in the area of polymer surface analysis.

DONG L. CHO is a guest scientist at the Institute for Surface Chemistry, Stockholm, Sweden. He received his Ph.D. degree from University of Missouri-Rolla. His main research interests are surface modification by plasma polymerization and plasma treatment, and he has 9 papers published or submitted for publication on these topics.

M. CHTAIB is with Facultes Universitaires Notre-Dame de la Paix-LISE-IRIS, Namur, Belgium.

GEORGE COLLINS is a Professor of Electrical Engineering at Colorado State University (CSU). Prior to joining CSU's College of Engineering, he was a Research Associate at Dunham Laboratory, Yale University, a Research Officer at Clarendon Laboratory, Oxford University, U.K., and a Visiting Scientist under the U.S.-Japan Cooperative Science Program at Nagoya University. He has served as a consultant to both industry and government. A graduate of Yale University, he is the author of more than 100 publications.

W.C. COWDEN is a senior technologist and 15 year veteran at Lawrence Livermore National Laboratory. For the past 7 years, he has been associated with the Electrodeposition Group and has worked on several projects dealing with electrodeposited gold, gold-copper alloys, silver diffusion bonding, and high precision copper plating. He has a BA degree in education from Colorado State College.

GREGORY DEE is member of the Polymer Physical Sciences Group of the Central Research and Development Department of the duPont Comapny in Wilmington, Delaware. He received a doctorate in Physics from Temple University. His research has concerned theoretical modelling of transport-coupled phase separation processes and equations of state for polymer melts.

N. JOHN DiNARDO is Assistant Professor of Physics at Drexel University which he joined in 1984. Before that he was a postdoctoral fellow (1981-1984) at IBM T.J. Watson Research Center, Yorktown Hts., N.Y. He receive his Ph.D. in Physics from the University of Pennsylvania in 1982. He was awarded Drexel University Research Scholar, 1986-1987. His research interests include application of scanning tunneling microscope to study local topology and spectroscopy at semiconductor interfaces, initial formation of metal-silicon and metal-gallium arsenide interfaces, and spectroscopic investigation of metal-polymer interfaces.

J.W. DINI is section leader, fabrication processes, at Lawrence Livermore National laboratory. He directs the activities of four groups: electroplating and metal finishing, vacuum processes, metal fabrication and plastics. He holds a degree in metallurgical engineering from Cleveland State University and is a national past president of the American Electroplaters and Surface Finishers Society (AESF).

WILLIAM V. DUMAS is a Chemist at General Electric Corporate R&D Center in Schenectady, N.Y. He received his Bachelor's degree in Chemistry from Union College, Schenectady, in 1984. He is involved in the metallization of high performance plastics.

FRANCIS EMMI is an advisory engineer and manager of a surface science department at the IBM Systems Technology Division, Endicott, NY. He received his Ph.D. in Analytical Chemistry from SUNY in 1984. His areas of interest include surface science, adhesion, metal-polymer interactions, and plasma etching chemistry. He has published on a number of topics.

ALAN B. ENTENBERG has been teaching in the Physics Department of Rochester Institute of Technology since 1983. Prior to 1983 he worked in laser-fusion research at the University of Rochester and in high energy particle research at the University of Pennsylvania. He received his Ph.D. in experimental particle physics from the University of Rochester in 1974. His current research interests are in the area of thin films. During the summer of 1985, he was a visiting faculty scientist at IBM Endicott where he studied the effects of RF plasmas to modify substrate surfaces for improved film adhesion.

MARK EYOLFSON was a physics student at Colorado State University working on his M.S.

FRANZ FAUPEL is a professor at University of Gottingen, W. Germany, where he is carrying out high T_c and metal/polyimide research. He received his Ph.D. degree in Metal Physics from University of Gottingen and spent a year at IBM T.J. Watson Research Center as a postdoctoral fellow studying metal/polyimide mechanical and diffusion properties.

LORETTA R. FENDROCK is currently working in magnetic resonance imaging at the University of Rochester Medical Center. She received B.S. degree in Physics from Rochester Institute of Technology in 1988. While an undergraduate, she was chosen to co-op in the RIT Thin Films Laboratory where she made measurements of stress and resistivity in sputtered copper films on polyimide.

G. FEYDER is in the Research Department of DuPont de Nemours (Lux.), Contern, Luxembourg.

DONALD F. FOUST is a Staff Chemist at General Electric Co. Corporate R&D Center in Schenectady, N.Y. He received his Ph.D. degree in Inorganic Chemistry from the University of Massachusetts, Amherst, in 1981. He is involved in the metallization of high performance plastics and has interest in the surface modification of polymers

WILLARD T. GRUBB was a physical chemist at GE Research and Development Center in Schenectady, N.Y. until his retirement in 1988. He is recognized for his contributions in the areas of electrochemistry, liquid metal embrittlement, trace element analysis, and electroless metal deposition. He has a Ph.D. degree.

HIROYUKI HIRAOKA is presently a research staff member of IBM Almaden Research Center, San Jose, CA. He joined T.J. Watson Research Center, Yorktown Hts., N.Y. in 1966 and moved to IBM San Jose Research Laboratory in 1971. Prior to joining IBM, he worked at Teijin Limited, Central Research Laboratory in Tokyo. He received his Ph.D. in Chemistry from Kyoto University. His main interest and study in IBM Research has been photochemistry and radiation chemistry of organic materials and applications to micro-lithographic fields. He has total 130 publications in scientific and technical journals or books, and 13 United States Patents issued. He received two IBM Outstanding Contribution Awards, a Research Division Award, and 7 Invention Awards. During 1987-1988, he was an adjunct professor at Colorado State University, Fort Collins, CO.

PAUL S. HO is Manager of the Interface Science Group in the Physical Science Department at the IBM T.J. Watson Research Center. He received a Ph.D. degree from Rensselaer Polytechnic Institute. He came to IBM in 1972 from Cornell University, where he had been an associate professor in the Department of Materials Science and Engineering. His current research is in the areas of thin films and interfaces in multilayered structures for VLSI applications. He has authored or coauthored many technical patents and has published extensively in the area of thin films and interfaces. He is a Fellow of the American Physical Society.

LARRY E. HOMSTAD is President of American Thin Film Laboratories, Inc., St. Paul, MN. which he founded in 1987 to provide high-quality sputtered films on flexible substrates for photolithographic and printed circuit applications. Previously, he established the thin film facility at APA Optics, Inc. He spent a five year sabbatical as Captain and owner of the sloop "NORSAGA" after 11 years with Honeywell in their Laser Gyro Mirror Development group. He holds a B.S. from the University of Minnesota.

SANG-KI HONG is currently doing doctoral work at Stevens Institute of Technology, Hoboken, N.J. He completed his M.S. in Materials Science and Engineering at Rochester Institute of Technology in 1988. His master's thesis research involved the dependences of stress and morphology of sputtered copper films on deposition conditions.

RONALD S. HORWATH has been since 1988 on IBM assignment at SEMATECH Corp. in Austin, TX, where he is presently a manager in the Pattern Transfer Group. As a senior engineer/manager from 1980 to 1987, he directed efforts in dry process development and implementation at the Packaging Laboratories of IBM in Endicott, N.Y. He joined IBM in East Fishkill in 1968 and in 1973 was promoted to manager of the Micron Device Technology Group which concentrated on the development of plasma and reactive ion etching processes. Before coming to IBM, he was employed (1956-1968) at several defense contractors, working in laser materials technology. He received his B.S. in Physics from Adelphi University in 1967.

RON P. HOWSON, is Reader in Physics at Loughborough University of Technology where he has a group investigating the preparation of thin films onto flexible polymer substrates. Based on their work this group was awarded the BP National and International Energy Prize in 1982. He is a member of the Organising Committee for the IPAT series of conferences and workshops. He is a member of the Vacuum Committee of the Institute of Physics and is on the Editorial Board of Solar Energy Materials, and Vacuum. He is author of a number of papers on the technology of the preparation of films and their properties. His current principal interest is the use of planar magnetron sputtering for the preparation of films for application on polymers. He has Ph.D. degree.

CHARLES JACKSON is member of the Polymer Physical Sciences Group of the Central Research and Development Department of the duPont Company in Wilmington, Delaware. He is a Research Technician who has contributed to a variety of programs at duPont including new polymerization chemistry, metal interlayer process development, and permeation measurements.

YOENG HUIE JENG is currently in the Physical Science Department at the IBM T.J. Watson Research Center in Yorktown Hts., N.Y. He received his Ph.D. degree in Chemical Engineering from Stanford University in 1988. He is presently working on mechanical properties and photoemission studies of metal-polymer interfaces.

GARY JORGENSEN is a senior optical materials scientist at the Solar Energy Research Institute. His most recent involvement has been with optical characterization of silvered polymer mirrors and optical/structural analyses of membrane dish concentrating collectors.

BRADLEY R. KARAS is a Staff Chemist at the General Electric Co. Corporate R&D Center in Schenectady, N.Y. He received his Ph.D. in Inorganic Chemistry from the University of Wisconsin-Madison in 1981. His research activities involve metallization of plastics and novel imaging techniques for 3-dimensional circuit boards.

DOV KATZ has been since 1968 with the Department of Materials Engineering at Technion, Haifa, Israel, first as Adjunct Professor and since 1973 as Professor of Polymer Engineering. He headed the Department of Materials Engineering during 1979-1980. He received his Ph.D. degree in 1960 from Hebrew University, Jerusalem and during 1960-1963 he worked as research associate with Prof. A.V. Tobolsky at Princeton University. He had had a number of positions including employment with the Scientific Department of the Israel Ministry of Defence and headed the Chemical Division during 1965-1973.

J. KRISHNASWAMY is a Research Assistant Professor in the Department of Materials Science and Engineering, North Carolina State University, Raleigh, NC. He received his Ph.D. degree from Nagoya University, Japan. His main interest is in the area of deposition and processing of thin films using particles and laser beams. He has serveral papers and patents in this area.

CHARLES J. KRYZAK joined the UNISYS GaAs Components Group in 1987 and is currently Project Manager of the Microwave Design, Packaging, and Testing Group. He joined UNISYS Corp. (Sperry Corp.) as a member of the Advanced Silicon Research Group in 1985. Before that he was with Hypres, Inc. He received his Ph.D. degree in solid state physics from Rensselaer Polytechnic Institute. He has published papers in the areas of MMIC packaging, laser processing, millimeter-wave transmission line design, and GaAs processing.

EDWARD J. LAMBY is a Chemist at the Corporate R&D Center of General Electric at Schenectady, N.Y. He received his M.S. degree from Union College, Schenectady, N.Y. in 1971. He is involved in the metallization of high performance plastics and photopatterning processes.

SOLOMON L. LEVINE is presently a Senior Chemist at IBM in Bromont, Quebec, Canada. Before that he held various positions in several IBM locations and had joined IBM in Burlington, VT. He has a Ph.D. degree. He has several patents and publications to his credit. His work has included analytical chemistry, instrument design and development, electrochemistry and plating, process engineering and process control.

MOSHE LEVY is on the faculty of the Department of Materials Science of the Weizmann Institute of Science in Rehovot, Israel. His contribution to the work described here was made during the course of sabbatical leave as Visiting Scientist with duPont (1984-1986). He is an organic chemist who has contributed to several areas of polymerization chemistry. More recently he has been active in the development of new technology for solar energy conversion.

ROBERT W. LEWIN is presently with INCO Selective Surfaces, Birmingham, U.K. Before that he worked in thin film research at Loughborough University of Technology for seven years. For the most part, he has been involved with the characterization of processes for, and the deposition and analysis of, films prepared by reactive d.c. magnetron sputtering. He received his B.Sc. in Physics from the University of Manchester in 1974.

LUMIN LI is a graduate student at Colorado State University from People's Republic of China.

VERN LINDBERG has been teaching in the Physics and Materials Science and Engineering Departments at Rochester Institute of Technology since 1979. He received his Ph.D. degree from Case Western Reserve University. He spent a sabbatical year at Eastman Kodak Research Laboratories working on metal film-polymer adhesion. He currently is Vice Chairman of the Upstate New York Chapter of the American Vacuum Society. His research interests include stress, adhesion and electrical conductivity of thin metallic films on polymers, surface modification of polymers, and optical properties of thin film conductors and dielectrics.

P. LUTGEN is in the Research Department of Du Pont de Nemours (Lux.), Contern, Luxembourg.

LEWIS MANRING is member of the Polymer Physical Sciences Group of the Central Research and Development Department of the duPont Company in Wilmington, Delaware. He received doctorate in Organic Chemistry from U.C.L.A. His recent research interests have included photochemical process in polymers and mechanisms of thermal depolymerization.

LUIS J. MATIENZO is an advisory engineer, IBM Corporation, Systems Technology Division in Endicott, NY. He received Ph.D. degree in Inorganic Chemistry from the University of Maryland in 1973. He has been a visiting professor of Chemistry at Universidad Nacional de Ingenieria in Lima, Peru, 1978. He has received several research awards. He has over sixty publications in scientific journals and books in the fields of surface science, adhesion, corrosion, materials science, advanced composites and surface modification.

TSUNETAKA MATSUMOTO is presently an emeritus professor of Kobe University, Kobe, Japan where he was professor from 1966 to 1985. He received Dr. Eng. degree from Kyoto University in 1962. In 1979 he received the award of the Chemical Society of Japan. He has been engaged in research on polymer colloids, polymer composites and magnetic composites.

STEPHEN MAZUR is member of the Polymer Physical Sciences Group of the Central Research and Development Department of the duPont Company in Wilmington, Delaware. he received doctorate in Organic Chemistry from U.C.L.A. His current research includes electrochemical processes of polymeric materials and molecular relaxations in amorphous bulk polymers.

KASHMIRI LAL MITTAL * is presently employed at the IBM US Technical Education in Thornwood, N.Y. He received his M.Sc. (First Class First) in 1966 from Indian Institute of Technology, New Delhi, and Ph.D. in Colloid Chemistry in 1970 from the University of Southern California. In the last 16 years, he has organized and chaired a number of very successful international symposia and in addition to this volume, he has edited 26 more books as follows: Adsorption at Interfaces, and Colloidal Dispersions and Micellar Behavior (1975); Micellization, Solubilization, and Microemulsions, Volumes 1 & 2 (1977); Adhesion Measurement of Thin Films, Thick Films and Bulk Coatings (1978); Surface Contamination: Genesis, Detection, and Control, Volumes 1 & 2(1979); Solution Chemistry of Surfactants, Volumes 1 & 2 (1979); Solution Behavior of Surfactants: Theoretical and Applied Aspects, Volumes 1 & 2 (1982); Adhesion Aspects of Polymeric Coatings, (1983); Physicochemical Aspects of Polymer Surfaces, Volumes 1 & 2 (1983); Surfactants in Solution, Volumes 1, 2 & 3 (1984); Adhesive Joints: Formation, Characteristics, and Testing (1984); Polyimides: Synthesis, Characterization and Applications, Volumes 1 & 2 (1984); Surfactants in Solution, Volumes 4, 5 & 6 (1986); Surface and Colloid Science in Computer Technology (1987); Particles on Surfaces 1: Detection, Adhesion and Removal, (1988); and Particles in Gases and Liquids 1: Detection, Characterization and Control (1989). Also he is Editor of the Series, Treatise on Clean Surface Technology, the premier volume appeared in 1987. In addition to these books he has published about 60 papers in the areas of surface and colloid chemistry, adhesion, polymers, etc. He has given many invited talks on the multifarious facets of surface science, particularly adhesion, on the invitation of various societies and organizations in many countries all over the world, and is always a sought-after speaker. He is a Fellow of the American Institute of Chemists and Indian Chemical Society, is listed in American Men and Women of Science, Who's Who in the East, Men of Achievement and many other reference works. He is or has been a member of the Editorial Boards of a number of scientific and technical journals, and is the Editor of the Journal of Adhesion Science and Technology, which made its debut in 1987.

KATSUHIKO NAKAMAE has been a professor in the Faculty of Engineering, Kobe University, Kobe, Japan since 1985 which he had joined in 1966. He received Dr. Eng. degree (Polymer Chemistry) from Kyoto University in 1967. In 1988 he received the award of the Society of Fiber Science and Technology, Japan. He has been engaged in research on polymer properties, surface properties of polymers and magnetic composites.

Y. NOVIS is Research Engineer at the "Institut de Recherche sur les Interfaces Solides", University of Namur, Belgium in the framework of a Belgian technological program of the Belgian Office for Science Policy (1984-1987) and an EEC BRITE program (1988-1991). He has collaborated with DuPont de Nemours (Luxembourg) and BASF (Fed. Rep. Germany). He received Diploma of Engineer in Physical Science from the University of Louvain (Belgium) in 1980, and during 1981-1983 was a science teacher in Mbarara, Uganda, East Africa. He is author or coauthor of about 8 papers on polymer surface and polymer surface modification.

KAZUKI OKA has been working since 1983 at UNITIKA Ltd., Japan on development of an ion gun, and is currently working in reactive magnetron sputtering for a doctorate. He received a M.Phil (Hons.) in Science from Nagoya University in 1983.

---

* As the editor of this volume.

THOMAS J. O'KEEFE is currently Curators Professor of Metallurgical Engineering and a Senior Investigator in the Graduate Center for Materials Research at UMR. He has held industrial positions with Dow Metal Products and Air Products and Chemicals, Metals System Division. He received a Ph.D. degree from the University of Missouri-Rolla in Metallurgical Engineering. His main areas of interest include chemical and process metallurgy, with emphasis in electrolytic processing and thin film deposition.

JAE M. PARK has been managing (since 1987) High Performance Materials Department at IBM System Technology Division in Endicott. He received his Ph.D. degree in Chemical Engineering/Polymer Science from University of Connecticut in 1982 and joined IBM T.J. Watson Research Center in Yorktown Hts., in 1983. His research interests include development of high performance materials and processes, metal-polymer interfacial phenomena, and adhesion, particularly for application to electronic packaging.

J.J. PIREAUX is with Facultes Universitaires Notre-Dame de la Paix-LISE-IRIS, Namur, Belgium.

SHIMON REICH is on the faculty of the Department of Materials Science of the Weizmann Institute of Science in Rehovot, Israel. His contribution to the work described here was made during the course of sabbatical leave as Visiting Scientist with duPont (1984-1986). He is a physicist whose research concerns polymer phase separation processes and physical properties of metal/polymer composites.

MALKA ROTEL is currently a research chemist at the Israel Institute of Metals, Haifa, Israel where she has been since 1983. Before that she was (1976-1983) a research chemist at the Environmental Centre at the Technion. She received her M.Sc. in Environmental Science in 1976. Her fields of interest are: laser material processes, laser enhanced deposition in liquid phase, and corrosion and electroplating.

PAUL SCHISSEL is a Principal Scientist at the Solar Energy Research Institute. As a member of the Materials Research Branch, he is involved in studies of polymeric materials for solar energy applications.

GENE R. SPARROW is President of Advanced R&D, Inc., St. Paul, MN, an Independent Research Laboratory which he formed in 1980 after 12 years of Analytical Service experience with 3M. He has been a recognized expert in Surface Instrumentation for the past 13 years and has authored numerous papers concerned with the principles and development of ion beam technology, primarily ISS and SIMS. He spent several years in the analytical application of High Resolution Mass Spectrometry and the development of computer data systems, and several years in Applied Surface Analysis. He holds a Ph.D. degree from Iowa State University.

ALARIC G. SPENCER has worked since 1983 at Loughborough University of Technology on magnetron sputtered thin films and is currently working in reactive magnetic sputtering for a doctorate. His published work include papers on magnetron design, and stability in reactive magnetron sputtering.

BYUNG K. SUN is currently studying for his Ph.D. degree in Metallurgical Engineering at University of Missouri-Rolla and is a graduate research assistant in the Graduate Center for Materials Research. He received his MS degree in Metallurgical Engineering from National Chonbuk University, Chonju, Korea. His present research interests are metallization of nonconducting substrates, and thin film deposition by plasma polymerization and electroplating.

SATOSHI TANIGAWA received M.Sc. degree (Polymer Chemistry) from Kobe University in 1987. He has been engaged in research on surface chemistry of magnetic composites.

W.J. VAN OOIJ is currently Senior Staff Scientist, Armco Research and Technology in Middletown, OH. Before that he was a Professor of Chemistry at the Colorado School of Mines. His research is in the areas of Surface Analysis of Polymers and Interfaces, Surface Modifications of Materials and Corrosion Control of Metals by Organic Coatings. He received his Ph.D. at the Technological University of Delft, The Netherlands in 1971, and worked for Akzo, a multinational Dutch chemical company, for 12 years. He is Editor of the Journal of Adhesion Science and Technology. He has about 90 publications and several patents on rubber additives, paint pigments and tire cords.

J. VOHS is with Facultes Universitaires Notre-Dame de la Paix-LISE-IRIS, Namur, Belgium.

T.A. WASH is a senior technologist in the vacuum processes group of the Materials Fabrication Division at Lawrence Livermore National Laboratory. He has been involved in various projects including fabrication of x-ray photoabsorption targets and other diagnostic devices. He has a BS degree in industrial technology from California State University, Fresno.

HIROTSUGU K. YASUDA is Professor and Chairman of the University of Missouri-Columbia Chemical Engineering Department and is also the Director of the Institute for Thin Films Processing Science. He received his Ph.D. in physical and polymer chemistry from the College of Environmental Science and Forestry in Syracuse, New York. His primary research interests are: polymer membrane technology; biomedical materials; thin-film technology; plasma polymerization; surface properties and modification of surfaces; adhesion of polymer coatings and control of interface and triboelectric properties of materials. He has written approximately 200 publications for scientific journals, books and encyclopedias.

JOSEPH ZAHAVI is currently the Head of the Corrosion, Surface Treatment and Laser Laboratories at the Israel Institute of Metals, Technion, Haifa, Israel where he received his D.Sc. degree in Materials Engineering in 1970. He spent three years at University of Illinois (1970-1973) and two years at IBM Poughkeepsie, N.Y. He has been the recipient of a number of prizes and awards, such as the Best Paper at the Interfinish 80, 10th World Congress on Metal Finishing, the Gold Medal for best paper published in Plating and Surface Finishing Journal in 1982 and 1983. He has more than 50 papers, many research reports and several American patents to his credit. His fields of interest are: laser material process, laser enhanced deposition in gaseous and liquid phase, and corrosion and electroplating.

INDEX

Acidity (surface) of Cr, Ni, Cu and their oxides investigated by extended Hückel type band calculations, 247-263
Adhesion
  at aluminum/poly(ethylene terephthalate), influence of polymer surface treatment, 193-204
  behavior of thin metal films on polyimide, 265-270
  of cobalt (ion-plated) PET film, 242-243
  of Cr, Ni, Cu to polymers containing basic functional groups, 247-263
  of iron film to PI surface 238-239
  of metallized PI and PET, effect of plasma treatment and ion-plating, 235-245
Adhesion energy
  vs Cr interfacial thickness in Cu/Cr/PI system, 269
  of metal-polymer structure, stretch deformation test for, 266
Adsorption of ammonia on clean metals, 250
Air (corona) treatment of PET, 197-199
Aluminum/poly(ethylene terephthalate) adhesion
  influence of PET surface treatment on, 193-204
Aluminum/poly(ethylene terephthalate) interface
  XPS and HREELS study of, 193-204
Ammonia
  adsorption on clean metals, 250

Band calculations (extended Hückel type) and investigation of surface acidity of Cr, Ni, Cu, and their oxides, 247-263

Chemical vapor deposition (laser stimulated) of metals on polyimide, 1-8
Chromium interfacial thickness and adhesion energy in Cu/Cr/PI system, 269
Cobalt (ion-plated) PET film, adhesion of, 242-243
Contact angle of water on polyimide
  effect of O_2 plasma exposure time, 239
Copper-ammonia interactions, 255
Copper/chromium/polyimide structure, 268
  adhesion energy vs Cr interfacial thickness in, 269
Copper coatings (thick) on glass, 93-101
Copper/polyimide system, 166
Copper (sputtered) films onto flexible polyimide substrates, 103-113
Corona (air, N_2) treatment of poly(ethylene terephthalate) 197-199

Deformation behavior of thin metal films on polyimide, 265-270
Deformation (stretch) test for adhesion energy of metal/polymer structure, 266
Durability of silvered polymer mirrors, 79-92

Electroless plating, 14-15
Electron beam (pulsed)
  metal-on-plastics patterning by, 69-78
Electronic excitations in polymers
  spectra of, 154
Electroplating solution
  UV laser-induced metallization of polyimide from, 55-68

Flexible polymer sheet
  high rate reactive sputtering onto, 45-54

Glass
  deposition of adherent, thick copper coating on, 93-101
Glow discharge plasma polymerization
  graded metallization of nonconducting substrates by, 9-27
Gold deposited on polyimide substrate, SEM micrographs of, 57-58
Graded metallization of nonconducting substrates by glow discharge plasma polymerization, 9-27

Heat distortion temperature for various plastics, 39
Heat of formation of various metal oxides, 94
HREELS spectra of PET, 201
HREELS study of the aluminum/poly (ethylene terephthalate) interface, 193-204
Hückel (extended) type band calculations and investigation of surface acidity of Cr, Ni, Cu and their oxides, 247-263

Interlayers (metal) in polymer films, 115-134
Ion-plating
  effect of, on adhesion of metallized PET, 235-245
Iron film (evaporated)
  adhesion to PI surface, 238-239
IR spectra of PET, 201
ISS study of interfacial chemistry of metal films on polymers, 205-219

Kapton®
  ESCA spectra of, 214
  SIMS spectra of, 175-176
Kapton-H®, SIMS spectra of, 177
Kapton-HN®, SIMS spectra of, 178
Kapton-V®, SIMS spectra of, 179

Laser-induced metallization of polyimide from electroplating solution, 55-68
Laser stimulated chemical vapor deposition of metals on polyimides, 1-8

Magnetron sputtering (reactive, planar), 45-54
Metal-containing polymeric film, deposition of, 13-14
Metal films on polyimide(s)
  adhesion and deformation behavior of, 265-270
  static SIMS study of interfaces between, 171-192
Metal films on polymers
  interfacial chemistry studied by XPS, ISS, SIMS, 205-220
Metal interlayers in polymer films, 115-134
Metallization of plastics (polymers) 1-8, 9-27, 29-44, 45-54, 55-68, 79-92, 103-113, 115-134, 223-233
  graded, by glow discharge plasma polymerization, 9-27
  by laser stimulated chemical vapor deposition, 1-8
  by reactive sputtering, 45-54
  with resistance heated sources, 29-44
  uses of, 30
  UV laser-induced, from electroplating solution, 55-68
Metallized plastics (see also, Metal films on polyimides, Metal films on polymers, and Metallization of plastics)
  surface spectroscopic techniques applied to, 137-170
Metallized polymer reflector construction, 80
Metal-on-plastics patterning by pulsed electron beam, 69-78
Metal oxides
  ammonia adsorption on, 256
  heat of formation of, 94
Metal-polymer interface study by surface spectroscopic techniques, 137-170

Nitrogen (corona) treatment of poly (ethylene terephthalate), 197-199

Optical performance of silvered polymer mirrors, 79-92
Oxygen plasma treatment of polyimide and adhesion of thin iron film, 238-239

Peel test, schematic representation of, 237
Plasma modification or treatment
  and adhesion of metallized polyimide, 235-245
  apparatus for, 237

Plasma polymerization (glow discharge)
graded metallization of nonconducting substrates by, 9-27
Plasma reactor system, 10
Plastics
chronology of, 32
heat distortion temperature for, 39
solubility of, 38
Polyetherimide
characterization of surface modifications during metallization of, 223-233
Poly (ethylene terephthalate)
cobalt ion plated, adhesion of, 242-243
corona treatment of, 197-199
metallized, 235-245
plasma surface treatment of, and effect on its adhesion to aluminum, 193-204
UV lamp treatment of, 195-196
XPS and RHEELS study of interface between aluminum and, 193-204
Polyimide(s)
commercial, SIMS spectra of, 173-184
contact angle of water on, effect of O_2 plasma exposure, 239
core level photoemission spectroscopy of, 144-147
EELS spectra of, 159
EELS spectra of surface vibrational losses of, 158
laser stimulated chemical vapor deposition of metals on, 1-8
MEIS spectra for copper-, 164
-metal interactions, theoretical models of, 149-152
metal interlayers in films of, 115-134
-metal interface, study of, 137-170
metallized, effect of plasma treatment on adhesion of, 235-245
metal patterning of, by pulsed electron beam, 69-78
plasma modification of, 235-245
sputtered copper films onto flexible, 103-113
static SIMS study of interface between evaporated metal films and, 171-192
thin metal films on, adhesion and deformation behavior of, 265-270
UV laser-induced metallization of, from electroplating solution, 55-68
valence photoemission spectra of, 152-154
water contact angle on O_2 plasma modified, 239
Polymer films
metal interlayers in, 115-134
Polymer mirrors (silvered)
optical performance and durability of, 79-92
Polymers
spectra of electronic excitations in, 154
Polymers containing basic functional groups
adhesion of Cr, Ni, Cu to, 247-263
Pulsed electron beam
metal-on-plastics patterning by, 69-78

Reactive sputtering onto flexible polymer sheet, 45-54
Resistance heated sources
metallization of plastics with, 29-44

Silvered polymer mirrors
optical performance and durability of, 79-92
Solubility of various plastics, 38
SIMS spectra of
APICAL®, 180
Kapton®, 175-176
Kapton-H®, 177
Kapton-HN®, 178
Kapton-V®, 179
Novax®, 181
Thermid-MC-600®, 182
Upilex-R®, 183
Upilex-S®, 184
SIMS spectra of commercial polyimide films, 173-184
SIMS (static) study of interfaces between evaporated metal films and polyimides, 171-192
SIMS study of interfacial chemistry of metal films on polymers, 205-219
Spectroscopic (surface) techniques applied to metallized plastics, 137-170
Sputtering (reactive) onto flexible polymer sheet, 45-54
Static SIMS study of interfaces between evaporated metal films and polyimides, 171-192

Stress
dependence of, on deposition conditions for sputtered Cu films onto flexible PI substrate, 103-113
Stretch deformation test for adhesion energy in metal-polymer structure, 265-270
Surface modification or treatment
and adhesion in Al/PET system, 193-204
of PET by air corona, 197-199
of PET by nitrogen corona, 197-199
of PI by oxygen plasma, 238-239
of PET by UV lamp, 195-196
Surface modifications during metallization of polyetherimide, characterization of, 223-233
Surface spectroscopic techniques applied to metallized plastics, 137-170
Surface spectroscopy experiments
general scheme for, 142

UV lamp treatment of poly(ethylene terephthalate), 195-196
UV laser-induced metallization of polyimide from electroplating solution, 55-68
Ultem
metallization of, characterizatio of surface modifications during, 223-233

Water contact angle on polyimide surface
effect of O_2 plasma exposure time 239

XPS of polyetherimide during metallization processes, 223-233
XPS study
of aluminum/poly(ethylene terephthalate) interface, 193-204
of interfacial chemistry of meta films on polymers, 205-219